Prostaglandins
and Thromboxanes

NATO ADVANCED STUDY INSTITUTES SERIES

A series of edited volumes comprising multifaceted studies of contemporary scientific issues by some of the best scientific minds in the world, assembled in cooperation with NATO Scientific Affairs Division.

Series A: Life Sciences

Recent Volumes in this Series

The series is published by an international board of publishers in conjunction with NATO Scientific Affairs Division

A	Life Sciences	Plenum Publishing Corporation
B	Physics	New York and London
C	Mathematical and Physical Sciences	D. Reidel Publishing Company Dordrecht and Boston
D	Behavioral and Social Sciences	Sijthoff International Publishing Company Leiden
E	Applied Sciences	Noordhoff International Publishing Leiden

Prostaglandins and Thromboxanes

Edited by

F. Berti

University of Milan
Milan, Italy

B. Samuelsson

Karolinska Institute
Stockholm, Sweden

and

G. P. Velo

University of Padua
Verona, Italy

PLENUM PRESS • NEW YORK AND LONDON
Published in cooperation with NATO Scientific Affairs Division

Library of Congress Cataloging in Publication Data

Nato Advanced Study Institute on Advances in Prostaglandins, Erice, Italy, 1976.
   Prostaglandins and thromboxanes.

   (Nato advanced study institutes series: Series A, Life Sciences; v. 13)
   Includes index.
   1. Prostaglandins–Congresses. 2. Thromboxanes–Congresses. I. Berti, Ferruccio.
II. Samuelsson, Bengt. III. Velo, G. P. IV. Title. V. Series.
QP801.P68N37 1976   591.1'924   77-5364
ISBN-13: 978-1-4684-2780-6   e-ISBN-13: 978-1-4684-2778-3
DOI: 10.1007/978-1-4684-2778-3

Lectures presented at the NATO Advanced Study Institute on Advances in
Prostaglandins held in the "Ettore Majorana" Center in Erice, Sicily,
October 4–15, 1976

© 1977 Plenum Press, New York

Softcover reprint of the hardcover 1st edition 1977

A Division of Plenum Publishing Corporation
227 West 17th Street, New York, N.Y. 10011

PREFACE

This volume presents lecture notes from the session
of the International School of Pharmacology, at the
"Ettore Majorana" Centre, on Advances in Prostaglandins
that took place in Erice (Sicily), October 4 to October
15, 1976. The School was a NATO Advanced Study Institute.

The aim of this international course was a comprehensive discussion by experts in various disciplines of
the present status of our knowledge of the biological
role of prostaglandins and thromboxanes.

The synthesis, metabolism, and function of prostaglandins have been evaluated at the cellular level, in
isolated tissues, and/or in the intact organism. The
mode of action of prostaglandins and their interactions,
particularly with cyclic nucleotides, in normal and
pathological conditions, has been discussed with the
aim of understanding the possible role of these compounds
in hormonal regulation and cell response.

The prostaglandin biosynthetic capacity of different
tissues, under various experimental conditions, in the
presence of specific precursors, and the inhibitory activity of different agents have been examined to ascertain the relationship between function and metabolism.

For these reasons the discussion has also been
extended to the methods (biological, immunological, and
spectrometric) available for the direct and specific
determination of prostaglandins and the evaluation of
their synthesis and metabolism.

The Editors hope that the book will be useful both
to beginners and to those with a long-time interest in
prostaglandins.

                                    F. Berti
                                    B. Samuelsson
                                    G. P. Velo

CONTENTS

CHEMISTRY OF PROSTAGLANDINS AND THROMBOXANES

Elisabeth Granström

Dept. of Chemistry, Karolinska Institutet

S-104 01   Stockholm 60, Sweden

The structures of several prostaglandins were determined during the early 1960's. First, two biologically active compounds were isolated from sheep vesicular glands; these compounds were called PGE_1 and $PGF_{1\alpha}$ (for a review, see Ref. 1). Reduction of PGE_1 with sodium borohydride gave a mixture of two epimeric compounds, $PGF_{1\alpha}$ and $PGF_{1\beta}$. Further studies on the structures of these compounds revealed that they were derivatives of a C_{20} carboxylic acid, later named prostanoic acid, which contained a five-membered ring (C-8 to C-12) with two side chanis. Prostaglandins of the E type were shown to have a keto group at C-9, whereas a hydroxyl group was found in this position in the F prostaglandins. The index α or β indicates the stereochemical position of this hydroxyl group. Common to all these prostaglandins and most others is the presence of an L-hydroxyl group at C-15 and a Δ^{13}-trans double bond.

The different classes of prostaglandins are further subgrouped according to the degree of saturation, and an index indicates the number of double bonds. Prostaglandins of the "1" type contain only the Δ^{13}-trans double bond, PG_2 contains also a Δ^5-cis double bond, and PG_3 an additional Δ^{17}-cis double bond. Thus PGE_2 e.g. is $11\alpha,15$-dihydroxy-9-ketoprost-5(cis),13(trans)-dienoic acid.

PGE_1, E_2, E_3, $F_{1\alpha}$, $F_{2\alpha}$ and $F_{3\alpha}$ were called the "primary" prostaglandins, and all except $PGF_{3\alpha}$ were also identified in human seminal plasma. From this source were later isolated eight additional prostaglandins. These were all dehydrated prostaglandins, formed by loss of the 11α hydroxyl group from a PGE compound, which resulted in the formation of a double bond either in the Δ^{10} position (PGA com-

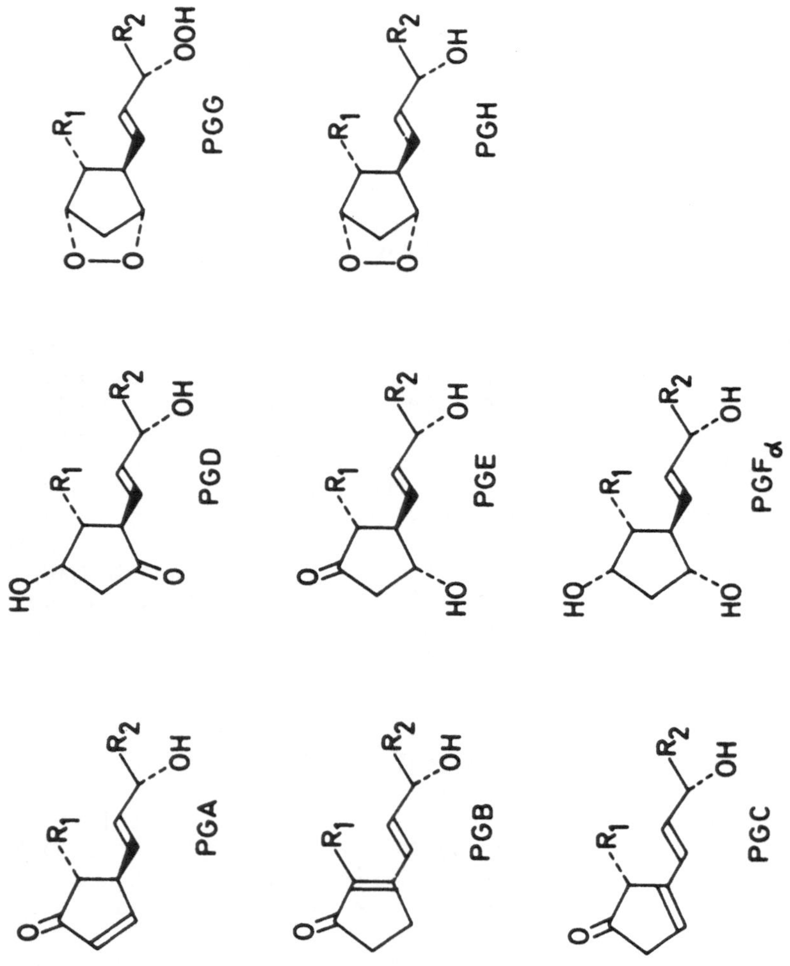

Fig. 1.   Nomenclature of prostaglandins.

pounds) or in the Δ^8 (12) position (PGB compounds). The eight additional prostaglandins found in human seminal plasma were PGA_1, PGA_2, PGB_1, PGB_2, and 19-hydroxylated derivatives of these compounds (2,3). Recently, however, several studies indicate that these dehydrated prostaglandins were artifacts and formed during storage of the seminal plasma (4,5). A different dehydrated prostaglandin, PGC, has later been identified (6). This prostaglandin has a double bond in the Δ^{11} position.

Prostaglandins of the D type are isomers of the PGE compounds, with a keto group at C-11 and an α-hydroxyl at C-9 (7,8). The nomenclature of prostaglandins according to the substituents in the ring and at C-15 is summarized in Fig. 1.

Prostaglandins are biosynthesized from certain polyunsaturated fatty acids. Arachidonic acid (5,8,11,14-eicosatetraenoic acid) is e.g. the precursor of prostaglandins of the "2" type. Studies on the conversion of 8,11,14-eicosatrienoic acid into PGE_1 showed that the two oxygens in the five-membered ring of PGE_1 were derived from the same oxygen molecule (1). This finding, together with the occurrence of 12L-hydroxy-8,10-heptadecadienoic acid and malondialdehyde as byproducts of the biosynthesis, strongly indicated the existence of an endoperoxide intermediate oxygenated at C-15 (Fig. 1) (7,9,10). This was also supported by the simultaneous formation of PGE, PGF_α and PGD compounds in certain systems.

A more direct evidence for the formation of endoperoxides during prostaglandin biosynthesis was later obtained. In these experiments, a compound reducible to the corresponding PGF_α compounds could be detected in short-time incubations of precursor acids with a prostaglandin synthetase preparation. This compound was isolated from incubations with arachidonic acid and found to be 15-hydroxyprostaglandin endoperoxide (PGH_2, 15-hydroxy-9α,11α-peroxidoprosta-5,13-dienoic acid (Fig. 1)). In subsequent studies, an additional endoperoxide was isolated from incubations carried out in the presence of p-mercuribenzoate, viz. 15-hydroperoxy-9α,11α-peroxidoprosta-5,13-dienoic acid, PGG_2 (Fig. 1). The structures of these unstable compounds were assigned mainly by a number of chemical transformations into previously known prostaglandins (Fig. 2) (11). Thus, treatment of either endoperoxide with stannous chloride or triphenylphosphine resulted in the formation of $PGF_{2\alpha}$. Treatment of PGG_2 with lead tetraacetate converted the hydroperoxy group at C-15 into a keto group, and after a subsequent reduction with triphenylphosphine, 15-keto-$PGF_{2\alpha}$ could be isolated. PGG_2 and PGH_2 were both unstable in aqueous medium: the main product formed from PGG_2 was 15-hydroperoxy-PGE_2, whereas PGH_2 was mainly converted into PGE_2.

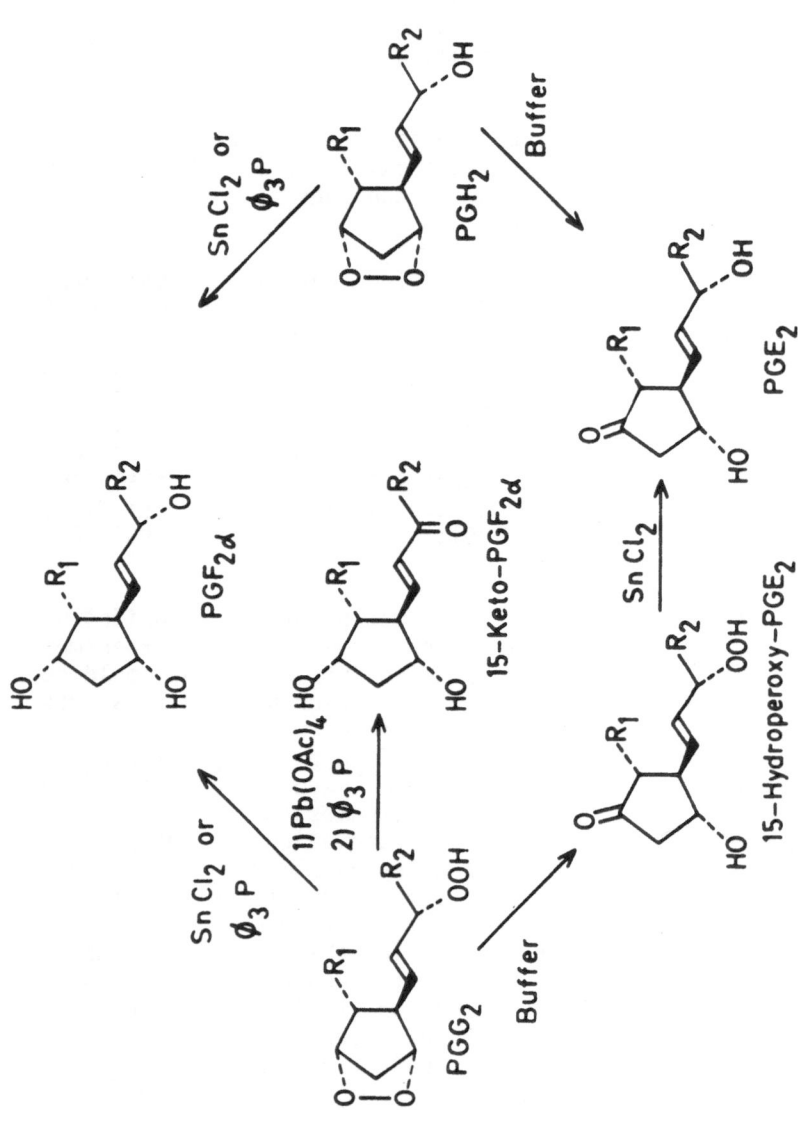

Fig. 2. Reactions carried out on PGG$_2$ and PGH$_2$. R$_1$: CH$_2$-CH=CH-(CH$_2$)$_3$-COOH. R$_2$: (CH$_2$)$_4$-CH$_3$. Ø: Phenyl.

The half-lives of these endoperoxides were about 4-5 min in aqueous medium. They were shown to have pronounced biological activity, both on several smooth muscle preparations (12) and as agents capable of inducing platelet aggregation (11).

In certain tissues or cells, e.g. lung tissue and platelets, arachidonic acid has been found to be converted only to a minor extent into PGE_2 and $PGF_{2\alpha}$. The major products formed were instead 12L-hydroxy-5,8,10,14-eicosatetraenoic acid (HETE), 12L-hydroxy-5, 8,10-heptadecatrienoic acid (HHT) and a novel compound, the hemiacetal derivative of 8-(1-hydroxy-3-oxopropyl)-9,12L-dihydroxy-5, 10-heptadecadienoic acid (thromboxane B_2 (TXB_2), earlier named PHD) (Fig. 3). The structure of this last-mentioned compound was assigned mainly by mass spectrometric analysis of a number of derivatives and by oxidative ozonolysis (13).

Indomethacin and aspirin (inhibitors of fatty acid cyclo-oxygenase) blocked the formation of HHT and thromboxane B_2 but increased the formation of HETE (14). It was also demonstrated that PGG_2 added to a suspension of human platelets was rapidly converted into HHT and TXB_2. These findings indicated that the latter two compounds are formed from PGG_2, whereas HETE is formed by a separate pathway.

Results of experiments in which arachidonic acid was incubated with platelets under $^{18}O_2$ suggested that TXB_2 was formed from PGG_2 by rearrangement and subsequent incorporation of one molecule of H_2O (15). It was postulated that the rearranged intermediate might be possible to trap in the presence of nucleophilic reagents. Short time incubations (30 sec) were interrupted by the addition of large volumes of methanol, ethanol or sodium azide, and in all cases products less polar than TXB_2 appeared (Fig. 4). These products were not seen when the incubation was interrupted after 5 min instead. The products formed were identified as mono-O-methyl TXB_2, mono-O-ethyl TXB_2 and an azido derivative of TXB_2, respectively (15).

These trapping experiments demonstrated the presence of a very unstable intermediate in the conversion of PGG_2 into TXB_2. The half-life of this intermediate in aqueous medium was determined using an isotope dilution method and bioassays (effect on rabbit aorta or platelet aggregation). In all experiments, a $t_{1/2}$ of about 30-40 sec was found.

The unstable intermediate has been assigned the structure given in Fig. 4 and is called thromboxane A_2. Two other alternate structures could be ruled out either from experiments with deuterium labeled precursor or from the results of the half-life determination.

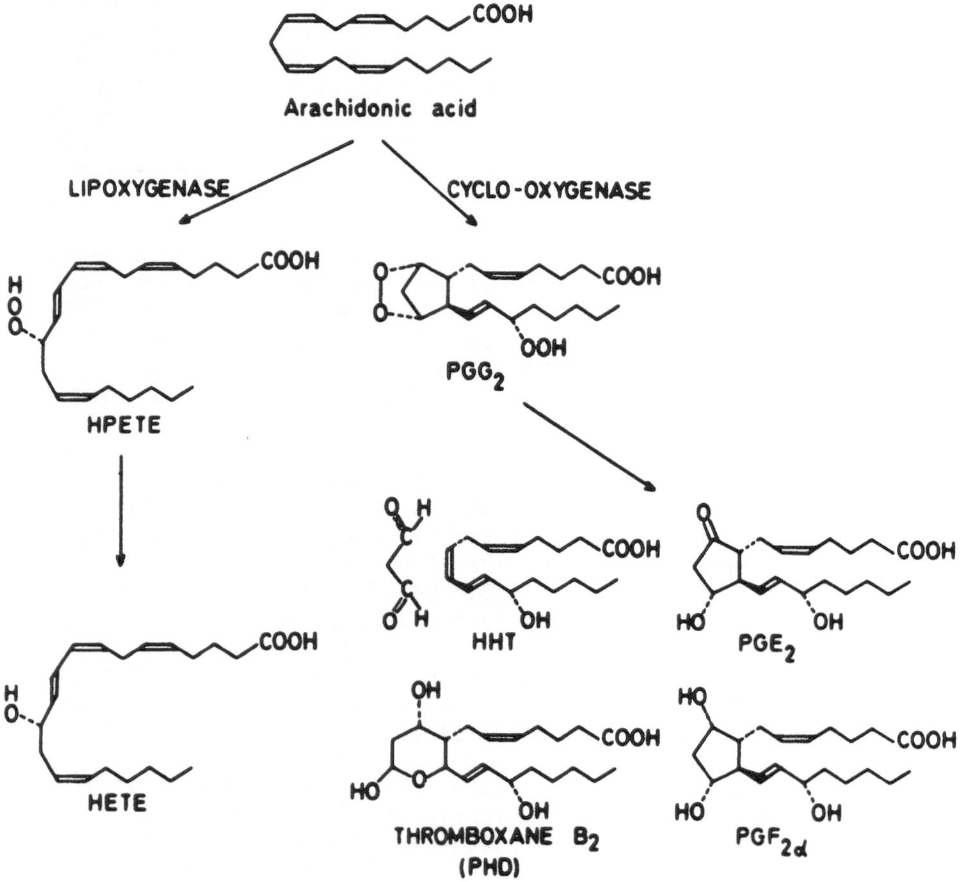

Fig. 3. Transformation of arachidonic acid in human platelets.

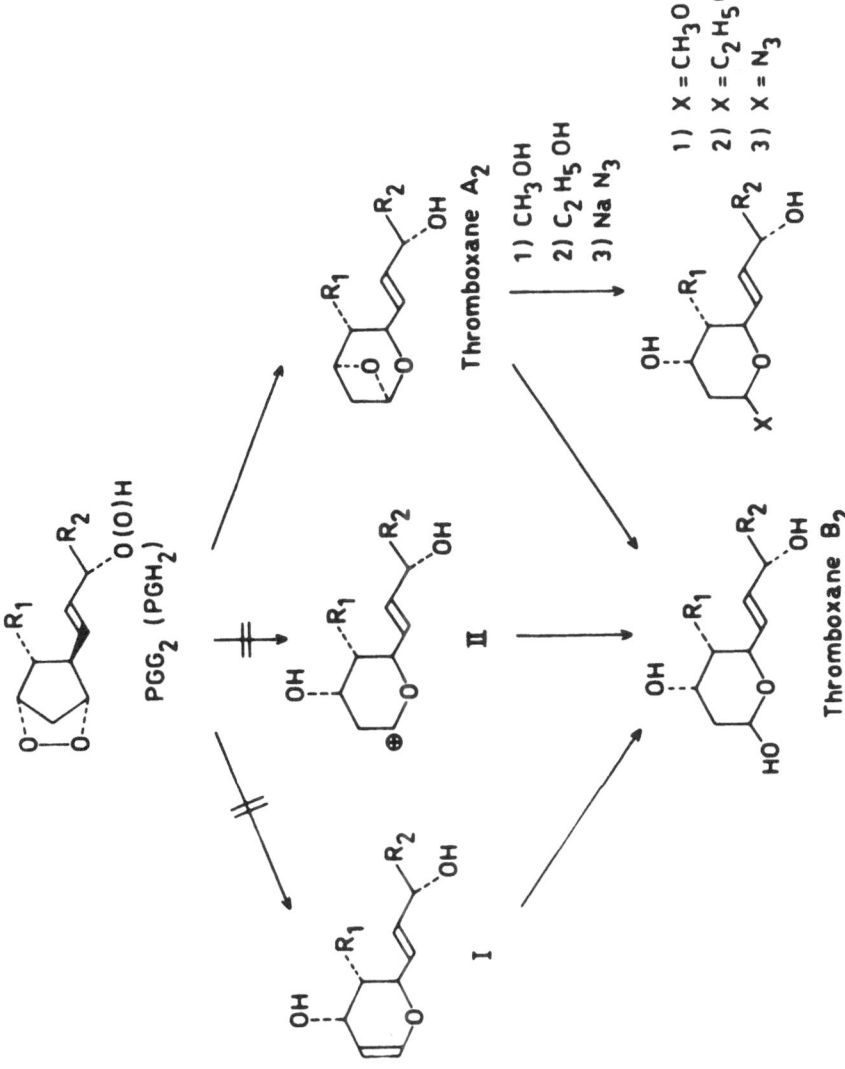

Fig. 4. Transformation of PGG₂ into thromboxanes.

In addition to the prostaglandin endoperoxides and the stable prostaglandins, the thromboxanes are postulated to be important mediators of the actions of certain polyunsaturated fatty acids.

REFERENCES

1. Samuelsson, B., In: Lipid Metabolism. Ed.: S.J. Wakil. Academic Press, Inc., New York, 1970, pp. 107-153.

2. Hamberg, M. and Samuelsson, B., J. Biol. Chem. 241 (1966) 257.

3. Hamberg, M., Eur. J. Biochem. 6 (1968) 147.

4. Taylor, P.L. and Kelly, R.W., Nature (Lond.) 250 (1974) 665.

5. Jonsson, H.T., Middleditch, B.S. and Desiderio, D.M., Science, 187 (1975) 1093.

6. Jones, R.L., Cammock, S., and Horton, E.W., Biochim. Biophys. Acta 280 (1972) 588.

7. Nugteren, D.H., Beerthuis, R.K. and van Dorp, D.A., Rec. Trav. Chim. Pays-Bas, 85 (1966) 405.

8. Granström, E., Lands, W.E.M. and Samuelsson, B., J. Biol. Chem. 243 (1968) 4104.

9. Hamberg, M. and Samuelsson, B., J. Am. Chem. Soc. 88 (1966) 2349.

10. Hamberg, M. and Samuelsson, B., J. Biol. Chem. 242 (1967) 5344.

11. Hamberg, M., Svensson, J., Wakabayashi, T., and Samuelsson, B., Proc. Nat. Acad. Sci. USA, 71 (1974) 345.

12. Hamberg, M., Hedqvist, P., Strandberg, K., Svensson, J., and Samuelsson, B., Life Sci. 16 (1975) 451.

13. Hamberg, M. and Samuelsson, B., Proc. Nat. Acad. Sci. USA, 71 (1974) 3400.

14. Hamberg, M., Svensson, J. and Samuelsson, B., Proc. Nat. Acad. Sci. USA, 71 (1974) 3824.

15. Hamberg, M., Svensson, J. and Samuelsson, B., Proc. Nat. Acad. Sci. USA, 72 (1975) 2994.

SOME SYNTHETIC ASPECTS OF PROSTAGLANDINS

Pierre Crabbé

Université Scientifique et Médicale

Grenoble (France)

A- INTRODUCTION

The prostaglandins (PG) form a class of natural products with many and potent biological activities. If their physiological role has not been clearly defined because they are intriguing biochemicals, their potency and activity in many apparently unrelated biological systems have awaken the interest of scientists of various disciplines (1-5).

Natural occurring PG may be regarded as derivatives of prostanoic acid (1), a C-20 organic acid with a substituted cyclopentane unit.

The natural PG are divided into three series, namely the first which presents only one double bond in the chains between positions 13 and 14, the second group has an additional olefinic bond between C-5 and C-6, and the third group has another double bond located between C-17 and C-18. Furthermore, natural PG form different families, (A, B, C, D, E, F, G, H, etc.), as shown in Charts 1 to 3. The six primary PG present an oxygen at position 9 (ketone in the E-series, hydroxyl in the F-series) and at C-11 and C-15 (α-hydroxyl groups), as well as a _trans_ double bond between C-13 and C-14.

The stereochemistry, _in extenso_ the configuration of the hydroxyls at C-9, C-11 (R) and C-15 (S), as well as that of the chains located at C-8 and C-12 and the chain double bonds (_cis_ or Z at C-5 and _trans_ or E at C-13), is important for biological activity.

Chart 1

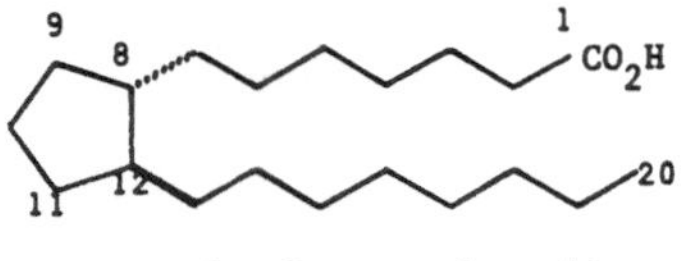

$\underline{1}$: Prostanoic acid

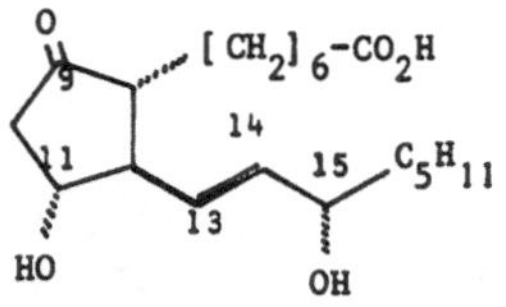

$\underline{2}$: PGE$_1$

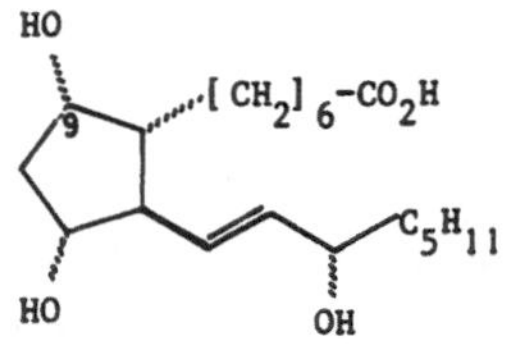

$\underline{3}$: PGF$_{1\alpha}$

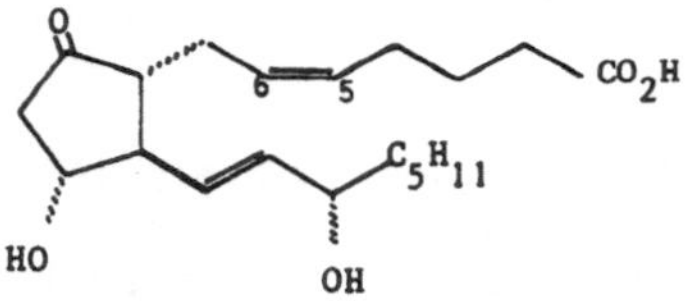

$\underline{4}$: PGE$_2$

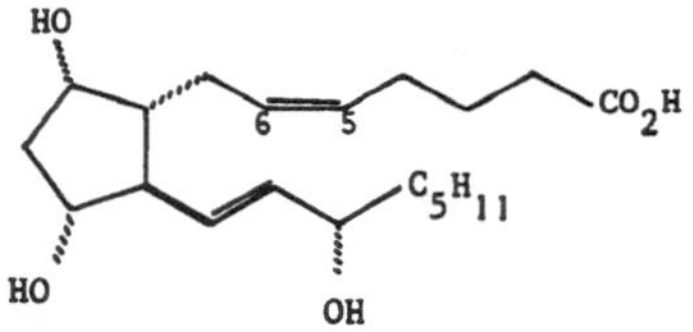

$\underline{5}$: PGF$_{2\alpha}$

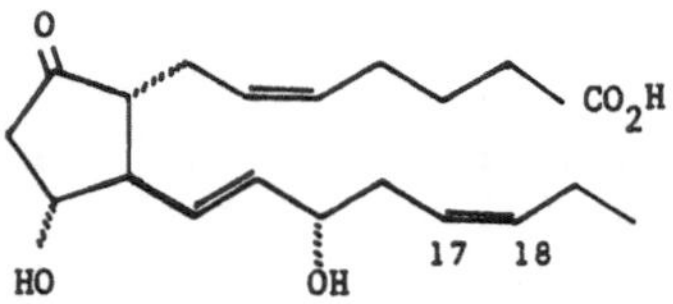

$\underline{6}$: PGE$_3$

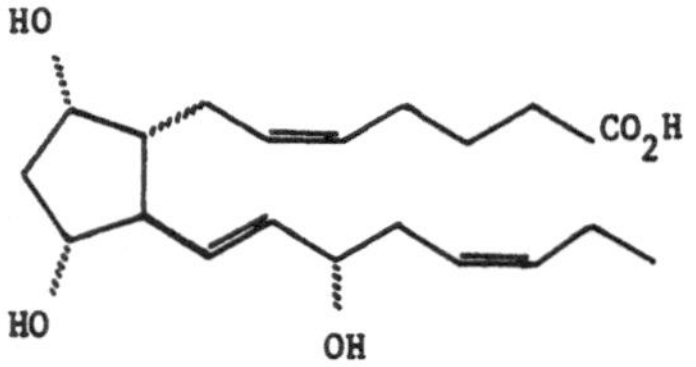

$\underline{7}$: PGF$_{3\alpha}$

Besides the primary PG, PGA_2 (8), PGB_2 (9), and PGC_2 (10) possess a carbonyl group at position 9, like the PG of the E family (Chart 2), but they differ by the position of the double bond in the five-member ring.

PG of the D family, such as PGD_2 (11), are 11-oxo-prostanoids, whereas PGG_2 (12), PGH_2 also called PGR_2 or LASS (labile aggregation-stimulating substance) (13) are biosynthetic endoperoxides intermediates. PGG_2 (12) and/or PGH_2' (13) readily metabolize to give newly isolated entities, in extenso thromboxane A_2 (14) and then thromboxane B_2 (PHD) (15). Interest in PG of the D family has been occasioned by the observation that PGD_2 (11) is a product formed during nonenzymatic decomposition of the endoperoxide PGH_2 (13), derived from arachidonic acid with microsomes of sheep vesicular gland and is believed to be biologically active. Finally, 19-hydroxy-PG (16) form one additional group of biologically important natural PG (see Chart 3).

The lack of an efficient biosynthetic process coupled with the therapeutic importance of PG have prompted the investigations of biosynthetic as well as chemical approaches to alleviate the scarcity of these biologically important compounds. In 1965, scientists in the Netherlands (6) and in Sweden (7), simultaneously discovered how to prepare relatively large quantities of PG by incubation of fatty acids with sheep glands. Moreover, because of the worldwide impetus in PG pharmacological and clinical research, several laboratories undertook the synthesis of these C-20 carboxylic acids, with the first report of a complete total synthesis appearing in 1967 (8). Furthermore the isolation of PGA_2 (8) from the marine corals Plexaura homomalla (9) provides scientists with an unexpected but welcome supply of material, used both for the preparation of primary PG, as well as of modified entities.

Although the total synthesis of PG constitutes a formidable challenge for organic chemists, numerous conceptually different synthetic routes have been explored successfully. A difficulty that one faces in their synthesis is that PG molecules present a number of functional groups of different nature. In addition, the stereochemistry of the PG, in extenso the geometry of the double bonds, the configuration of the chains at C-8 and C-12 and the configuration of hydroxyls at positions 9, 11 and 15, is critical for the bioactivity. Five asymmetric centers are present in many natural PG and their incorporation into a total synthetic scheme constitutes a difficult objective. One should also mention that the PG molecules are not easy to handle technically, since most of them are unstable in acidic or alkaline medium and are sensitive to air and heat.

It was only shortly after the final proof of the structure and the determination of the stereochemistry of primary PG that the

Chart 2

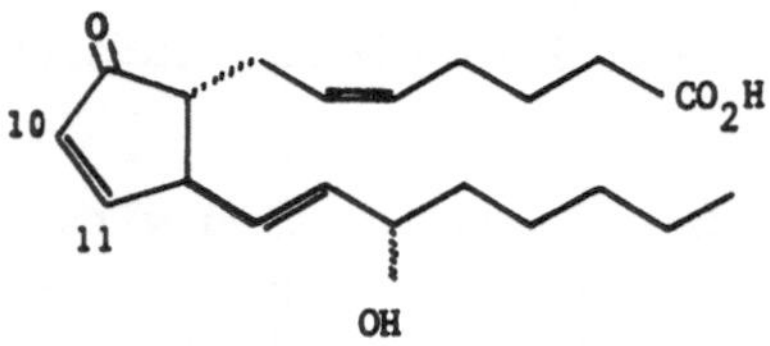

8 : PGA$_2$

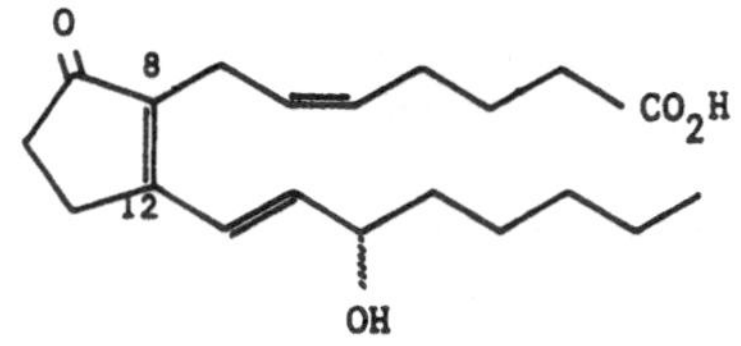

9 : PGB$_2$

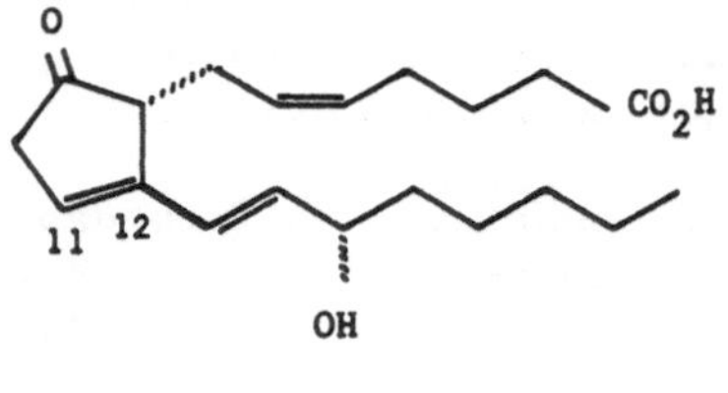

10 : PGC$_2$

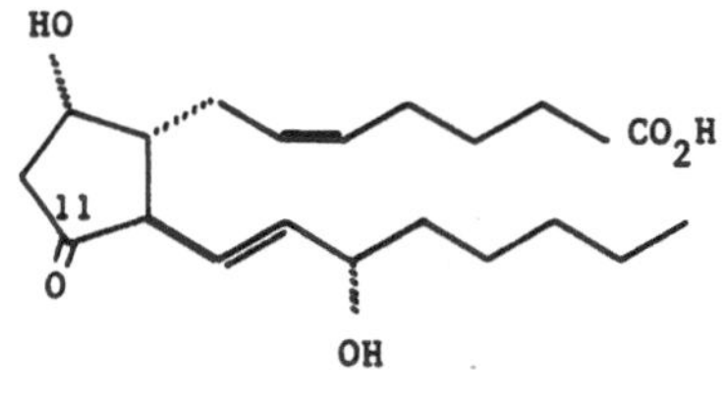

11 : PGD$_2$

first total synthesis of PG was published (8). This was followed
by a number of imaginative, unusual, and elegant approaches.

The motivation given to chemists to achieve this goal has been
extremely beneficial to organic synthesis in general. A number of
novel synthetic procedures has been reported and new reagents have
been designed for the synthesis of PG. Furthermore, many research
groups have elaborated rather sophisticated, sometimes extremely
clever strategies, to accomplish their task. It is not the purpose
of this chapter either to review systematically all the novel
reagents which have been reported or to discuss in detail the new
methodology which has been used.

Rather, the object of this chapter is to review some aspects
of the chemistry which has been reported. Excellent reviews of the
chemical work have already appeared (5,10), so we have opted to
mention only four different syntheses and realize that such a choice
is rather arbitrary. However, these approaches should serve to
illustrate the potential the chemists have now at their disposal
to build sophisticated molecules of biological importance.

B- THE BICYCLIC LACTONE ROUTE

One of the goals of the synthesis by Corey and co-workers (11)
by the γ-lactone approach, was the preparation of an intermediate,
with functional groups of such a nature as to allow the building
of the final fragments to complete the work.

The starting material is cyclopentadiene sodium (17a), which
was alkylated with chloromethyl methyl ether to afford the intermediate (18a) as shown in Chart 4.

This intermediate is immediately submitted to a Diels-Alder
reaction with 2-chloro-acrylonitrile. The bicyclic derivative (19a),
sometimes contaminated with its isomer (19c), was hydrolyzed with
base to provide the β,γ-unsaturated ketone (20). Baeyer-Villiger
oxidation of ketone (20) with m-chloroperbenzoic acid (MCPBA) in
the presence of sodium bicarbonate, or with hydrogen peroxide and
sodium hydroxide (12), gave almost quantitatively the lactone (21).
Base hydrolysis of the lactone group furnished the acid (22), which
by iodolactonization produced the key-intermediate (23), which
presented both the correct stereochemistry as well as the necessary functionality to complete the synthesis.

As indicated in Chart 4, acetylation of the secondary alcohol
group produced the acetate (24a), deiodination with tributyltin
hydride gave (25a), and cleavage of the methyl ether group with
boron tribromide afforded the primary alcohol (26a). Collins

Chart 3

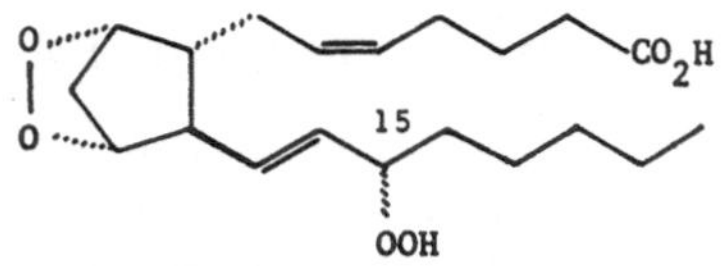

12 : PGG$_2$

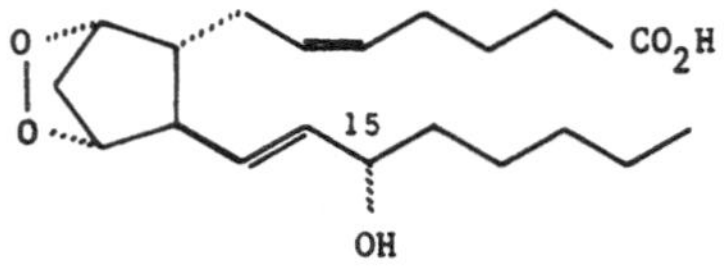

13 : PGH$_2$ = R$_2$ = LASS

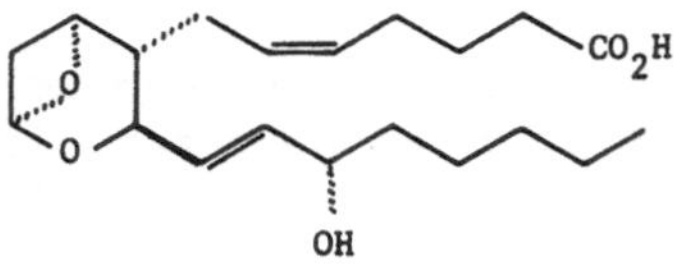

14 : Thromboxane A$_2$

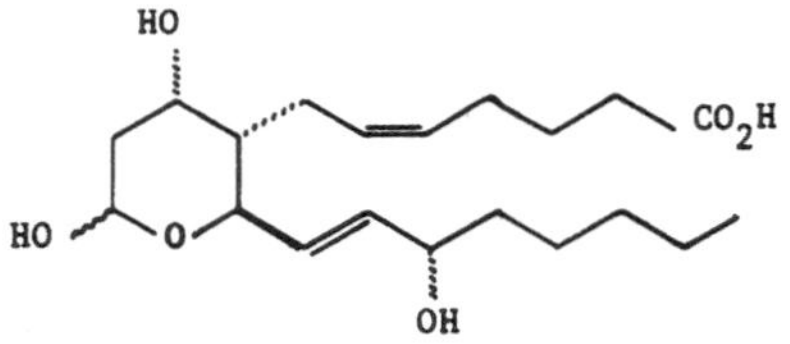

15 : Thromboxane B$_2$ = PHD

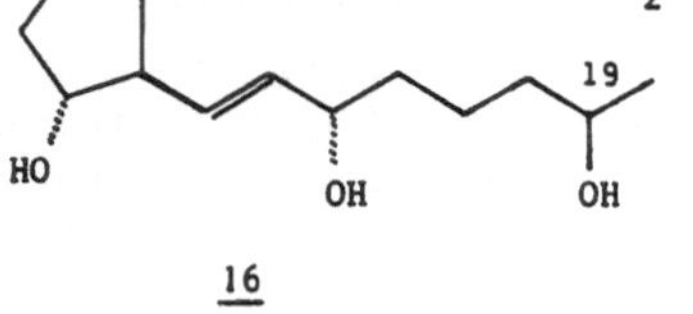

16

oxidation then gave the corresponding aldehyde (27a), suitable for
an Emmons-Horner reaction (13) to build the chain at position 12.

    One of the drawbacks of the original work was the formation
of the side-reaction compound (19c) as the result of a prototropic
rearrangement before or during the Diels-Alder addition. This side
reaction was prevented by using cyclopentadienyl thallium (17b)
(14). Other improvements consisted in the alkylation of the salt
(17b) with chloromethyl benzyl ether to give (18b) and in the ester-
ification of the iodo-lactone (23b) with p-phenylbenzoyl chloride
to provide the intermediate (24b). The alcohol (26b) was then obtained in good yield by hydrogenolysis. The p-phenylbenzoate group
gave crystalline intermediates (36). Another improvement consisted
in using 2-chloro-acrylic chloride (CH_2=CCl-COCl) as the dienophile
in the reaction with the diene (18b), thus yielding the dichloro-
derivative (28) further converted to lactone (21b) by treatment
with sodium azide, followed by aqueous acetic acid (15).

    The PG synthesis was completed by construction of both chains
from the bicyclic intermediate (27b). The side chain at position
12 was formed by an Emmons-Horner reaction (13) of the aldehyde
(27b) with the sodium salt of dimethyl 2-oxo-heptylphosphonate (29),
obtained by reaction of ethyl hexanoate with dimethyl α-lithiome-
thane phosphonate (16). This furnished the enone (30) with the carbonyl at position 15 to give access to the corresponding secondary
hydroxyl group. Treatment of the ketone at C-15 with zinc borohydride or aluminum isopropoxide (17) gave a mixture of 15(R)- and
15(S)- isomers, separated by preparative chromatography. The non-
natural 15(R)-epimer could be recycled by oxidation to the enone
(30) with dichloro-dicyano-benzoquinone (DDQ). Alkaline hydrolysis
of the ester group at position 11 (acetate, p-phenylbenzoate, etc.)
in the synthetic intermediate (31) afforded the corresponding 11,
15-diol (32a). The hydroxyls at C-11 and C-15 were protected as the
bis-tetrahydropyranyl ether (32b) and the lactone group was then
reduced with diisobutyl aluminum hydride (DIBAL) in toluene solution
(18), thus yielding the hemi-acetal (33).

    Reaction between the intermediate (33) and the di-sodium salt
of triphenylphosphoniopentanoic acid (34) in dimethylsulfoxide
(DMSO) provided the bis-tetrahydropyranyl ether derivative of $PGF_{2\alpha}$
(35). Acid hydrolysis of the ether groups in (35) afforded $PGF_{2\alpha}$
(36). Jones oxidation (19) of the 9-hydroxy group of the di-ether
(35) followed by acid treatment yielded PGE_2 (37).

    Several attempts have been made to improve the stereoselectiv-
ity in the reduction of the enone (30) to the 15(S)-isomer (31). A
good result was obtained by reduction of ketone (30a) with 2,9-
thexyllimonylborohydride (TLBH), which provided a 15(S):15(R) ratio
of 4.5 to 1. In the case of the 11α-p-phenylphenylurethane (30b),

Chart 4

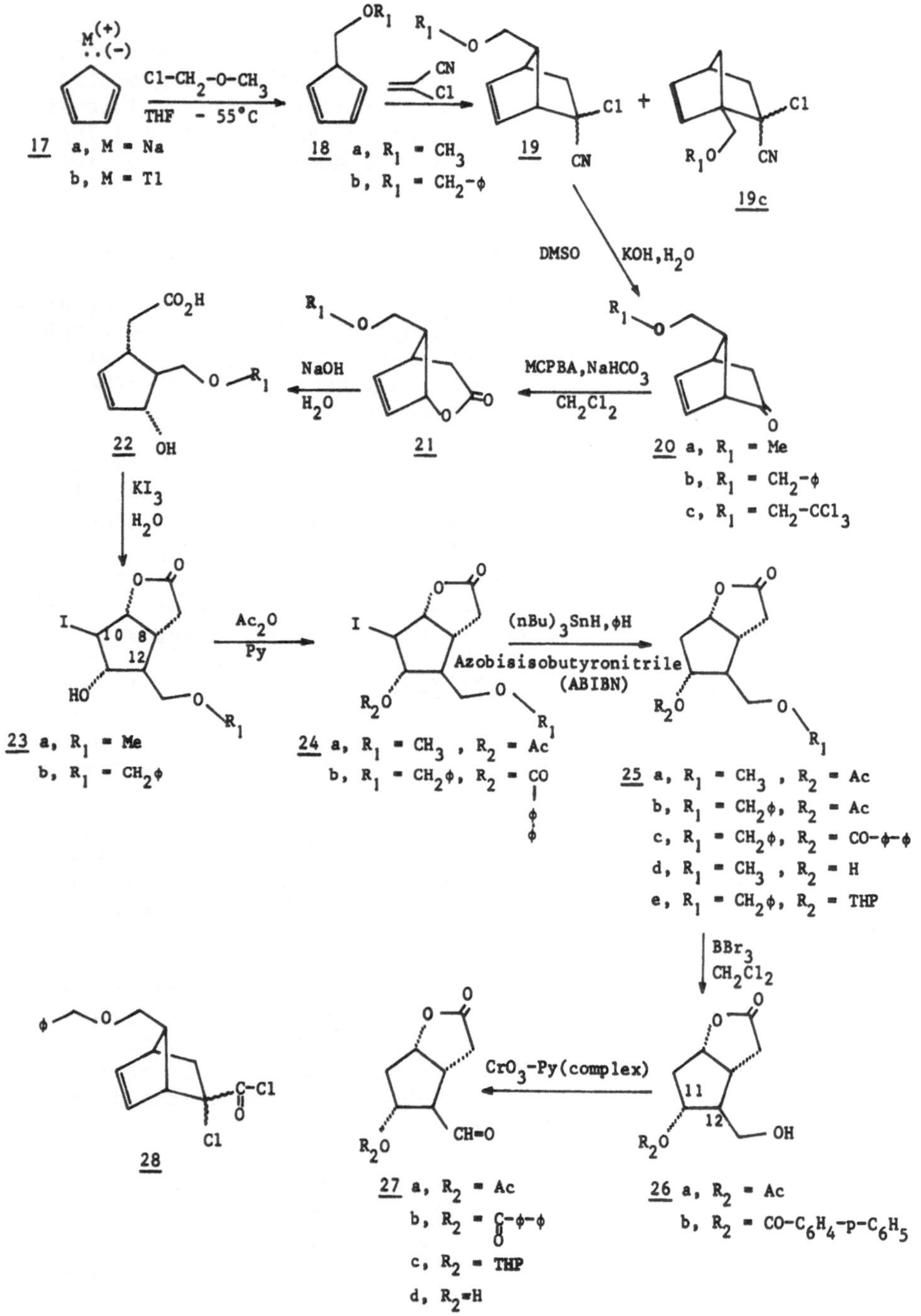

M(+)
..(-)
Cl-CH2-O-CH3
THF - 55°C
17 a, M = Na
b, M = Tl
18 a, R1 = CH3
b, R1 = CH2-ϕ
CN
Cl
19
19c
DMSO
KOH,H2O
20 a, R1 = Me
b, R1 = CH2-ϕ
c, R1 = CH2-CCl3
MCPBA,NaHCO3
CH2Cl2
21
NaOH
H2O
22
KI3
H2O
23 a, R1 = Me
b, R1 = CH2ϕ
Ac2O
Py
24 a, R1 = CH3 , R2 = Ac
b, R1 = CH2ϕ, R2 = CO-ϕ-ϕ
(nBu)3SnH,ϕH
Azobisisobutyronitrile
(ABIBN)
25 a, R1 = CH3 , R2 = Ac
b, R1 = CH2ϕ, R2 = Ac
c, R1 = CH2ϕ, R2 = CO-ϕ-ϕ
d, R1 = CH3 , R2 = H
e, R1 = CH2ϕ, R2 = THP
BBr3
CH2Cl2
26 a, R2 = Ac
b, R2 = CO-C6H4-p-C6H5
CrO3-Py(complex)
27 a, R2 = Ac
b, R2 = C(O)-ϕ-ϕ
c, R2 = THP
d, R2=H
28

in which the attack of the hydride on the carbonyl at C-15 occurs
from the side opposite to that occupied by the 8,9-lactone and 11-
substituent groups in the appropriate conformation, the (S):(R)
ratio was 92 to 8. The p-phenylphenylurethane group also provides
compounds that are crystalline and absorb in ultraviolet (UV) ;
moreover this group is also easily removed.

Another potential way to increase the yield of the desired
15(S)-epimer is to use an efficient conversion of the undesired
15(R)-PG intermediate to its 15(S)-isomer (31c) by nycleophilic
displacement. Thus, reaction of the 11-acetoxy 15(R)-alcohol with
methanesulfonyl chloride in the presence of triethylamine produced
the corresponding mesylate, which was isolated and treated with a
solution of potassium superoxide (KO$_2$) and 18-crown-6 in dimethyl-
sulfoxide-dimethylformamide-dimethoxyethane (DMSO-DMF-DME) solution.
The desired 15(S)-dihydroxy-lactone (31c) was obtained (20).

PGF$_{2\alpha}$ (36) and E$_2$ (37) have been prepared in the optically
active form by resolution of the hydroxy-acids (22) with (+)-ephe-
drine or (+)-amphetamine (11,12,21a). An improved method for the
preparation of the key intermediate (23b) (Chart 4) in optically
pure form without resolution has appeared (21b). The process util-
izes a Diels-Alder reaction between 5-benzyloxymethylcyclopentadie-
ne (18b) and an optically pure acrylate, prepared from (-)-pulegone.

C- THE BICYCLO[3.1.0] HEXANE ROUTE

The ring strain in the bicyclo [3.1.0] hexane system makes it
very reactive affording rearranged products under appropriate conditions.

Since PG may be considered as products resulting from such a
rearrangement occurring at C-11 and C-12, this solvolysis has been
used by Just et al. (8,22) for the synthesis of the lower part of
the PG molecule.

Chart 6 mentions the total synthesis of PGF$_{1\alpha}$ (47) (22,23).
The bicyclo [3.1.0] hexane system was obtained from 3-cyclopentenol
(38) by copper-catalyzed addition of carbethoxy-carbene to the
THP-derivative. After base equilibration to give the exo-isomer
(39), lithium aluminum hydride reduction provided the alcohol (40).
Oxidation to the aldehyde was followed by a Wittig reaction which
furnished a mixture of olefins (41). Acid hydrolysis of the THP
protecting group, followed by Jones oxidation gave the ketone (42).
Treatment of this intermediate with methyl 7-iodoheptanoate in the
presence of potassium t-butoxide afforded the alkylated cyclopenta-
nones (43) and (44), separated by column chromatography. Reduction
of the carbonyl group provided the mixture of alcohols (45) and

Chart 5

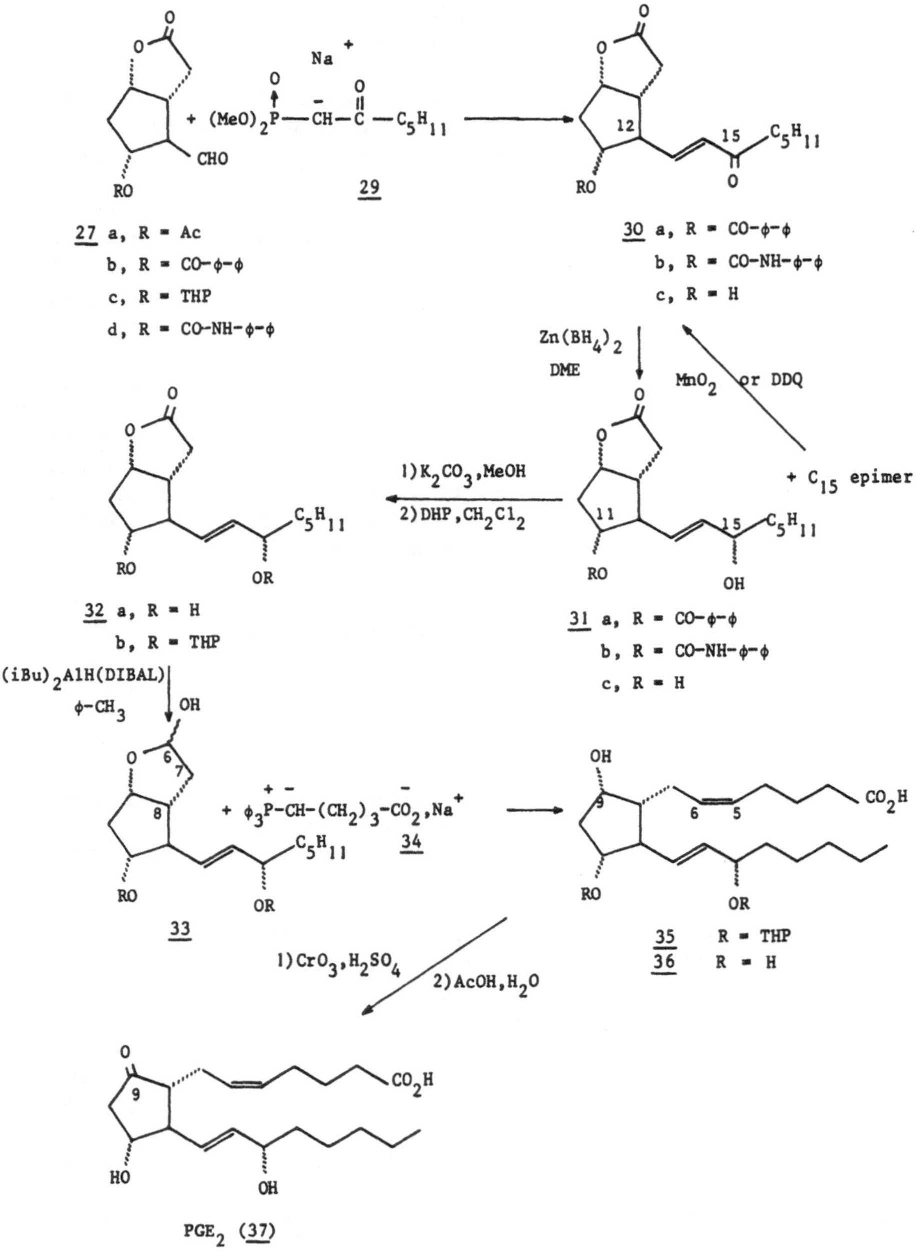
Na +
(MeO)2P—CH—C—C5H11
29
27 a, R = Ac
b, R = CO-ϕ-ϕ
c, R = THP
d, R = CO-NH-ϕ-ϕ
CHO
RO
12
15
C5H11
30 a, R = CO-ϕ-ϕ
b, R = CO-NH-ϕ-ϕ
c, R = H
Zn(BH4)2
DME
MnO2 or DDQ
+ C15 epimer
1) K2CO3, MeOH
2) DHP, CH2Cl2
11
OH
31 a, R = CO-ϕ-ϕ
b, R = CO-NH-ϕ-ϕ
c, R = H
OR
32 a, R = H
b, R = THP
(iBu)2AlH(DIBAL)
ϕ-CH3
6
7
8
+ ϕ3P-CH-(CH2)3-CO2,Na+
34
33
9
5
CO2H
35 R = THP
36 R = H
1) CrO3, H2SO4
2) AcOH, H2O
HO
PGE2 (37)

(46). After separation and epoxidation, the corresponding epoxides were solvolyzed. Alkaline hydrolysis of the resulting formates and chromatography then furnished PGF$_{1\alpha}$ (47), F$_{1\beta}$ (48), and isomers, as their methyl esters (22,23).

Numerous modifications have been published, so that the bicyclo[3.1.0]hexane route has been substantially improved recently (24,25) and is now much more practical.

D- THE 1,4-ADDITION ROUTE

Since PG of the E family can be converted more easily to those belonging to other families than vice versa, PGE have been the goal of most synthetic approaches (26).

Chart 7 outlines the synthetic route followed by Sih and coworkers (27,28). The acid chloride (51) was transformed to the methyl-ketone (52) by treatment with the ethoxy-magnesium salt of diethyl malonate, followed by hydrolysis and decarboxylation. The trione (53) was then prepared by a known procedure (29). The selective asymmetric reduction of trione (53) was achieved both microbiologically and catalytically (27,28). Treatment of the triketone (53) with Dipodascus uninucleatus gave stereospecifically the 11-(R)-alcohol (54) in high yield. Catalytic hydrogenation of compound (53) in the presence of (1,5-cyclooctadiene)bis(0-anisylcyclohexylmethylphosphine) rhodium (I) tetrafluoroborate yielded the same alcohol (54). The conversion of the 1,3-dione system (54) to the enone (57) could be achieved without racemization of the alcohol group both by a known sequence (30) as well as by a modification. The basic method consisted of the formation of the enol ether (55a) of the 1,3-dicarbonyl system, followed by reduction with an appropriate hydride and acid hydrolysis. In the modification, the enol benzoate (55b) was used instead of the enol ether (55a).

The selectivity observed in the transformation of the diketone (54) to its enol (55) is remarkable. The preferred formation of enol (55) instead of its isomer (56) seems to be attributable to steric factors. This selectivity was improved by increasing the size of the group R in the enol (55), as evicenced with the mesityl sulfonyl ester (55c) (28). The synthesis was completed by a 1,4-addition of a lithium dialkylcopper reagent.

A noteworthy feature of this synthesis is the combination of chemical and microbiological methods.

A similar synthesis has been reported (31), in which in place of a microbiological reduction, a resolution of the hydroxy-cyclopentenone (57) was achieved via the (R)-2-aminoxy 4-methyl valeric acid derivative (59). Regeneration of the carbonyl group was

Chart 6

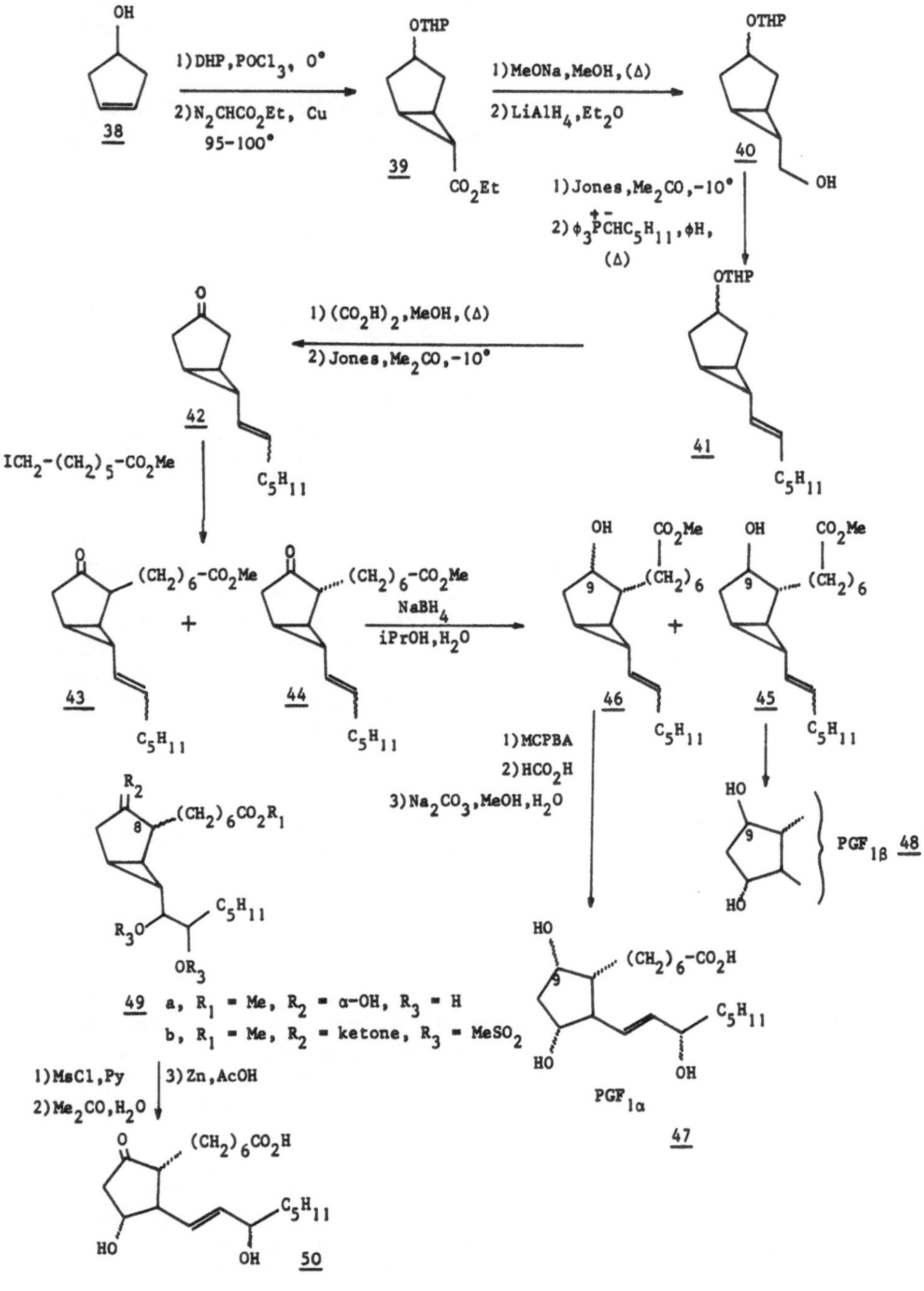
OH
38
1)DHP,POCl3, 0°
2)N2CHCO2Et, Cu
95-100°
OTHP
39
CO2Et
1)MeONa,MeOH,(Δ)
2)LiAlH4,Et2O
OTHP
40
OH
1)Jones,Me2CO,-10°
2)φ3PCHC5H11,φH,
(Δ)
OTHP
41
C5H11
1)(CO2H)2,MeOH,(Δ)
2)Jones,Me2CO,-10°
O
42
C5H11
ICH2-(CH2)5-CO2Me
(CH2)6-CO2Me
43
C5H11
44
C5H11
NaBH4
iPrOH,H2O
OH
CO2Me
(CH2)6
9
46
C5H11
45
1)MCPBA
2)HCO2H
3)Na2CO3,MeOH,H2O
HO
PGF1β 48
R2
(CH2)6CO2R1
8
C5H11
R3O
OR3
49 a, R1 = Me, R2 = α-OH, R3 = H
b, R1 = Me, R2 = ketone, R3 = MeSO2
1)MsCl,Py
2)Me2CO,H2O
3)Zn,AcOH
(CH2)6-CO2H
PGF1α
47
(CH2)6CO2H
HO
OH
50

effected by reduction of the oximino-derivative with titanium tri-
chloride.

E- THE TROPOLONE ROUTE

The last synthetic approach which will be discussed here not
only leads potentially to natural PG, but also is a useful entry
to modified entities not readily available by other total synthetic
schemes (32).

Photochemical irradiation of tropolone methyl ether (60) in
methanol solution afforded 7-methoxy-3,6-bicyclo [3.2.0] heptadiene-
2-one (61) in high yield. The double bonds of the intermediate (61)
display a selective chemical reactivity. Indeed, on the one hand,
ozonolysis of compound (61) in pyridine-methanol-methylene chloride
solution takes place selectively at the enol ether bond to provide
the ozonide, which was cleaved with distilled sulfur dioxide to
produce the dimethyl acetal (62). This substituted cyclopentenone
can be transformed to natural PG by the methodology known in PG
chemistry (26). On the other hand, as shown in Chart 8, Michael
addition of cyanide ion to the enone (61) afforded the 11-cyano
derivative (65), which can also be converted to natural PG, such
as PGE$_1$ (64), by known procedures.

In addition, catalytic hydrogenation of the substituted cyclopentenone (61) takes place selectively at the conjugated double
bond, yielding the cyclopentanone (66). Ozonolysis of the enol ether
bond followed by treatment of the ozonide with liquid sulfur dioxide
in methanol solution provided the dimethyl acetal (67). Alkylation
of this keto-ester at position 8 was performed by reaction with
potassium hydride in DMSO, followed by treatment with ethyl 7-iodo-
heptanoate, producing the diester (68). Acid cleavage of the dimethyl acetal gave the corresponding aldehyde (69), which was submitted to the classical Emmons-Horner reaction (13), thus affording
the enedione (70). Reduction of the 15-keto group in (70) was performed with L-Selectride, thus providing regioselectively a mixture
of (15R) and (15S)-alcohol isomers of which the latter (71) was separated by column chromatography. Decarboxylation of the β-keto-
ester group and alkaline hydrolysis of the ester at position 1 provided 11-desoxy PGE$_1$ (72).

11-Desoxy PG are of interest not only because of their intrinsic biological properties, such as antagonists of the PG belonging
to the E and F-series, but also since they can be transformed
chemically (33) to PGA such as (63), and hence to 11-hydroxylated
entities, i.e. (64).

Chart 7

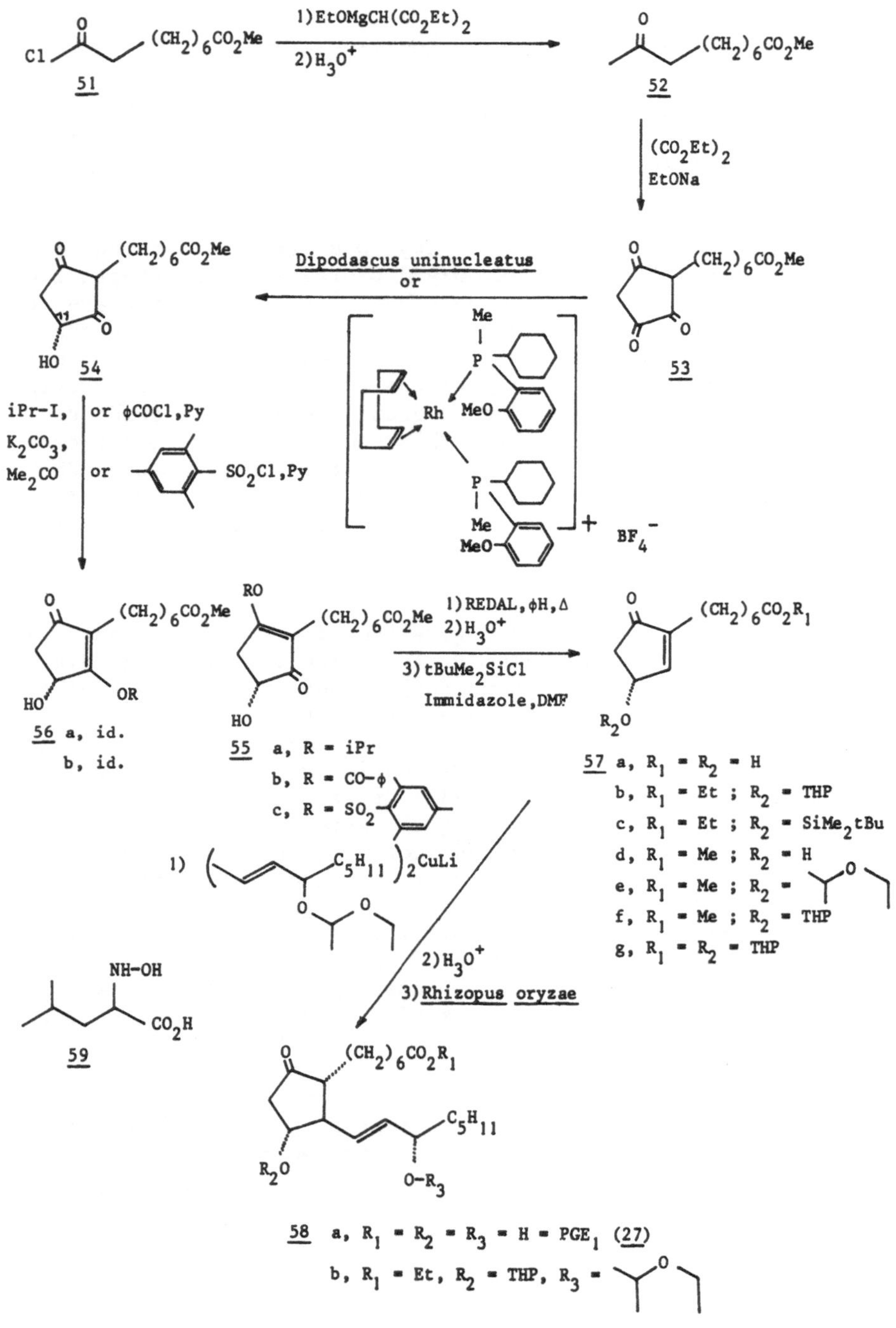

51 → (1) EtOMgCH$(CO_2Et)_2$, 2) H_3O^+ → 52

$(CO_2Et)_2$ EtONa

Dipodascus uninucleatus or

54 53

iPr-I, K_2CO_3, Me_2CO or φCOCl,Py or ── SO_2Cl,Py

$+ BF_4^-$

56 a, id. b, id.

55 a, R = iPr
b, R = CO-φ
c, R = SO_2

1) REDAL,φH,Δ
2) H_3O^+
3) tBuMe$_2$SiCl Immidazole,DMF

57 a, $R_1 = R_2 = H$
b, $R_1 = Et$; $R_2 = THP$
c, $R_1 = Et$; $R_2 = SiMe_2tBu$
d, $R_1 = Me$; $R_2 = H$
e, $R_1 = Me$; $R_2 = $
f, $R_1 = Me$; $R_2 = THP$
g, $R_1 = R_2 = THP$

1) $\left(\quad C_5H_{11} \right)_2 CuLi$

2) H_3O^+
3) Rhizopus oryzae

59

58 a, $R_1 = R_2 = R_3 = H = PGE_1$ (27)
b, $R_1 = Et, R_2 = THP, R_3 = $

Chart 8

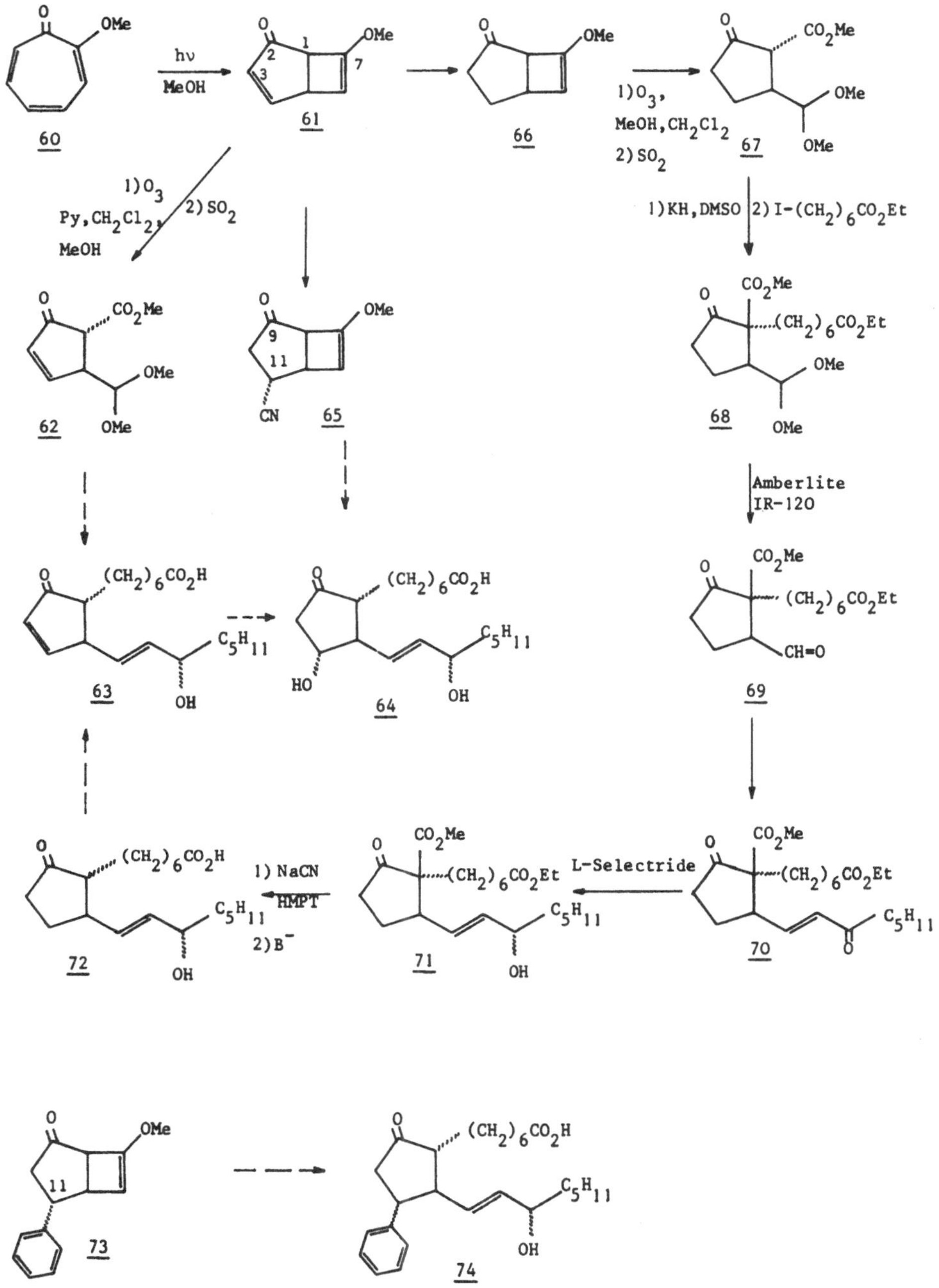

60
hν
MeOH
61
66
1) O_3, MeOH, CH_2Cl_2
2) SO_2
67
1) O_3
Py, CH_2Cl_2, MeOH
2) SO_2
1) KH, DMSO 2) I-$(CH_2)_6CO_2Et$
62
65
68
Amberlite IR-120
63
64
69
72
1) NaCN HMPT
2) B^-
71
L-Selectride
70
73
74

A number of alkyl and aryl groups have been added to enone (61). In particular, conjugate addition of lithium diphenylcopper to the cyclopentenone (61) proceeded regioselectively to afford mainly the 11-substituted cyclopentanone (73). Further transformations, as described above, afforded 11-phenyl 11-desoxy PGE_1 (74).

Of course, other synthetic routes to natural and modified PG have been reported (26) and many changes have been introduced in the PG molecule. Indeed, some modified PG are either more active or have a better selectivity than their natural occuring counterparts (26).

Finally, it should also be mentioned that PGA_2 (8) from the marine corals (9) has been converted into natural as well as modified PG. These chemical transformations are well documented (5,26).

The aforementioned synthetic pathways should serve to illustrate some approaches which have been followed for the preparation of PG. In view of their intriguing biological properties, no doubt that numerous other novel and imaginative routes will appear in the future.

REFERENCES

1. U.S. von Euler, and R. Eliasson, Prostaglandins, Medicinal Chemistry. A series of Monographs. Vol.8, Academic Press, New York (1967).
2. S. Bergström, and B. Samuelsson, (Eds.), Prostaglandins, Nobel Symposium 2, Interscience, New York (1967).
3. S. Bergström, Science, 157, 382 (1967) ; S. Bergström, and B. Samuelsson, Endeavour, 27, 109 (1968).
4. B. Samuelsson, and R. Paoletti, (Edit.), Advances in Prostaglandin and Thromboxane Research, Vol. I, II, Raven Press, New York (1976).
5. W.P. Schneider, in Prostaglandins : Chemical and Biochemical Aspects, S.M.M. Karim (Edit.), MTP, Edinburgh (1976).
6. D.A. van Dorp, R.K. Beerthuis, D.H. Nugteren, and H. Vonkeman, Nature, 203, 839 (1964) ; id., Biochim. Biophys. Acta, 90, 204 (1964).
7. S. Bergström, H. Danielsson, and B. Samuelsson, Biochim. Biophys. Acta, 90, 207 (1964) ; S. Bergström, H. Danielsson, D. Klenberg, and B. Samuelsson, J. Biol. Chem., 239, PC 4006 (1964).
8. G. Just, and Ch. Simonovitch, Tetrahedron Letters, 2093 (1967).
9. A.J. Weinheimer, and R.L. Spraggins, Tetrahedron Letters, 5183 (1969).
10. Inter alia : a) U.F. Axen, Ann. Rep. Med. Chem., 290 (1967) ; b) J.F. Bagli, Ann. Rep. Med. Chem., 170 (1969) ; c) G.L. Bundy, Ann. Rep. Med. Chem., 137 (1970) ; d) M.P.L. Caton, in

Progress in Medicinal Chemistry, G.P. Ellis, and G.B. West, (Edit.), Vol.8, Butterworths, London (1971) ; e) U.F. Axen, J.E. Pike, and W.P. Schneider, in _Progress in Total Synthesis of Natural Products_, J.W. Apsimon, (Edit.), p.81, J. Wiley, New York (1973) ; f) R. Clarkson, in _Progress in Organic Chemistry_, W. Carruthers, and J.K. Sutherland, (Edit.), Vol.8, Butterworths, London (1973) ; g) P.H. Bentley, _Chem. Soc. Rev._, 2, 29 (1973).

11. E.J. Corey, N.M. Weinshenker, T.K. Schaaf, and W. Huber, _J. Amer. Chem. Soc._, 91, 5675 (1969).

12. N.M. Weinshenker, and R. Stephenson, _J. Org. Chem._, 37, 3741 (1972).

13. a) W.S. Wadsworth, and W.D. Emmons, _J. Amer. Chem. Soc._, 83, 1733 (1961) ; b) L. Horner, H. Hoffmann, W. Klink, H. Ertel, and V.G. Toscano, _Chem. Ber._, 95, 581 (1962) and references therein.

14. a) E.J. Corey, U. Koelliker, and J. Neuffer, _J. Amer. Chem. Soc._, 93, 1489 (1971) ; b) E.J. Corey, S.M. Albonico, U. Koelliker, T.K. Schaaf, and R.K. Varma, _J. Amer. Chem. Soc._, 93, 1491 (1971) ; c) See also : N.M. Weinshenker, _Prostaglandins_, 3, 219 (1973).

15. a) E.J. Corey, T. Ravindranathan, and S. Tirashima, _J. Amer. Chem. Soc._, 93, 4326 (1971) ; b) E.J. Corey, and T. Ravindranathan, _J. Amer. Chem. Soc._, 94, 4013 (1972) ; c) E.J. Corey, and P.L. Fuchs, _J. Amer. Chem. Soc._, 94, 4014 (1972).

16. a) E.J. Corey, and G.T. Kwiatkowski, _J. Amer. Chem. Soc._, 88, 5654 (1966) ; b) E.J. Corey, and E. Hamanaka, _J. Amer. Chem. Soc._, 89, 2758 (1967).

17. J. Bowlery, K.B. Mallion, and R.A. Raphael, _Synth. Comm._, 4, 211 (1974).

18. L.I. Zakharkin, and I.M. Khorlina, _Tetrahedron Letters_, 619 (1962).

19. K. Bowden, I.M. Heilbron, E.R.H. Jones, and B.C.L. Weedon, _J. Chem. Soc._, 39 (1946).

20. E.J. Corey, K.C. Nicolaou, M. Shibasaki, Y. Machida, and C.S. Shiner, _Tetrahedron Letters_, 3183 (1975).

21. a) E.J. Corey, T.K. Schaaf, W. Huber, U. Koelliker, and N.M. Weinshenker, _J. Amer. Chem. Soc._, 92, 397 (1970) ; b) E.J. Corey, and H.E. Ensley, _J. Amer. Chem. Soc._, 97, 6908 (1975) ; c) see also : E.J. Corey, H.E. Ensley, and J. Suggs, _J. Org. Chem._, 41, 380 (1976) ; d) K.G. Paul, F. Johnson, and D. Favara, _J. Amer. Chem. Soc._, 98, 1285 (1976).

22. G. Just, and C. Simonovitch, _Canad. J. Chem._, 19, 41 (1967).

23. a) W.P. Schneider, U. Axen, F.H. Lincoln, J.E. Pike, and J.L. Thompson, _J. Amer. Chem. Soc._, 90, 5895 (1968) ; b) G. Just, C. Simonovitch, F.H. Lincoln, W.P. Schneider, U. Axen, G.B. Spero, and J.E. Pike, _J. Amer. Chem. Soc._, 91, 5364 (1969) ; c) W.P. Schneider, U. Axen, F.H. Lincoln, J.E. Pike, and J.L. Thompson, _J. Amer. Chem. Soc._, 91, 5372 (1969).

24. a) R.C. Kelly, V. van Rheenen, I. Schletter, and M.D. Pillai,
     J. Amer. Chem. Soc., 95, 2746 (1973) ; b) R.C. Kelly, and V.
     van Rheenen, Tetrahedron Letters, 1709 (1973).
25. a) D.R. White, Tetrahedron Letters, 1753 (1976) ; b) V. van
     Rheenen, R.C. Kelly, and D.Y. Cha, Tetrahedron Letters, 1973
     (1976), and references therein.
26. P. Crabbé (Edit.), Prostaglandin Research, Academic Press, in
     print.
27. C.J. Sih, J.B. Heather, G.P. Peruzzotti, P. Price, R. Sood,
     and L.F. Hsu Lee, J. Amer. Chem. Soc., 95, 1676 (1973) ; see
     also H.W. Whitlock, J. Amer. Chem. Soc., 98, 3225 (1976).
28. a) C.J. Sih, R.G. Salomon, P. Price, R. Sood, and G. Peruz-
     zotti, J. Amer. Chem. Soc., 97, 857 (1975) ; b) C.J. Sih, J.
     B. Heather, R. Sood, P. Price, G. Peruzzotti, L.F. Hsu Lee,
     and S.S. Lee, J. Amer. Chem. Soc., 97, 865 (1975).
29. J. Katsube, and M. Matsui, Agr. Biol. Chem., 33, 1078 (1969).
30. W.F. Gannon, and H.O. House, "Organic Syntheses", Vol. Coll.
     V, 294, 539.
31. a) R. Pappo, P. Collins, and C. Jung, Ann. N.Y. Acad. Sci.,
     180, 64 (1971) ; b) R. Pappo, and P.W. Collins, Tetrahedron
     Letters, 2627 (1972) ; c) R. Pappo, P. Collins, and C. Jung,
     Tetrahedron Letters, 943 (1973).
32. a) A. Greene, and P. Crabbé, Tetrahedron Letters, 2215 (1975) ;
     b) P. Crabbé, A. Cruz, J.P. Deprès, M.C. Meana, and A. Greene,
     Heterocycles, in press ; c) P. Crabbé, E. Barreiro, H.S. Choi,
     A. Cruz, J.P. Deprès, G. Gagnaire, A.E. Greene, M.C. Meana,
     and L. Williams, Bull. Soc. Chim. France, in press.
33. G. Stork, and S. Raucher, J. Amer. Chem. Soc., 98, 1583 (1976).

PROSTAGLANDIN BIOASSAY

S.H. FERREIRA

Department of Pharmacology, Faculty of Medicine

of Ribeirão Preto, S. Paulo, BRAZIL

> "New chemical methods have been
> described in recent years which are
> a great improvement on older methods
> and there is little doubt that as
> time goes on biological methods of
> assay will be less used, but they
> are still important and chemical
> methods will only inspire universal
> confidence if they are shown to
> give the same results as the
> biological methods".
> Sir JOHN GADDUM, 1959.

Starting from the very discovery of prostaglandins
itself bioassay has provided crucial information on the
role of lungs in the removal of circulating prostaglandins
(FERREIRA & VANE, 1967), on the participation of
prostaglandins in inflammatory reaction (WILLIS, 1969 a,
b), on the contribution of prostaglandins in the autoregulation and maintenance of resting blood flow to the
kidney (HERBACZYNSKA-CEDRO & VANE, 1973; LONIGRO et al.,
1973), on the inhibitory effect of aspirin-like drugs
upon the synthesis of prostaglandins (VANE, 1971; SMITH
& WILLIS, 1971; FERREIRA et al., 1971), on the mediation
of pyrogen fever by prostaglandins (FELDBERG et al.,
1973), and on the release of rabbit aorta contracting
substance, RCS, from lungs during anaphylaxis (PIPER &
VANE, 1969), etc.
GADDUM'S statement, which was made about
catecholamines (see VANE, 1966), also applies to the

prostaglandins field. Today, a series of new and more
sophisticated chemical methods (gas chromatography,
radio-immunoassay, mass spectrometry, mass fragmentometry,
competitive binding, etc.) are being developed and
perfected. In this mini-review I shall analyse some
experiments based on bioassay in order to illustrate its
applicability and limitations as well as to discuss its
usefulness in current prostaglandin research.

ASSAY TISSUES

Although the definition of a "prostaglandin unit"
was based on the lowering of rabbit's blood pressure
(Von EULER, 1939) its isolation and identification was
based upon isolated rabbit duodenum (SAMUELSSON, 1963).
As stimulants of smooth muscle preparations, prostaglandins A_1 and A_2 are much less potent than prostaglandins E_1 and E_2. The rat fundus is the most sensitive
but 100-200 ng of A_2 and 1-2 μg A_1 are necessary.
However, by using cat blood pressure as an assay (spinal
or anesthetized) it is possible to detect 5-10 ng of A_2
and 25-50 ng A_1 (HORTON & JONES, 1969).

Since prostaglandins are capable of contracting
almost any segment of the intestinal tract, pieces of
gut from many species were used for their detection and
quantification (Table I).

Today, the most commonly used assay tissue is a strip
from rat stomach probably because it is easily available,
it possesses high sensitivity and is quite reliable,
being little influenced by seasonal variations, hormonal
state and feeding. The rat stomach strip was ranked
second when compared with gerbil colon, guinea pig ileum,
and rat duodenum (WEEKS et al., 1968). These preparations
responded to threshold concentrations of the order of
1-10 ng; gerbil colon, however, showed a steeper dose-
-response curve and a more rapid recovery. The hamster
stomach fundus has been also described as very sensitive
to prostaglandins and having a recovery time smaller
than the rat stomach strip (UBATUBA, 1973). Preparations
with rapid recovery are, in fact, very valuable when a
great number of samples have to be quantitated.

In my opinion, the worst drawback of the bioassay
is the period necessary for stabilization of the tissues
before an assay can be made. This time varies from
preparation to preparation, depending on the presence
of antagonists in the bathing fluid and on the regularity
of agonist addition. On a practical level this is
reflected by the stability of the baseline. I don't
know of any systematic study having been made on this
factor. I suspect that the lack of such studies, as well

as of those comparing the performance of several
isolated preparations in different bathing solutions
or with different types of transducers, makes the choice
of the assay tissue a very personal matter. It is quite
possible that a laboratory which uses a force transducer
with "isotonic springs" (Grass) will obtain different
dose-response curves from those using a rotary motion
transducer (Havard) fitted with an auxotonic lever
(PATON, 1957). It is well known that the stability and
sensitivity of the preparation depend upon the bathing
solution; however, the type of transducer is important
for the degree and maintenance of sensitivity in an
isolated preparation. The rat stomach strip, for example, does not perform well with isometric transducers.
The best advice to a neophyte is to reproduce as closely
as possible the conditions described by those authors
claiming the best results. I found it rather interesting
that sometimes, after many unsuccessful attempts, a visit
to the mother laboratory and the learning of a few tricks
resulted in a much improved assay. Possibly the best
assay tissue is the one the researcher knows best how
to handle.

ORGAN BATH AND SUPERFUSION TECHNIQUE

MAGNUS, 1903, introduced the idea of suspending an
isolated fragment of smooth muscle in a chamber containing
a nutrient fluid. The organ baths used today are slightly
modified versions of that used by DALE, 1912 (see
Pharmacological Experiments on Isolated Preparations,
1970).

The basic objectives guiding the evolution of the
assay methods were to increase its sensitivity and
specificity. This has been done either: a) by keeping
to a minimum the amount of fluid bathing the tissues;
b) by using tissues sensitive only to the substance under
study; c) if no such tissue exists, by using a
combination of several tissues which present a
characteristic pattern of response to the substance; d)
by using antagonists to other substances which might
interfere in the assay; e) if there are no methods for
stabilizing the substance, the assay should be done as
soon as the perfusate or exudate is obtained.

GADDUM (1953) applied the experimental design
developed by FINKLEMAN (1930) to the assay of minute
amounts of biologically active substances.

His technique was denominated "superfusion" (in
contrast to "infusion") and consisted of bathing an
assay tissue with a stream of fluid which was stopped
at the moment of addition of the test substance. VANE
(1969) introduced the idea of superfusing several

tissues in cascade (generally up to 4). This arrangement
permits analysis by several different assay organs of
the material present in a fluid stream taken from the
outflow of a perfused organ or of blood continuously
sampled from an animal (blood bathed organ technique).

Furthering the concept of keeping to a minimum the
amount of fluid bathing an isolated assay tissue, we
have superfused assay tissues immersed in mineral oil.
In VANE'S technique, tissues are superfused at a rate
of 10-5 ml/min. At a slower rate than that, it is
difficult to keep the preparation moist and at a
constant temperature, the result usually being either
loss of sensitivity or an unstable baseline. In our
technique, the tissues are superfused at very low rates
(0.2 ml/min). In the classical superfusion technique,
the rat stomach strip gives a graded response to
injections of 1-10 ng. In our laminar flow technique,
quantification is made in a range of 100-500 pg
(FERREIRA & SOUZA COSTA, 1976). This is the range
generally attained by chemical methods such as radioimmunoassay. However, it is too soon to make a fair
judgment of the usefulness of this method. Furthermore,
since it is not of a cascade type, although the amount
of material needed is very small, several samples have
to be used for a parallel assay.

IMMEDIATE DIRECT BIOASSAY

In this assay, the sample is tested without any
chemical treatment. This type of bioassay permits only
to define a "prostaglandin-like substance". This is done
by means of a combination of three or more assay tissues
treated with a mixture of specific antagonists to block
unwanted substances (FERREIRA & VANE, 1967; GILMORE et
al., 1968).

Pharmacological parallel assay was taken by GADDUM
(1959) and VANE (1969) as strong evidence for the
identity of an endogenous substance. Table II shows
that a concentration of prostaglandin such as is likely
to be found in inflammatory exudates or in blood
during a physiological event (1-20 ng) contracts the
rat stomach strip, rat colon and chick rectum; this
response pattern is peculiar to prostaglandins E_2, E_1,
$F_{1\alpha}$ and $F_{2\alpha}$. It is difficult to analyse the
composition of a mixture of prostaglandins by bioassay,
but, because of the greater sensitivity of the rat colon
to PGF's one can define the predominant type of
prostaglandin in a sample (FERREIRA et al., 1973).

As pointed out above, the specificity of the bioassay
can be further increased by the use of antagonists. The

interference of 5-HT, which contracts the rat stomach
strip and rat colon, can be avoided by the use of a
specific antagonist such as methylsergide maleate (2 x
10^{-7} g/ml). The interference of histamine, adrenaline,
noradrenaline and acetylcholine can also be avoided by
a mixture of antihistaminics (mepyramine maleate, 10^{-7}
g/ml), adrenergic α and β blocking agents (phenoxybenzamine
hydrochloride, 10^{-7} g/ml and propranolol hydrochloride,
12 x 10^{-6} g/ml) and by atropine-like agents (hyoscine
hydrobromide, 10^{-7} g/ml) (for details, see GILMORE et
al., 1968). Non-steroidal anti-inflammatory agents are
generally added to the bathing solution in order to
inhibit intramural generation of prostaglandins from
arachidonic acid which might also be present in the
sample (indomethacin, 1-2 µg/ml).

The prostaglandin antagonists so far available are
not capable of inhibiting all the pharmacological
effects of prostaglandins, but they are quite effective
inhibitors of their spasmogenic effects. In some
instances, they may be helpful in the identification
of prostaglandin-like material. From the three groups
of inhibitors, the dibenzoxazine derivatives (especially
SC-19220) seem to be more reliable inhibitors than the
7-oxa-prostaglandin analogs or phosphorylated polymers
of phloretin (see SANNER, 1974, for discussion).

There are three experimental designs in which
immediate direct assay has proved to be of value: a)
when using pure synthetic prostaglandins and measuring
their disappearance during passage through a vascular
bed; b) for correlating a physiological event and
appearance of prostaglandin-like substances in the
tissue outflow; c) to detect substances whose half-life
is very short and for which there is no available
stabilizing method.

By continuously monitoring the arterial concentration
of prostaglandin with the blood bathed organ technique,
FERREIRA & VANE, 1967, demonstrated that more than 95%
of prostaglandin concentration disappeared during a single
passage through cat lung.

They compared the responses of the assay tissues to
infusion of prostaglandin into the right atrium and in
the ascending aorta. Inactivation of prostaglandins by
blood was negligible since no loss of activity was
observed after incubation up to 3 min. Our results were
confirmed by McGIFF et al., 1969, who further showed
that PGA_1 and PGA_2 survived the passage through the lungs.
PIPER et al., 1970, using isolated guinea pig lungs and
measuring prostaglandin activity directly on the effluent,
confirmed lung ability to inactivate prostaglandins.

The contribution of prostaglandins to several

physiological functions has been established by measuring the variable under study and the concomitant release of prostaglandin-like material directly assayed in the organ outflow. This is illustrated by the demonstration of the importance of the release of prostaglandins for the nociceptive effect of bradykinin injected into the splenic artery (FERREIRA et al., 1973).

COLLIER (1972) has developed a system in which, instead of bathing the assay tissue directly with blood, the blood is dialyzed against Krebs solution which is then used to superfuse the assay tissues. Using this device, it is possible to detect PGE_2 concentrations of the order of 15-10 ng/ml in the circulating blood. One of the advantages of this technique is that the treatment of isolated tissues with antagonists does not reach the blood donor.

The immediate direct assay contributed markedly to the knowledge of biological activity of unstable intermediates of the oxidation of arachidonic acid. PIPER & VANE (1969), described a new substance released from lungs during anaphylaxis capable of contracting rabbit aorta strips. They also showed that aspirin-like drugs block this release. The half-life of this substance was so short that almost all activity was lost by increasing the delay for the effusate to reach the assay tissues.

VARGAFTIG & DAO (1971), also using a direct biossay, related RCS to the prostaglandin family by demonstrating its generation upon injection of arachidonic acid into perfused guinea pig lungs. Initially, it was thought that the activity of RCS was due to the unstable cyclic endoperoxide intermediates of the synthesis of prostaglandins (GRIGLEWISKI & VANE, 1972). Recently, HAMBERG et al. (1975). suggested that the major component of the rabbit aorta contracting activity released from lung was due to another active material generated by oxidation of arachidonic acid, thromboxane A_2. This conclusion was strengthened by the demonstration that, like RCS, thromboxane A_2 contracts strips of rabbit coeliac artery, while PGG_2 causes relaxation (BUNTING et al., 1976). PGE_2 also relaxes the coeliac artery.

Rabbit aorta strip seems to be a good tissue to detect fatty acid peroxides for it contracts the peroxides derived from linolenic, linoleic and arachidonic acid generated by soybean lipoxygenase. Fatty acid peroxides also contract rat stomach strip (FERREIRA & VARGAFTIG, 1974).

BIOASSAY AFTER STABILIZATION AND PARTIAL

Bioassay specificity can be improved when there is

enough material and an adequate method for stabilization
and purification of an active material. Prostaglandin
extraction with ethyl acetate or with chloroform is the
most popular at present. Addition of a small amount of
radioactive prostaglandins is becoming routine, because
it permits to estimate the recovery. This is particularly
important because, depending on the nature of the sample,
solvent extraction causes intense protein denaturation
and consequent PG loss by adsorption. There are
advantages to solvent extraction: a) the bioassay can
be made independently from the experiment; b) it
desalts and concentrates the material, which is an
important step for chromatography; c) it increases the
specificity of the bioassay because some contaminants
against which there is no good antagonist are excluded
by the extraction (kinins, for example).

   Assay of an extracted material does not give much
more information than the immediate direct assay. At
best, its bioassay can indicate the predominant
prostaglandin present in the mixture.

   Assay after chromatographic fractionation (GRÉEN &
SAMUELSSON, 1964; ÄNGGÅRD & SAMUELSSON, 1964) permits a
better identification and quantification of the different
types of prostaglandins present in the mixture. However,
one must be aware of the limitations of chromatographic
fractionation of the prostaglandin family, because there
are metabolites capable of contracting assay tissue which
have an Rf similar to that of intact prostaglandins.
CRUTCHLEY & PIPER (1975, 1976) showed that the pulmonary
metabolites of PGE_2 and $PGF_{2\alpha}$, in high doses, mimic the
effect of parent compounds. In most of the assay
tissues, the ratio of potency between parent and
metabolites is roughly the same. Furthermore, using the
GRÉEN & SAMUELSSON A II system, they showed the Rf
similarity between PGE_1 and 13,14-dihydro-PGE_2 (Rf 0.8)
as well as between PGE_2 and 13,14-dihydro-$PGF_{2\alpha}$ (Rf 0.5).
These limitations are important when analysing perfusates
from tissues with high metabolic capacity as is the case
for lung.

   However, prostaglandin bioassay, as we have discussed,
is still very useful in many experimental situations, its
main virtue residing in the ability to show an unexpected
aspect of a system. I have mainly discussed here the
bioassay performed with smooth muscle preparations, but
chances are that new facets of the prostaglandin field
will be discovered using other types of bioassay, such
as platelet aggregation or bronchoconstriction.

   I thank Wellcome and FAPESP for research grants.

TABLE I

SMOOTH MUSCLE PREPARATIONS USED IN PG BIOASSAY

(THRESHOLD SENSITIVITY 1-10 ng/ml)

PREPARATION	PROSTAGLANDIN	REF.
Rat stomach fundus	$E_2 > E_1 > F_{2\alpha} > F_{1\alpha}$	1,7
Rat colon	$F_{2\alpha} > E_2 > E_1 > F_{1\alpha}$	9,8
Rat uterus	$E_2 > E_1 > F_{1\alpha} > F_{2\alpha}$	3,9,12
Gerbil colon	$E_1 > F_{1\alpha} > F_{2\alpha} > E_2$	3,5,7
Chick rectum	$E_2 > E_1 > F_{2\alpha} > F_{1\alpha}$	2,5,11
Hamster colon	$F_{2\alpha} > E_2 > E_1 > F_{1\alpha}$	10,12
Hamster stomach fundus	$E_2 > F_{2\alpha}$	13
Guinea pig ileum	$E_2 > E_1 > F_{2\alpha} > F_{1\alpha}$	3,7,10,12
Guinea pig uterus	$E_1 > F_{1\alpha}$	3,11
Rabbit jejunum	$F_{2\alpha} > F_{1\alpha} > E_2 > E_1$	3,10,12
Rabbit aorta strip	$TxA_2 > G_2 \sim H_2$	14
Rabbit coeliac artery	$G_2 \sim H_2 < E_2 \ (\downarrow); \ TxA_2 \ (\uparrow)$	14
Rabbit mesenteric artery	$G_2 \sim H_2 < E_2 \ (\downarrow); \ TxA_2 \ (\uparrow)$	14

1- COCEANI, F. & WOLFE, 1966; 2- FERREIRA, S.H. & VANE, 1967; 3- BERGSTRÖM, S. et al., 1968; 4- GILMORE, N. et al., 1968; 5- DUNHAM, E.W. & ZIMMERMAM, 1970; 6- FERREIRA, S.H. et al., 1973; 7- WEEKS, J.R. et al., 1968; 8- GAGNON, D.J. & SINOIS, 1972; 9- VANE, J.R. & WILLIANS, 1979; 10- HORTON, E.W. & MAIN, 1963; 11- BERGSTRÖM, S. et al., 1939; 12- HORTON, E.W. & MAIN, 1965; 13- UBATUBA, F.B., 1973; 14- BUNTING, S. et al., 1976.

TABLE II

RESPONSE OF ISOLATED ASSAY TISSUES TO MEDIATORS

	Ad	Nad	Ang. II	BK	Hist.	PG'S
RAT STOMACH STRIP	↓	↓	↑	↑	o	↑
CHICK RECTUM	↓	o	o	o	o	↑
RAT COLON	↓	↓	↑	o	o	↑
CAT JEJUNUM	o	o	o	↑	o	o
CAT ILEUM	↓	↓	o	↑	↑	o
GUINEA-PIG ILEUM	o	o	↑	↑	↑	↑
CONCENTRATION 0.5 – 20ng/ml						

After VANE , 1969

REFERENCES

Ånggård, E. and Samuelsson, B., 1964, Metabolism of prostaglandin
    E_1 in guinea-pig lung: the structures of two metabolites, J.
    Biol. Chem. 239: 4097.

Bergström, S., Eliasson, R.V.S. von and Sjovall, J., 1959, Some
    biological effects of two crystaline prostaglandin factors,
    Acta Physiol. Scand. 45: 133.

Bergström, S., Carlson, L.A. and Weeks,J.R., 1968, The prostaglandins:
    A family of Biologically active lipids, Pharmacol. Rev. 20: 1.

Bunting, S., Moncada, S. and Vane, J.R., 1976, The effects of
    prostaglandin endoperoxides and thromboxane A_2 on strips of
    rabbit coeliac artery and certain other smooth muscle
    preparations, Brit. J. Pharmacol. 57: 462.

Coceani, F. and Wolfe, L.S., 1966, On the action of prostaglandin
    E_1 and prostaglandins from brain on the isolated rat stomach,
    Canad. J. of Physiol. and Pharmac. 44: 933.

Collier, J.G., 1972, New dialysis technique for the continuous
    measurement of the concentration of vasoactive hormones, Br. J.
    Pharmacol. 44: 383.

Crutchley, D.J. and Piper, P.J., 1975, Comparative bioassay of
    prostaglandin E_2 and its three pulmonary metabolites, Br. J.
    Pharmac. 54: 397.

Crutchley, D.J. and Piper, P.J., 1976, Bioassay and thin-layer
    chromatography of prostaglandins and their pulmonary metabolites,
    Br. J. Pharmacol. 57: 463.

Dale, H.H., 1912, The anaphylatic reaction of plain muscle in the
    guinea pig, J. Pharmacol. Exptl. Therap. 4: 167.

Department of Pharmacology University of Edinburg, 1970,
    Pharmacological experiments on isolated preparations, E. & S.
    Livingstone, London.

Duham, E.W. and Zimmerman, B.G., 1970, Release of prostaglandin-
     like material from dog kidney during nerve stimulation, Am. J.
     of Physiol. 219: 1279.

Euler, U.S. von, 1939, Weitere untersuchungen über prostaglandin,
     die physiologisch aktive substanz gewisser genitaldrüsen, Skand.
     Arch. Physiol. 81: 65.

Feldberg, W., Gupta, K.P., Milton, A.S. and Wendlandt, S., 1973,
     Effect of pyrogen and antipyretics on prostaglandin activity
     in cisternal CSF of unanaesthetized cats, J. of Physiol. 234:
     279.

Ferreira, S.H. and Vane, J.R., 1967, Prostaglandins: their
     disappearance from and release into the circulation, Nature
     216: 868.

Ferreira, S.H., Moncada, S. and Vane, J.R., 1971, Indomethacin and
     aspirin abolish prostaglandin release from the spleen, Nature,
     New Biol. 231: 237.

Ferreira, S.H., Moncada, S. and Vane, J.R., 1973, Some effects of
     inhibiting endogenous prostaglandin formation of the responses
     of the cat spleen, Brit. J. Pharmacol., 47: 48.

Ferreira, S.H., Moncada, S. and Vane, J.R., 1973, Prostaglandins
     and the mechanism of analgesia produced by aspirin-like drugs,
     Brit. J. Pharmacol. 49: 86.

Ferreira, S.H. and Vargaftig, B.B., 1974, Inhibition by non-steroid
     anti-inflammatory agents of rabbit aorta contracting activity
     generated in blood by slow reacting substance C, Br. J. Pharmacol.
     50: 543.

Ferreira, S.H. and Costa, F.S., 1976, A laminar flow superfusion
     technique with much increased sensitivity for the detection of
     smooth muscle stimulating substances, Europ. J. Pharmacol.,
     accepted for publication.

Finkleman, B., 1930, On the nature of inhibition in the intestine, J. Physiol., Lond. 70: 145.

Gaddum, J.H., 1953, The technique of superfusion, Br. J. Pharmac. Chemotherap. 8: 321.

Gaddum, J.H., 1959, Biossay procedures, Pharmac. Rev. 11: 241.

Gagnon, D.J. and Sirois, P., 1972, The rat isolated colon as a specific assay organ for angiotensin, Br. J. Pharmacol. 46: 89.

Gilmore, N., Vane, J.R. and Wyllie, J.H., 1968, Prostaglandins released by the spleen, Nature 218: 1135.

Green, K. and Samuelsson, B., 1964, Thin-layer chromatography of prostaglandins, J. Lipid Res. 5: 117.

Gryglewski, R. and Vane, J.R., 1972, Rabbit aorta contracting substance (RCS) may be a prostaglandin precursor, Brit. J. Pharmacol. 43: 420.

Hamberg, M., Svensson, J. and Samuelsson, B., 1975, Thromboxanes: A new group of biologically active compounds derived from prostaglandin endoperoxides, Proc. Nat. Acad. Sci. U.S.A. 72: 2994.

Herbaczynska-Cedro, K. and Vane, J.R., 1973, Contribution of intra-renal generation of prostaglandin to autoregulation of renal blood flow in the dog, Circ. Res. 33: 428.

Horton, E.W. and Main, I.H.M., 1963, A comparison of the biological activities of four prostaglandins, Brit. J. Pharmacol. Chemotherap. 21: 182.

Horton, E.W. and Main, I.H.M., 1965, Actions of prostaglandins $F_{2\alpha}$ and E_1 on smooth muscle, Brit. J. Pharmacol. Chemotherap. 24: 470.

Horton, E.W. and Jones, R.L., 1969, The biological assay of prostaglandins A_1 and A_2, J. Physiol. (London) 200: 56.

Lonigro, A.J., Itskovitz, H.D., Crowshaw, K. and McGiff, J.C., 1973, Dependency of renal blood flow on prostaglandin in synthesis in the dog, Circ. Res. 32: 712.

Lonigro, A.J., Terragno, N.A., Malik, K. U. and McGiff, J.C., 1973, Differential inhibition by prostaglandins of the renal actions of pressor stimuli, Prostaglandins 3: 595.

Magnus, R., 1903, Pharmakologie der magen und darmbewegungen, Ergebn. Physiol. 2: 637.

MacGiff, J.C., Terragno, N.A., Strand, J.C., Lee, J.B., Lonigro, A. J. and Ng, K.K.F., 1969, Selective passage of prostaglandins across the lung, Nature 223: 742.

Paton, W.D.M., 1957, A pendulum auxotonic lever, J. Physiol., Lond., 137: 35.

Piper, P.J. and Vane, J.R., 1969, Release of additional factors in anaphylaxis and its antagonism by anti-inflammatory drugs, Nature 223: 20.

Piper, P., Vane, J.R. and Wyllie, J.H., 1970, Inactivation of prostaglandins by lungs, Nature 225: 600.

Samuelsson, B., 1963, Isolation and identification of prostaglandins from human seminal plasma, J. Biol. Chem. 238: 3229.

Sanner, J.H., 1974, Substances that inhibit the actions of prostaglandins, Arch. Intern. Med. 133: 133.

Smith, J.B. and Willis, A.L., 1971, Aspirin selectively inhibits prostaglandin production in human platelets, Nature New Biol. 231: 235.

Ubatuba, F.B., 1973, The use of the hamster stomach in vitro as an assay preparation for prostaglandins, Br. J. Pharmacol. 49: 662.

Vane, J.R., 1966, B. The estimation of Catecholamines by biological assay, Pharmacol. Rev. 18: 317.

Vane, J.R., 1969, The release and fate of vaso-active hormones in the circulation, Br. J. Pharmacol. 35: 209.

Vane, J.R., 1971, Inhibition of prostaglandin synthesis as a mechanism of action for aspirin-like drugs, Nature New Biol. 231: 232.

Vane, J.R. and Williams, K.I., 1973, The contribution of prostaglandin production to contractions of the isolated uterus of the rat, Br. J. Pharmacol. 48: 629.

Vargaftig, B.B. and Dao Hai, N., 1971, Release of vasoactive substances from guinea-pig lungs by slow reacting substance C and arachidonic acid, Pharmacology 6: 99.

Weeks, J.R., Schultz, J.R. and Brown, W.E., 1968, Evaluation of smooth muscle bioassays for prostaglandins E_1 and F_1 , J. of Appl. Physiol. 25: 783.

Willis, A.L., 1969a, Parallel assay of prostaglandin-like activity in rat inflammatory exudate by means of cascade superfusion, J. Pharm. Pharmac. 21: 126.

Willis, A.L., 1969b, Release of histamine, kinin and prostaglandin during carrageenin-induced inflammation in the rat, in: Prostaglandins, peptides and amines (P. Mantegazza and E. W. Horton, eds.), pp. 31-38, Academic Press, London.

RADIOIMMUNOASSAYS OF PROSTAGLANDINS AND THROMBOXANES

Elisabeth Granström

Dept. of Chemistry, Karolinska Institutet

S-104 01 Stockholm 60, Sweden

Radioimmunoassay is based on a competition between labeled and unlabeled molecules of a certain compound for the binding sites on an antibody directed against the same compound. The presence of a large amount of unlabeled compound will lead to an extensive displacement of labeled molecules from the binding sites, thus, to a high radioactivity in the unbound fraction. The absolute amounts of the compound assayed are obtained from comparison with a standard curve with known amounts.

In the field of prostaglandin assay, radioimmunoassay has in the past few years gained a widespread use (for a review, see Ref. 1). The method has a large capacity in comparison with the more time consuming bioassays or mass spectrometric methods. The sensitivity is also generally high: the limit of detection in most published assays generally ranges from only a few picograms to about 50 picograms, which often allows determination of prostaglandins in very small samples.

However, these radioimmunological methods are not entirely specific for the compounds they are designed to measure. The antiserum generally cross-reacts to a variable extent with other compounds with related structures, and this cross-reactivity may even be essentially complete, e.g. in the case of antisera against certain urinary metabolites (see below). However, the cross-reactivity of the antiserum may be studied and a certain correction for the presence of other prostaglandins or metabolites is often possible.

Of far greater importance is the non-specific influence by other, entirely unknown and unrelated compounds upon the binding between antigen and antibody. It is important to realise, that the standards and the samples are never analysed under identical conditions. Any factor that inhibits the antigen-antibody binding will naturally increase the radioactivity of the unbound fraction and thus lead to a too high "prostaglandin" value. These non-specific factors may be impurities from the solvents used in an extraction of the sample, or unrelated compounds present in the sample in the case of assays without extraction. No doubt these unknown factors constitute part of the explanation for the wide variation in reported results (2) and for the highly conflicting data obtained from different laboratories, even when identical samples were measured (3).

The development of a radioimmunoassay involves several steps: raising of the antiserum, preparation of a labeled ligand with a high specific activity, development of a method for separation of the antibody-bound and the free fraction, processing of samples, and evaluation of the method by estimating its sensitivity, specificity, precision and accuracy.

Antibodies to prostaglandins are generally raised by injection of a prostaglandin-protein conjugate into a suitable animal. A convenient method to prepare the conjugate is the coupling between the carboxyl group of the prostaglandin and an amino group on the protein molecule. The antibodies produced will then preferrably recognize the more distant parts of the molecule, e.g. the ring, which renders them highly specific for prostaglandins. As a coupling reagent carbodiimide (4) or carbonyldiimidazole (5) are generally employed. In the case of the dioic acid metabolite of PGF_α, a selective coupling to the ω-carboxyl may be obtained after the compound has been converted into its δ-lactone form (6).

The labeled ligand should be of a specific activity as high as possible to allow a high sensitivity of the assay. Recently, commercial preparations of several primary prostaglandins with specific activities around 100 Ci/mmole have become available. These are labeled with 2-8 tritium atoms per molecule. The metabolites, e.g. the 15-keto-13,14-dihydro compounds or the 11-keto-tetranorprosta-1,16-dioic acids, are not commercially available yet but have to be prepared by chemical or enzymatic methods. If enzymatic methods, either in vivo or in vitro, using crude enzyme preparations are employed, it is important to include also inhibitors of prostaglandin biosynthesis to prevent dilution with endogenous material (1).

One of the most critical steps in the radioimmunoassay is the separation of the antibody-bound from the unbound fraction. A large number of methods exist. In the prostaglandin field, the most employed ones involve precipitation of the globulin fraction with either second antibody (antiserum directed against the γ-globulin of the species used for production of the prostaglandin antibody (7)), or with more non-specific protein precipitation methods, e.g. polyethylene glycol (8) or ammonium sulphate. Dextran-coated charcoal on the other hand absorbs the unbound molecules and leaves the antigen-antibody complex in solution (4). Other methods include adsorption on nitrocellulose membranes, the use of a solidified first antibody, etc.

The processing of samples in most cases involves an extraction step followed by a silicic acid chromatography (4,9). The advantages of this method are the high purity obtained of the sample and the group separation of the various prostaglandins. However, the possible introduction of interfering substances from the solvents or from other steps in the procedure, the considerable reduction in capacity by these time-consuming steps, and furthermore, the need for addition of labeled compounds for estimating the recovery, together render this processing less suitable in many types of studies. In direct assays of biological fluids without purification, however, only very small volumes can be assayed, 0.1-0.5 ml generally, and the possible presence of interfering substances is obvious. These problems have to be carefully considered in each study. When e.g. prostaglandin metabolite levels in blood or urine are followed during a long time in a large number of samples in the same individual, assay of unextracted samples may be suitable.

Of great importance in this field is also the choice of the compound to measure. The earliest prostaglandin radioimmunoassays were all aimed at primary prostaglandins or prostaglandins of the A and B type, and for several years these methods were employed quite uncritically, e.g. for measurements of prostaglandin levels in the peripheral circulation (1). This is unfortunately still the case in many places. It is however well known that the measured levels of these prostaglandins do not reflect the endogenous levels (for a discussion, see Refs. 1-3). Prostaglandins are formed to a considerable extent during the sample collection, from platelets and white blood cells and even from non-enzymatic cyclization of the precursor acid. Furthermore, the half-life of the primary prostaglandins is very short, probably less than half a minute (1), and the true endogenous plasma levels of e.g. $PGF_{2\alpha}$ have been calculated as 2 pg/ml or less (2). For all these reasons, it is obvious that for studies on endogenous release of prostaglandins, the major plasma metabolites, the 15-keto-13,14-dihydro compounds, are the compounds of choice. These metabolites are not formed when

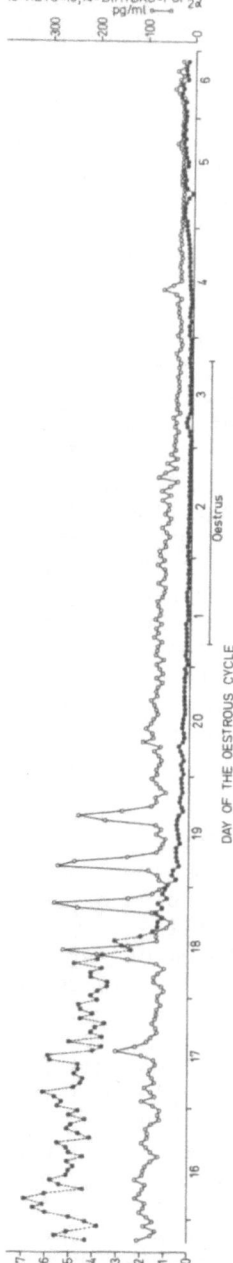

Fig. 1. Peripheral plasma levels of progesterone and 15-keto-13,14-dihydro-PGF$_{2\alpha}$ in a heifer. Blood samples were collected every hour.

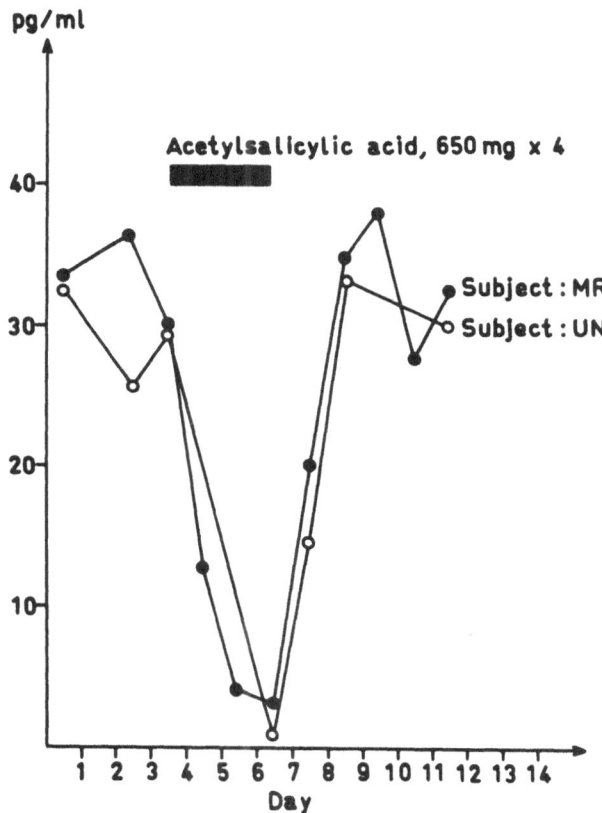

Fig. 2.   Inhibition of prostaglandin biosynthesis in the human.
Plasma levels of 15-keto-13,14-dihydro-PGF$_{2\alpha}$ were measured before,
during and after administration of aspirin, 2.6 g per day.

the blood sample is taken, they occur in 20-70 fold higher concentrations, and they have considerably longer half-lives (around 8 min). In later years a number of radioimmunoassays for 15-keto-13,14-dihydro-PGF$_{2\alpha}$ have been developed (see Ref. 6 and references therein).

These assays for the major metabolite of PGF$_{2\alpha}$ in the circulation have been employed in a number of studies. Great interest has recently been taken in the role of PGF$_{2\alpha}$ as the naturally occurring uterine luteolytic substance in several species (see Ref. 10 and references therein). When only assays for PGF$_{2\alpha}$ were available, it was necessary to study the release of PGF$_{2\alpha}$ from the uterus by obtaining samples from the uterine vein. With assays for the major metabolite of this prostaglandin, however, it has been possible to carry out detailed studies using peripheral plasma samples. Thus, the above-mentioned difficulties have been avoided and the samples obtained under considerably more physiologic conditions for the animal. Fig. 1 shows the results of measurements of progesterone and 15-keto-13,14-dihydro-PGF$_{2\alpha}$ in peripheral plasma of a heifer during the last days of the oèstrus cycle. The pulsative release of the prostaglandin is seen to occur coinciding with the decrease of progesterone.

A different kind of study is illustrated in Fig. 2. In this experiment the effects of anti-inflammatory drugs on prostaglandin biosynthesis (1) was studied in human subjects by monitoring the plasma levels of 15-keto-13,14-dihydro-PGF$_{2\alpha}$ before, during and after ingestion of aspirin (Fig. 2), indomethacin or naproxen. Almost undetectable values of the metabolite were obtained on the second day of administration of the drug. The levels again returned to normal two days after the administration was stopped (6).

Since it is known that prostaglandin release often occurs intermittently (e.g. as exemplified in Fig. 1), even the half-life of the 15-keto-13,14-dihydro compounds in the circulation may be too short, if the blood sampling is not frequent enough. Thus, in some studies measurement of the urinary metabolites of prostaglandins may be more reliable, since a major part of prostaglandins that reach the blood stream is eventually excreted into urine as shorter compounds (1). In recent years, a few radioimmunoassays for the main urinary metabolites of PGF$_{2\alpha}$ in the human or other species, viz. 11-ketotetranorprostanoic acid derivatives, have been published (see Refs. 6 and 11 and references in these papers).

In the case of the human metabolite, 5α,7α-dihydroxy-11-keto-tetranorprosta-1,16-dioic acid, the method of choice has been to couple the compound selectively at the ω-carboxyl group to the protein molecule prior to immunization. The resulting antibodies did then not recognize the ω-end of the molecule but were instead

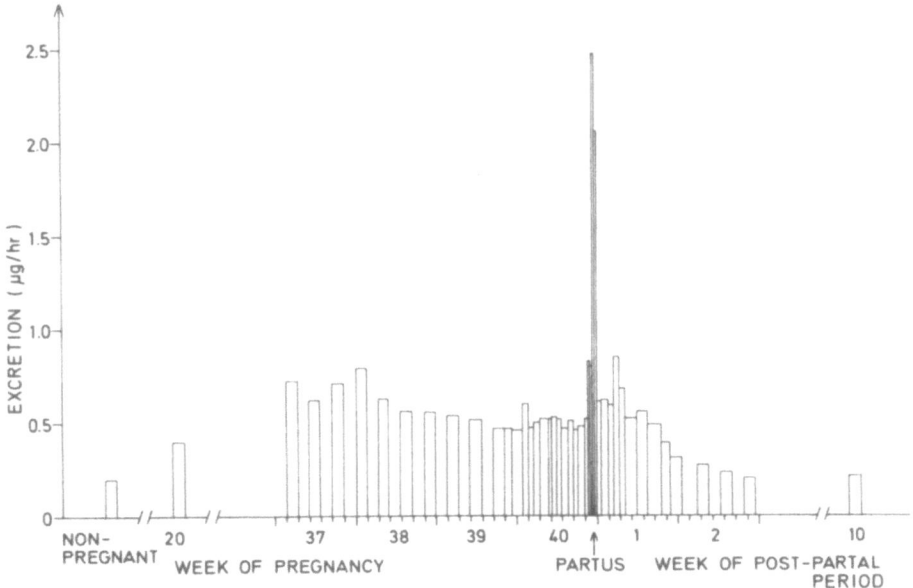

Fig. 3. Excretion of 5α,7α-dihydroxy-11-ketotetranorprosta-1,16-
-dioic acid in a pregnant human female. From week 36 urine was collected in 24 hr portions every second day. From 8 days before delivery all urine was collected; during the day of parturition in portions.

very specific for 11-ketotetranor compounds. Thus the same
antiserum could be used also for measurements of the main urinary
metabolite in the guinea pig, 5α,7α-dihydroxy-11-ketotetranorprosta-
noic acid (11).

     Fig. 3 shows the results obtained in a study of the urinary
excretion of 5α,7α-dihydroxy-11-ketotetranorprosta-1,16-dioic acid
in the human during pregnancy. The normal basal levels in this
subject were around 5 μg/day and increased about three-fold during
the last weeks of pregnancy. Very high levels were seen during labour and immediately after delivery.

     Fig. 4 shows a study on the role of $PGF_{2\alpha}$ during the oestrus
cycle in the guinea pig (cf. the bovine study in Fig. 1). In the
guinea pig it is obviously impossible to obtain frequent blood
samples during long periods for assay of 15-keto-13,14-dihydro-
$PGF_{2\alpha}$. Monitoring 5α,7α-dihydroxy-11-ketotetranorprostanoic acid,

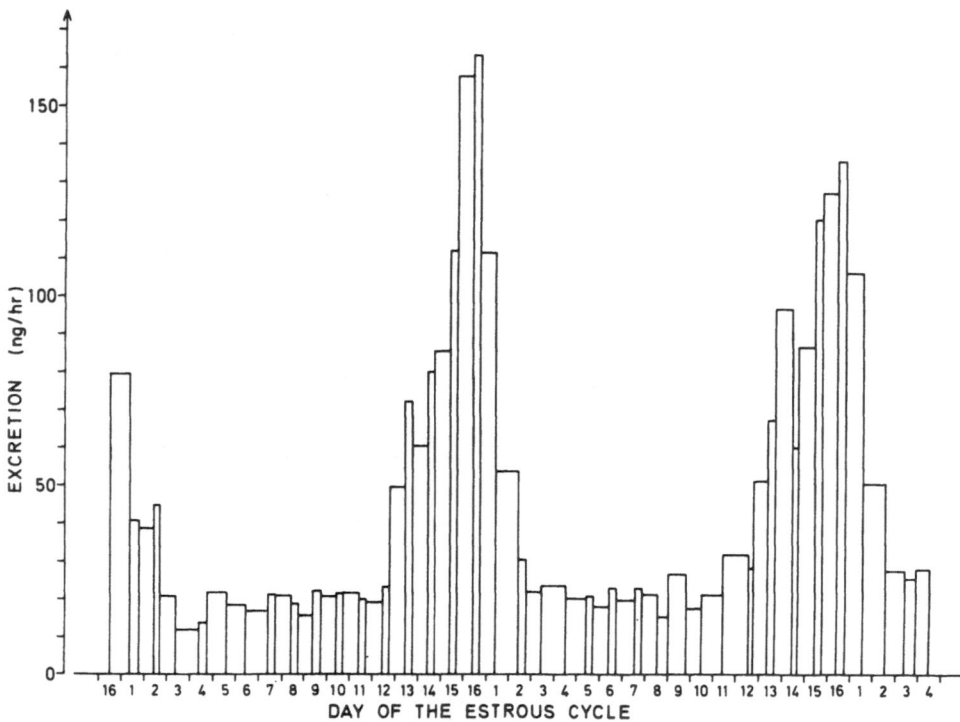

Fig. 4. Excretion of 5α,7α-dihydroxy-11-ketotetranor-prostanoic acid into urine of a guinea pig during two consecutive oestrus cycles. Day 1 of the cycle was the day of heat.

the main urinary metabolite, however, allows the collection of samples under completely physiological conditions for an unlimited time period. Also in this species a low basal prostaglandin production was seen throughout most of the cycle. A few days prior to oestrus, however, the $PGF_{2\alpha}$ synthesis was increased severalfold. This was reflected by the great increase in the excretion of 5α, 7α-dihydroxy-11-ketotetranorprostanoic acid into urine.

It has recently been shown that in certain tissues and cells, e.g. platelets and lung tissue, the prostaglandin endoperoxides PGG_2 and PGH_2 are only to a minor extent converted into the stable prostaglandins PGE_2 and $PGF_{2\alpha}$ (12,13). The major part of the endoperoxides is instead transformed into the so called thromboxanes and related structures (14). For the studies of various physiolo-

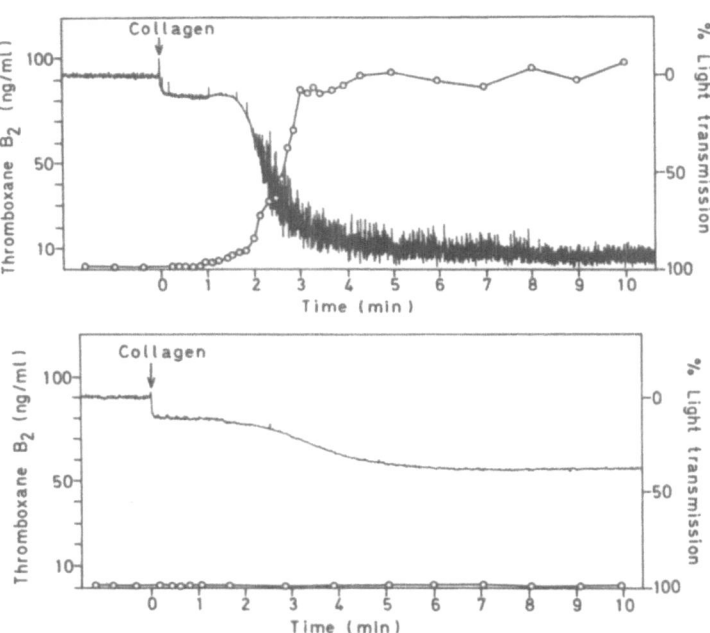

Fig. 5. Release of TXB$_2$ after addition of collagen to human PRP. Upper panel: PRP without indomethacin. Lower panel: PRP containing 1.4 x 10^{-4} M indomethacin. Light transmission recorded simultaneously in an aggregometer. Arrows indicate addition of collagen.

gic mechanisms involving the very potent compounds TXA_2 and the prostaglandin endoperoxides, it is thus not reliable to measure PGE_2 and $PGF_{2\alpha}$ in these systems. Thromboxanes and thromboxane metabolites must be monitored instead. Recently, a radioimmunoassay for TXB_2 has been published (15).

Platelet aggregation has been studied using this radioimmunoassay. Fig. 5 shows an experiment where TXB_2 was determined at different times in a sample of human platelet rich plasma (PRP) before and after addition of collagen. Normal aggregation started about 1 min after the addition (upper panel). A rise in the TXB_2 level could be seen to start shortly before the aggregation was noticeable and continued for ca 2 min to a level of about 90 ng/ml. Collagen was also added to PRP pretreated with indomethacin (lower panel). In this experiment no aggregation occurred, and no TXB_2 increase was seen during the experiment.

Prostaglandin radioimmunoassays have thus in many cases been found to be a useful tool in the studies of the roles of prostaglandins in various physiological and pathological conditions. It is however of utmost importance that these methods are employed critically; that the proper compound is monitored in each study; that the investigator is aware of the many pit-falls that are inherent in this kind of methodology; and, if possible, that the data obtained are regularly checked by the use of an independent quantitative method, preferrably a mass spectrometric method.

REFERENCES

1. Samuelsson, B., Granström, E., Grēen, K., Hamberg, M., and Hammarström, S., Ann. Rev. Biochem. 44: 669 (1975).

2. Samuelsson, B., Adv. Biosci. 9: 7 (1973).

3. Samuelsson, B., Axen, U., Behrman, H., Granström, E., Grēen, K., Jaffe, B.M., Kirton, K., Levine, L., Skarnes, R.C., Speroff, L., and Wolfe, L.S., Adv. Biosci. 9: 121 (1973).

4. Caldwell, B.V., Speroff, L., Brock, W.A., Auletta, F.J., Gordon, J.W., Andersen, G.G., and Hobbins, J.C., J. Reprod. Med. 9: 361 (1972).

5. Axen, U., Prostaglandins 5: 45 (1974).

6. Granström, E., and Kindahl, H., In: Advances in Prostaglandin and Thromboxane Research. Eds. B. Samuelsson and R. Paoletti. Raven Press, New York, Vol. 1, p. 81 (1976).

7. Cornette, J.C., Kirton, K.T., Barr, K.L., and Forbes, A.D., J. Reprod. Med. 9: 355 (1972).

8. Van Orden, D.E., and Farley, D.B., Prostaglandins 4: 215 (1973).

9. Jaffe, B.M., Behrman, H.R., and Parker, C.W., J. Clin. Invest. 52: 398 (1973).

10. Kindahl, H., Edqvist, L.-E., Bane, A., and Granström, E., Acta Endocrin. (Kbh.) 82: 134 (1976).

11. Granström, E., and Kindahl, H., Prostaglandins, in press (1976).

12. Hamberg, M. and Samuelsson, B., Proc. Nat. Acad. Sci. USA, 71: 3400 (1974).

13. Hamberg, M., and Samuelsson, B., Biochem. Biophys. Res. Commun. 61: 942 (1974).

14. Samuelsson, B., Hamberg, M., Malmsten, C., and Svensson, J., In: Advances in Prostaglandin and Thromboxane Research. Eds. B. Samuelsson and R. Paoletti. Raven Press, New York, vol. 2, p. 737 (1976).

15. Granström, E., Kindahl, H., and Samuelsson, B., Anal. Lett. 9: 611 (1976).

VAPOR-PHASE METHODS FOR QUANTITATIVE EVALUATION OF PROS-
TAGLANDINS AND RELATED COMPOUNDS IN BIOLOGICAL SAMPLES

S. Nicosia and G. Galli

Institute of Pharmacology and Pharmacognosy,

Laboratory of Applied Biochemistry, University

of Milan, Milan, Italy

Since their discovery, many procedures have been
developed for prostaglandin (PG) analysis, based on different principles: biological assay, radioimmunoassay
(which will be discussed in other sections of this book),
gas chromatographic and mass spectrometric methods.
PGs are a class of compounds all showing very similar
structure, but different functions and effects, which are
present in biological tissues and fluids in very low concentration.

A method suitable for quantitative PG evaluation in
biological samples should present both a high degree of
specificity and a high sensitivity; in addition it should
allow the quantitative evaluation of more than one PG at
the same time, possibly on a routinary base.
Gas chromatographic analysis can reach high sensitivity if a proper detector is used (e.g., electron capture
detector) or when it is coupled with mass spectrometry.
The mass spectrometer can also provide the required specificity: different PGs can be distinguished and can be
monitored contemporaneously. The evaluation of PGs in
biological samples, using gas chromatographic techniques,
requires extraction from the tissue or fluid and some
purification to eliminate not only interfering substances
but also impurities with long retention time.
Gas phase analysis also requires derivatization of
PGs, since they are not enough thermally stable.

The most common methods for purification and derivative formation of PGs and related compounds will be discussed here, with particular emphasis on massfragmentographic methods.

EXTRACTION AND PURIFICATION

It has been shown that rapid PG biosynthesis occurs during handling of biological tissues (1,2). If the in situ levels of PGs have to be measured, it is important to prevent this phenomenon by denaturing the PG synthetase as soon as possible. This is usually accomplished by homogenizing the tissue immediately after dissection either in ethanol (1) or in a medium at a pH (3.0-3.5) that does not modify PG chemical structure. In the case of cerebral tissue, the use of a focused microwave oven is the method of choice (3,4), since it allows inactivation of enzymes within a few seconds before dissection. As discussed by Samuelsson et al. (5), PG biosynthesis occurs in. platelets during blood drawing and handling. Therefore, the levels of primary PGs which can be measured in plasma or serum are not correlated with the total production of PGs in the body, but the concentration of the metabolites should be taken as a better indication.

Methods for PG extraction and purification from plasma and urine involve either an ether-ethyl acetate extraction at pH 3 (6) or a chromatography of the fluid, after adjustment at pH 3, through an Amberlite XAD-2 column (7-9). Plasma PGs are further purified by reversed-phase partition chromatography and silicic acid chromatography with increasing concentrations of ether in petrol ether (7).
   For the analysis of arachidonic acid metabolites different from PGs in platelets, ethyl ether extraction at pH 3 and thin layer chromatography with ethyl acetate: trimethylpentane:water as eluent have been described (10).
   The method set up in our laboratory for PG extraction and purification from cerebral tissues (3,11) has been modified, and is shown in Fig. 1.
   Very recently, high performance liquid chromatography has been proposed (12,13) as a method to purify PGs from biological samples prior to the mass fragmentographic analysis.

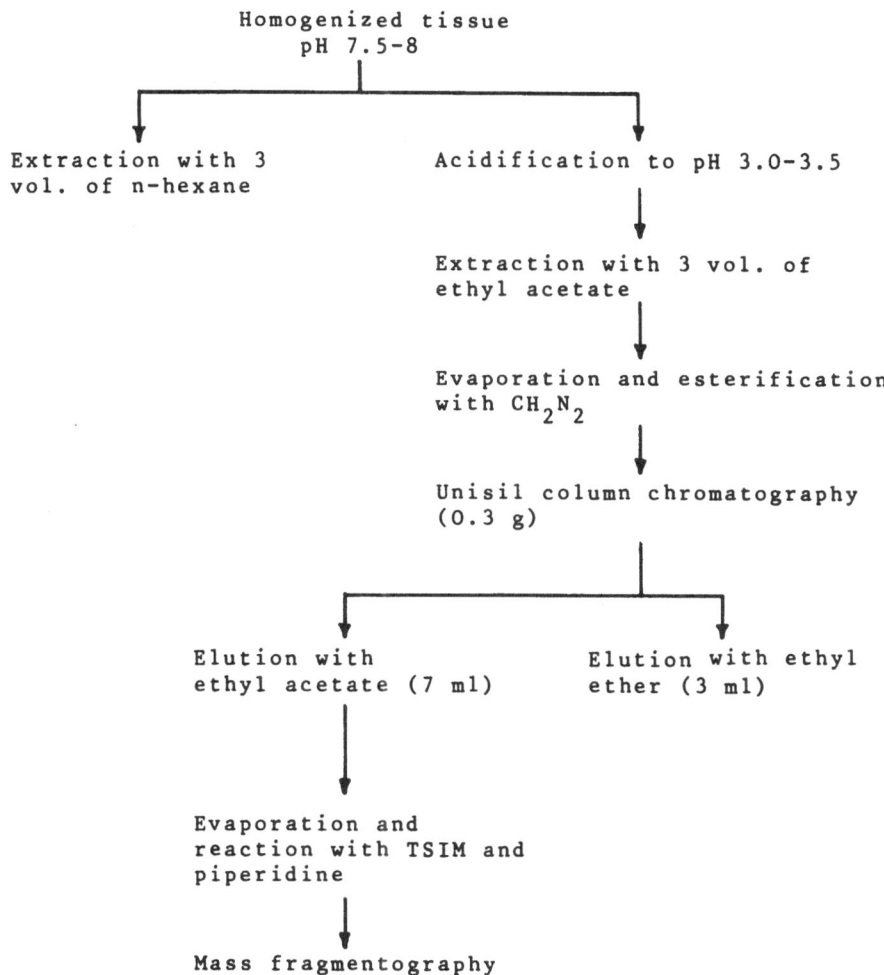

Fig. 1 - Extraction and purification of PG from cerebral
         tissue.

INTERNAL STANDARD AND MASS FRAGMENTOGRAPHY

An internal standard should be added at the earliest
stages of the analytical procedure, not only to make the
quantitative assay independent from possible losses
during the procedure itself, but also to act as a carrier
of the very minute amounts of natural PGs and to prevent
adsorption on the gas-chromatographic column. To achieve
these goals, the internal standard should be as similar
as possible to the compound to be evaluated. A deuterated
analog of the natural compound is the standard of choice
if the detector can discriminate between the protium and
deuterium forms. A mass spectrometer can be used for this
purpose, both in mass fragmentography (or multiple ion
detection, MID) and in repetitive magnetic scanning. The
former method consists in setting the mass spectrometer
so that, through an accelerating voltage alternator, it
records only fragments characteristic of the compound to
be determined and the corresponding fragments of the
deuterated standard.

One of the first applications of MID with deuterated
standard to quantitative problems has been the evaluation
of PGE_1 at nanogram level (14).

The second approach consists in scanning the magnetic
field repetively over a narrow mass range, including ions
carachteristic of the natural substance and of the deuterated standard. The intensities of these ions are plotted
through the aid of a computer, as a function of the scan
number, to yield a mass chromatogram. Repetitive magnetic
scanning avoids the minor technical problems connected
with ion focusing, but a computer is needed for collection
and interpretation of data, and the sensitivity is usually
lower than in MID. Applications to PG quantitation have
been published (15,16).

The synthesis of many deuterated PGs and related
compounds has been described (7,8,10,17).

To increase the specificity of the mass fragmentographic method it is advisable to monitor more than one
ion for each compound; in this way the product is iden-
tifiable not only by the retention time and by the
presence in the mass spectrum of one fragment, but also
by the presence of the second fragment and by the ratio
of these two fragment heights.

DERIVATIVE FORMATION

A derivative suitable for vapor-phase analysis should meet the following requirements: it should (a) be chemically and thermally stable, (b) be obtained in high yield and (c) as rapidly as possible; (d) show good gas chromatographic properties. Moreover, if a mass fragmentographic method is used, the mass spectrum of the derivative should present a prominent peak (representing a high percentage of the total ion current) at a high mass value.

Mass Fragmentography

The most commonly used derivatives for mass fragmentography of PGs are listed in Table 1. All the reactions involve functionalization of the carboxyl and hydroxyl groups, and in some instances of the keto group. The double bonds are not involved in the derivative formation, and therefore the reactions described for PGE_2 and $PGF_{2\alpha}$ may apply also to PGE_1 and $PGF_{1\alpha}$.

As far as the carboxyl group is concerned, methyl esters (ME) are usually prepared by addition of diazomethane before derivatization of hydroxyl and keto groups.

Both tris-trimethyl-silyl ether (TMS) (18, reaction No. 1), and tris-acetate (Ac) (17, reaction No. 2) have been proposed as derivatives for $PGF_{2\alpha}$. TMS has been described as being formed with N,O-bis-trimethylsilyl-trifluoroacetamide (BSTFA) in 3 hrs at room temperature, but the same product can be obtained with other reagents in shorter time as discussed later. Acetate formation is accomplished with acetic anhydride and pyridine overnight at room temperature, and extraction of the product from the reaction mixture is required.

Reaction of $PGF_{2\alpha}$ with butyl-boronic acid (6,19, reaction No. 3) generates a cyclic derivative involving the hydroxyl groups in position 9 and 11. Functionalization of the 15 hydroxyl group is then achieved with BSTFA (2 hrs, room temperature). The mass spectrum of $PGF_{2\alpha}$-ME-butylboronate-TMS presents a very prominent peak at m/e 435 suitable for mass fragmentography; however this derivative can be prepared only on prostaglandins of the F type.

PGEs contain a keto group, which can be transformed into the corresponding methyloxime (MO) (20,21) by reaction with methyl-hydroxylamine hydrochloride, overnight at room temperature (reaction No. 4 and 5).

TABLE 1

No.	PG	DERIVATIVE	REACTION CONDITIONS AFTER ESTERIFICATION	SENSITIVITY AND PRECISION ($\pm$% S.D.)[a]
1	$PGF_{2\alpha}$	$PGF_{2\alpha}$-ME-TMS	BSTFA, 3 hrs	0.05 ng[b]
2	$PGF_{2\alpha}$	$PGF_{2\alpha}$-ME-Ac	Ac_2O, Py overnight, extraction	0.1 ng[c] (0.1 ng $\pm$ 25% 0.5 ng $\pm$ 7%)
3	$PGF_{2\alpha}$	$PGF_{2\alpha}$-ME-Butyl-boronate-TMS	1.But-B(OH)$_2$,60°C 500 mm Hg,20 min 2.BSTFA, 2 hr	0.2 ng
4	PGE_2	PGE_2-ME-MO-TMS	1.Methoxyamine-HCl,Py overnight 2.TMCS:HMDS, 1 hr	
5	PGE_2	PGE_2-ME-MO-Ac	1.Methoxyamine-HCl,Py overnight,extraction 2.Ac_2O, Py overnight,extraction	0.25 ng (1 ng $\pm$ 5.9%)
6	PGE_2	PGE_2-ME-BO-TMS	1.Benzoxyamine-HCl 2.BSA	1 ng $\pm$ 30%
7	PGE_2	PGA_2-ME-TMS	BSA+Py overnight	0.09 ng
8	PGE_2	PGB_2-ME-TMS	1.KOH MeOH,15-60 min 2.H$^+$, extraction 3.CH_2N_2 4.BSA or BSTFA	---
9	PGE_2+ $PGF_{2\alpha}$	PGB_2-ME-TMS $PGF_{2\alpha}$-ME-TMS	TSIM+Piperidine instantaneous	0.07 ng (0.25 ng$\pm$17%)

[a] In brackets the amount for which the precision has been determined.
[b] From: Sweetman et al., Prostaglandins, 3, 385 (1973)
[c] From: Green et al., Anal. Biochem., 54, 434 (1973)

The hydroxyl group in position 15 is converted either
directly into the TMS by reaction with trimethylchloro-
silane (TMCS) and hexamethyl-disilazane (HMDS) for 1
hr (No. 4), or, after extraction from the reaction mixture, into the acetate (No. 5). The latter has to be
extracted before injection into the gaschromatograph.

The formation of PGE_2-ME-MO-TMS or PGE_2-ME-MO-Ac
has the drawback that a series of two reactions after
the methylation is required; the overall reaction time
is therefore fairly long and in the case of acetylation
some work up is required after both reactions. In addition, the reaction gives rise to both the syn and the
anti isomers of the MO, which appear as two peaks in
many gas chromatographic conditions (20,21). Benzylo-
xymes (BO) have been also described (16, reaction No. 6),
but their formation presents the same disadvantages
described for MO, and in addition the response is not
linear below 10 ng.

PGEs can be converted into the corresponding PGBs
or PGAs (2,22), thermally more stable since they lack
of the labile β-ketol system. Sweetman et al. (23) described the formation of PGA_2-ME-TMS (reaction No. 7) by
treatment of PGE_2-ME with bis-trimethylsilylacetamide
(BSA) and pyridine overnight. However, PGA_2-ME-TMS is
unstable due to its conversion to PGB_2-ME-TMS in the
presence of even traces of alkali. Therefore, to avoid
decomposition, all the glassware and the gas chromatographic column have to be carefully silanyzed. PGB_2-ME-
TMS, on the contrary, is fairly stable but the series of
reactions proposed by Sweetman et al. (23)(treatment of
PGE_2 with methanolic KOH for 15-60 min., acidification
and extraction, methylation, treatment with BSA or BSTFA
overnight, reaction No. 8) is quite involved and can give
rise to by-product formation (24). On the other hand,
the same derivative can be easily obtained by a single,
instantaneous reaction, performed at room temperature
with trimethylsilylimidazol (TSIM) and piperidine (25,
reaction No. 9)· PGB_2-ME-TMS mass spectrum presents a
prominent peak at m/e 321, which makes this derivative
suitable for mass fragmentography.

All the derivatives described show good gaschromato-
graphic properties, are formed in good yield and are
chemically and thermally stable, with the exceptions that
have been discussed. However, for some of them such as
PGE_2-ME-MO-Ac and $PGF_{2\alpha}$-ME-Ac, a great excess (100-1000
fold) of deuterated analog is described to be present in
the injected mixture (7). The rationale behind the use

of a large amount of the deuterated form might be the
protection of the natural compound from absorption in the
gas chromatograph-mass spectrometer. On the other hand,
a larger amount of deuterated PG causes a higher back-
ground on the fragment focused for the protium form,
which is a drawback when extremely low amounts of natural
PG have to be quantitated.

All the methods described, with the exception of
No. 9, involve different reactions carried out separately
on PGs of the E and F series. On the contrary, both PGEs
and PGFs have been determined contemporaneously after
derivatization with reaction No. 9.

Table 1 shows the limit of detection and the precision of the different methods. The sensitivity is always
very high, ranging from 0.05 to 0.25 ng, with the exception of reaction No. 6.

The same reactions discussed for primary PGs have
been applied to quantitative determination of plasma and
urinary metabolites (7,13).

As mentioned in other sections of this book, during
the last few years new metabolites of arachidonic acid
have been identified in platelets (26); these metabolites,
namely 12L-hydroxy-5,8,10,14-eicosatetraenoic acid (HETE),
12L-hydroxy-5,8,10-heptadecatrienoic acid (HHT) and 8-
(1-hydroxy-3-oxopropyl)-9,12L-dihydroxy-5,10-heptadeca-
dienoic acid (PHD or Thromboxane B_2), have been quantitated
by MID of their ME-TMS (10). Both HHT and HETE mass spectra
show a prominent ion at m/e 295 and therefore they have
been quantitated contemporaneously, with octadeutero-HETE
as internal standard. Correspondingly, octadeutero-PHD
has been used for PHD analysis.

Electron Capture

PGBs, which can be obtained by alkaline treatment of
PGEs, possess intrinsic electron capturing properties, due
to the presence in the molecule of a highly conjugated
system; on the basis of these observations, a method was
developed by Jouvenaz et al. (1,2) for PGE_1 and E_2 evaluation by electron capture (EC) gas chromatography. On
the other hand, PGFs, in order to be evaluated by EC,
were transformed into halogenated derivatives such as
bromomethyldimethylsilyl ethers-ME (2) obtained by treatment with bromomethyldimethylsilyl chloride and diethyla-
mine (1 hr. at room temperature) in anhydrous conditions.
Omega-nor-PGE_2 and ω-homo-PGE_1 were used as internal
standards and the limit of detection is around 1 ng.

Later, Sweetman et al. (23) reported the detection of
200 pg of PGB$_2$ with EC.
    The use of heptafluorobutyrate-ME for PGFs has been
described (27,28) but their stability has been questioned
(27).
    More recently, a method for the conversion of PGF$_{2\alpha}$
into pentafluorobenzyl ester has been published (29).
The electron capturing properties of this derivative allow
the detection of 12.5 pg, and the reaction is performed
with a solution of pentafluorobenzyl bromide and di-iso-
propylethylamine in acetonitrile, 5 min at 40°C.
    EC gas chromatography presents some drawbacks,
namely that a more extensive purification is required,
since some of the impurities can also be converted into
compounds with electron trapping properties. On the other
hand, an EC detector is much less expensive than a mass
spectrometer, which makes this method of great usefulness
for many laboratories.

    Vapor-phase analysis has proved to be very useful in
quantitative determination of PGs in biological samples:
if a proper purification method and derivatization are
choosen, very high sensitivity, in the sub-nanogram range,
can be reached; the specificity of the technique allows
to evaluate different PGs in the same sample with high
reliability.

                        References

1. JOUVENAZ, G.H., NUGTEREN, D.H., BEERTHUIS, R.K. and
   VAN DORP, D.A. (1970) A sensitive method for the
   determination of prostaglandins by gas chromatography
   with electron-capture detection. Biochim. Biophys.
   Acta, 202, 231-234.
2. JOUVENAZ, G.H., NUGTEREN, D.H. and VAN DORP, D.A.
   (1973). Gas chromatographic determination of nanogram
   amounts of prostaglandins E and F. Prostaglandins,
   3, 175-187.
3. GUIDOTTI, A., CHENEY, D.L., TRABUCCHI, M., DOTEUCHI,
   M., WANG, C. and HAWKINS, R.A. (1974) Focussed Microwave Radiation: a Technique to Minimize Post-Mortem
   Changes of Cyclic Nucleotides, DOPA and Choline and to
   Preserve Brain Morphology, Neuropharmacol., 13, 1115-
   1122.
4. BOSISIO, E., GALLI, C., GALLI, G., NICOSIA, S., SPA-
   GNUOLO, C. and TOSI, L. (1976) Correlation between
   release of free arachidonic acid and prostaglandin
   formation in brain cortex and cerebellum. Prostaglandins, 11, 773-781.

5. SAMUELSSON, B., GRANSTRÖM, E., GREEN, K., HAMBERG, M. and HAMMARSTRÖM, S. (1975) Prostaglandins, Ann. Rev. Biochem., $\underline{44}$, 669-695.

6. KELLY, R.W. (1973) Method for the Measurement of Prostaglandin $F_{2\alpha}$ in Biological Fluids by Gas Chromatography-Mass Spectrometry. Anal. Chem., $\underline{45}$, 2079-2082.

7. GREEN, K., GRANSTRÖM, E., SAMUELSSON, B. and AXEN, U.(1973) Methods for Quantitative Analysis of $PGF_{2\alpha}$, PGE_2, 9α,11α-Dihydroxy-15-Keto-Prost-5-Enoic Acid and 9α, 11α,15-Trihydroxy-Prost-5-Enoic Acid from Body Fluids Using Deuterated Carriers and Gas Chromatography-Mass Spectrometry. Anal. Biochem., $\underline{54}$, 434-453.

8. HAMBERG, M. (1973) Quantitative Studies on Prostaglandin Synthesis in Man. II. Determination of the Major Urinary Metabolite of Prostaglandins $F_{1\alpha}$ and $F_{2\alpha}$. Anal. Biochem., $\underline{55}$, 368-378.

9. SWEETMAN, B.J., WATSON, J.T., CARR, K., OATES, J.A. and FROLICH, J.C. (1973) Quantitative vapor-phase analysis of prostaglandin $F_{2\alpha}$ in female human urine. Prostaglandins, $\underline{3}$, 385-387.

10. HAMBERG, M., SVENSSON, J. and SAMUELSSON, B. (1974) Prostaglandin Endoperoxides. A New Concept Concerning the Mode of Action and Release of Prostaglandins, Proc. Nat. Acad. Sci. USA, $\underline{71}$, 3824-3828.

11. NICOSIA, S. and GALLI, G. (1975) A mass fragmentographic method for the quantitative evaluation of brain prostaglandin biosynthesis. Prostaglandins, $\underline{9}$, 397-403.

12. CARR, K., SWEETMAN, B.J. and FRÖLICH, J.C. (1976) High performance liquid chromatography of prostaglandins: biological applications. Prostaglandins, $\underline{11}$, 3-14.

13. HUBBARD, W.C. and WATSON, J.T. (1976) Determination of 15-keto-13,14-dihydro-metabolites of PGE_2 and $PGF_{2\alpha}$ in plasma using high performance liquid chromatography and gas chromatography-mass spectrometry. Prostaglandins, $\underline{12}$, 21-35.

14. SAMUELSSON, B., HAMBERG, M. and SWEELEY, C.C. (1970). Quantitative Gas Chromatography of Prostaglandin E_1 at the Nanogram Level: Use of Deuterated Carrier and Multiple-Ion Analyzer. Anal. Biochem., $\underline{38}$, 301-304.

15. BACZYNSKYJ, L., DUCHAMP, D.J., ZIESERL, J.F., Jr. and AXEN, U. (1973) Computerized Quantitation of Drugs by Gas Chromatography-Mass Spectrometry. Anal. Chem., $\underline{45}$, 479-482.

16. AXEN, U., BACZYNSKYJ, L., DUCHAMP, D.J., KIRTON, K.T. and ZIESERL, J.F., Jr. (1973) Differentiation between endogenous and exogenous (administered) prostaglandins in biological fluids. In: Advances in the Biosciences, vol. 9, Bergström, S. Ed., Pergamon Press., pp. 109-116.

17. AXEN, U., GREEN, K., HÖRLIN, D. and SAMUELSSON, B.
    (1971) Mass spectrometric determination of picomole
    amounts of prostaglandins E_2 and $F_{2\alpha}$ using synthe-
    tic deuterium labeled carriers. Biochem. Biophys.
    Res. Comm., 45, 519-

18. THOMPSON, C.J., LOS, M. and HORTON, E.W. (1970).
    The separation, identification and estimation of
    prostaglandins in nanogram quantities by combined
    gas chromatography-mass spectrometry. Life Sci.,
    9, 983-988.

19. PACE-ASCIAK, C. and WOLFE, L.S. (1971) N-Butylboronate
    derivatives of the F prostaglandins. Resolution of
    prostaglandins of the E and F series by gas-liquid
    chromatography. J. Chromatogr., 56, 129-133.

20. GREEN, K. (1969) Gas chromatography-mass spectrometry
    of O-methyloxime derivatives of prostaglandins. Chem.
    Phys. Lipids, 3, 254-272.

21. VANE, F. and HÖRNING, M.G. (1969) Separation and
    characterization of the prostaglandins by gas chro-
    matography and mass spectrometry. Anal. Letters, 2,
    357-371.

22. HAMBERG, M. (1968) Metabolism of prostaglandins in rat
    liver mitochondria. Europ. J. Biochem., 6, 135-146.

23. SWEETMAN, B.J., FRÖLICH, J.C. and WATSON, J.T. (1973)
    Quantitative determination of prostaglandins A, B
    and E in the sub-nanogram range. Prostaglandins, 3,
    75-87.

24. Unpublished observations from this laboratory.

25. NICOSIA, S. and GALLI, G. (1974) A rapid gas chromato-
    graphic-mass spectrometric method for prostaglandin
    analysis at picomole levels. Anal. Biochem., 61,192-199.

26. HAMBERG, M. and SAMUELSSON, B. (1974) Prostaglandin
    endoperoxide. Novel transformations of arachidonic
    acid in human platelets. Proc. Nat. Acad. Sci. USA,
    71, 3400-3404.

27. LEVITT, M.J., JOSIMOVICH, J.B. and BROSKIN K.D. (1972)
    Analysis of prostaglandins by electron-capture gas
    chromatography. I. Thermal decomposition of hepta-
    fluorobutyrate methyl esters. Prostaglandins, 1, 121-
    131.

28. MIDDLEDITCH, B.S. and DESIDERIO, D.M. (1972) Gas
    chromatography of prostaglandin heptafluorobutyrate
    methyl esters. Prostaglandins, 2, 195-198.

29. WICKRAMASINGHE, A.J.F. and SHAW, R.S. (1974) An elec-
    tron-capture gas-liquid-chromatographic method for the
    determination of prostaglandin $F_{2\alpha}$ in biological fluids
    Biochem. J., 141, 179-187.

BIOSYNTHESIS OF PROSTAGLANDINS

Elisabeth Granström

Department of Chemistry
Karolinska Institutet
S-104 01 Stockholm 60, Sweden

The conversion of arachidonic acid into PGE_2 by homogenates of sheep vesicular glands was demonstrated in 1964 by two groups (1,2). Analogous formation of PGE_1 and PGE_3 from all-cis 8,11,14-eicosatrienoic acid and all-cis 5,8,11,14,17-eicosapentaenoic acid, was subsequently described (3,4). The parallel production of corresponding PGF compounds from the same precursors was demonstrated later.

The prostaglandin synthesizing system has been found in many other tissues and organs in a large number of species. That the biosynthesis of prostaglandins was not restricted to accessory genital glands was first demonstrated with guinea pig lung (5); later, the prostaglandin synthesizing system was found to have an almost ubiquitous distribution in mammals, and prostaglandin production was demonstrated in the kidney, spleen, gastro-intestinal tract, uterus, central nervous system, thymus, iris, skin, platelets, etc., as well as in a number of tissues of lower animals (for a recent review, see (6)).

A major clue to the elucidation of the mechanism of the prostaglandin biosynthesis was provided by isotope experiments, where the precursor acid was incubated with an enzyme preparation in the presence of $^{18}O_2$. It was found that all the three oxygen atoms of the prostaglandin molecule were derived from molecular oxygen, and later that the two oxygens in the ring originated in the same oxygen molecule. This last-mentioned finding was the result of an experiment where eicosatrienoic acid was converted into prostaglandins in an atmosphere consisting of 50% $^{16}O_2$ and 50% $^{18}O_2$. Mass spectrometric analysis of the products demonstrated the presence of equal amounts of compounds with two $^{18}O_2$ atoms in the ring and compounds with two $^{16}O_2$ atoms in these positions; "mixed" compounds were virtually

absent (7). From this study evolved the concept of an endoperoxide
intermediate in prostaglandin biosynthesis.

During the earlier experiments with 8,11,14-eicosatrienoic acid
and ovine vesicular gland, a large number of oxygenated products were
found (8,9). Apart from the expected PGE_1 and $PGF_{1\alpha}$, an isomer of
PGE_1 was also found, viz. PGD_1. Several monohydroxy fatty acids occurred among the products: 11-hydroxy-8-cis,12-trans,14-cis-eicosatri-
enoic acid, 15-hydroxy-8-cis,11-cis,13-trans-eicosatrienoic acid,
and a C_{17} monohydroxy acid that had been formed by loss of a three-
carbon fragment (C-9 to C-11) from the molecule, viz. 12-hydroxy-8-
trans,10-trans-heptadecadienoic acid. The three-carbon fragment was
also found and identified as malondialdehyde. Further products isolated were 15-hydroperoxy-PGE_1 and 15-keto-PGE_1.

The occurrence of some of these products, as well as the results
of certain isotope experiments, indicated that the biosynthesis of
prostaglandins initially proceeded via a lipoxygenase-like pathway.
The use of stereospecifically tritium labelled precursor acids as
substrate demonstrated that the initial step was the removal of the
L-hydrogen at C-13. This reaction was followed by introduction of
oxygen at C-11 to give 11-peroxy-8,12,14-eicosatrienoic acid (10).
This unstable intermediate is subsequently transformed into an endoperoxide by the following reactions: addition of oxygen at C-15,
isomerization of the Δ^{12} double bond, formation of the carbon-carbon
bond between C-8 and C-12, and attack by the oxygen radical at C-9.

The existence of an endoperoxide intermediate was postulated
already in 1965 (7), based on the findings from the $^{18}O_2/^{16}O_2$ experiments. The occurrence of a $9\alpha,11\alpha$-peroxidoprostanoate derivative,
oxygenated at C-15, was strongly supported by the isolation of 12-
hydroxy-8,10-heptadecadienoic acid plus malonaldehyde as byproducts
in the conversion of 8,11,14-eicosatrienoic acid, and was given
further support by the parallel formation of PGE, PGF_α and PGD compounds in certain systems.

Direct evidence for the formation of endoperoxides was obtained
later. When precursor acids were incubated with an enzyme preparation
from ovine vesicular gland, products were detected during the very
earliest stages which could be reduced to PGF_α compounds with $SnCl_2$
or $Na_2S_2O_4$ (11,12). These compounds could not be detected in incubations that had proceeded for longer periods. The unstable intermediates were isolated and were shown to consist of two products, both
prostaglandin endoperoxides (12,13). The isolation was facilitated
by including p-mercuribenzoate in the incubation medium, since it
had been shown that SH blocking agents reduced the rate of disappearance of the intermediates. The structures of the endoperoxides were
elucidated by chemical conversion of the compounds into known prosta-

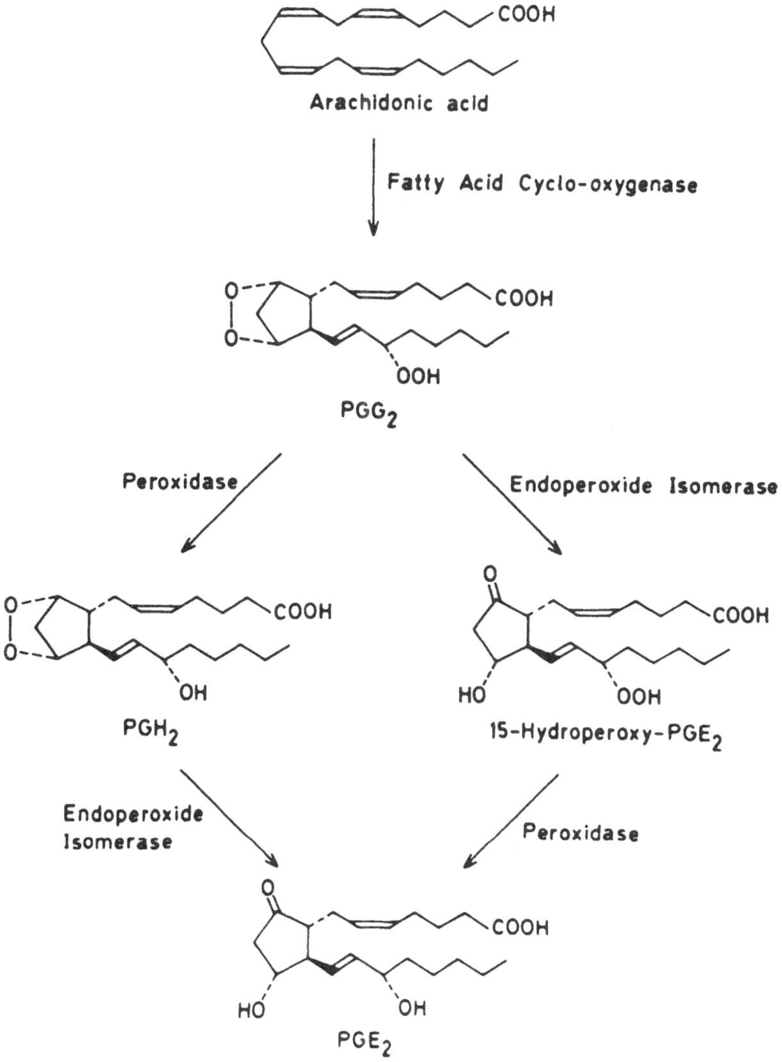

Fig. 1. Pathways in the biosynthesis of PGE2 from arachidonic
acid.

glandins. Thus, the endoperoxides formed from arachidonic acid were
15-hydroperoxy-9α,11α-peroxidoprosta-5,13-dienoic acid (PGG₂) and
15-hydroxy-9α,11α-peroxidoprosta-5,13-dienoic acid (PGH₂).

The steps involved in the biosynthesis of prostaglandins from
the precursor acid are thus first the formation of a PGG₂ compound,
catalyzed by the enzyme fatty acid cyclo-oxygenase. The endoperoxide
structure can then be isomerized into a β-hydroxy ketone, either
into a PGE or a PGD compound, and these reactions are catalyzed by
endoperoxide isomerases. A peroxidase catalyzes the reduction of the
hydroperoxy group at C-15 to a hydroxy group. Both the endoperoxide
isomerases and the peroxidase are stimulated by reduced glutathione,
and the peroxidase might be identical with glutathione peroxidase.
The preferred pathway in the biosynthesis of PGE compounds in the
vesicular gland seems to be PGG → 15-hydroxy-PGE → PGE (14) (Fig. 1).

The formation of PGF compounds involves reduction of the endoperoxide structure. A number of factors favouring this reduction
have been identified, e.g. copper-dithiol complexes (15), L-epi-
nephrine (16), ferrihaem and glutathione (12). It has been suggested
that the formation of PGF compounds from the endoperoxides is a
non-enzymatic process (17,18), due to the ease and efficiency with
which cyclic endoperoxides are reduced by cofactors and/or metallic
ions. This is in contrast to the established enzymic formation of
PGF from PGE compounds, catalyzed by the enzyme prostaglandin E
9-keto reductase. It has also been suggested that the stimulatory
effect of Cu²⁺ on PGF formation is inhibition of the endoperoxide
isomerase, thereby inhibiting PGE biosynthesis and thus promoting
PGF formation.

Many attempts have been made to purify the prostaglandin synthetase, which is mainly a microsomal enzyme system. Due to the
complex nature and instability of this enzyme system, great difficulties have been encountered, and this was particularly the case
when the over-all conversion from precursor acid into prostaglandins was used as enzyme assay. As the details of the prostaglandin
biosynthesis were elucidated, it became clear that the synthetase
consists of several components. The number of separate enzymes
making up the synthetase system is not known. Several reports on
the solubilization and partial purification of the prostaglandin
synthetase have been published. Detergents are generally used.
Sometimes a separation of the enzyme system into at least two components has been achieved during the purification procedure, viz.
the fatty acid cyclo-oxygenase and the endoperoxide isomerase
(18,19). A different approach is the use of oxygen consumption as
enzyme assay, which allows the selective study of the fatty acid
oxygenase (20).

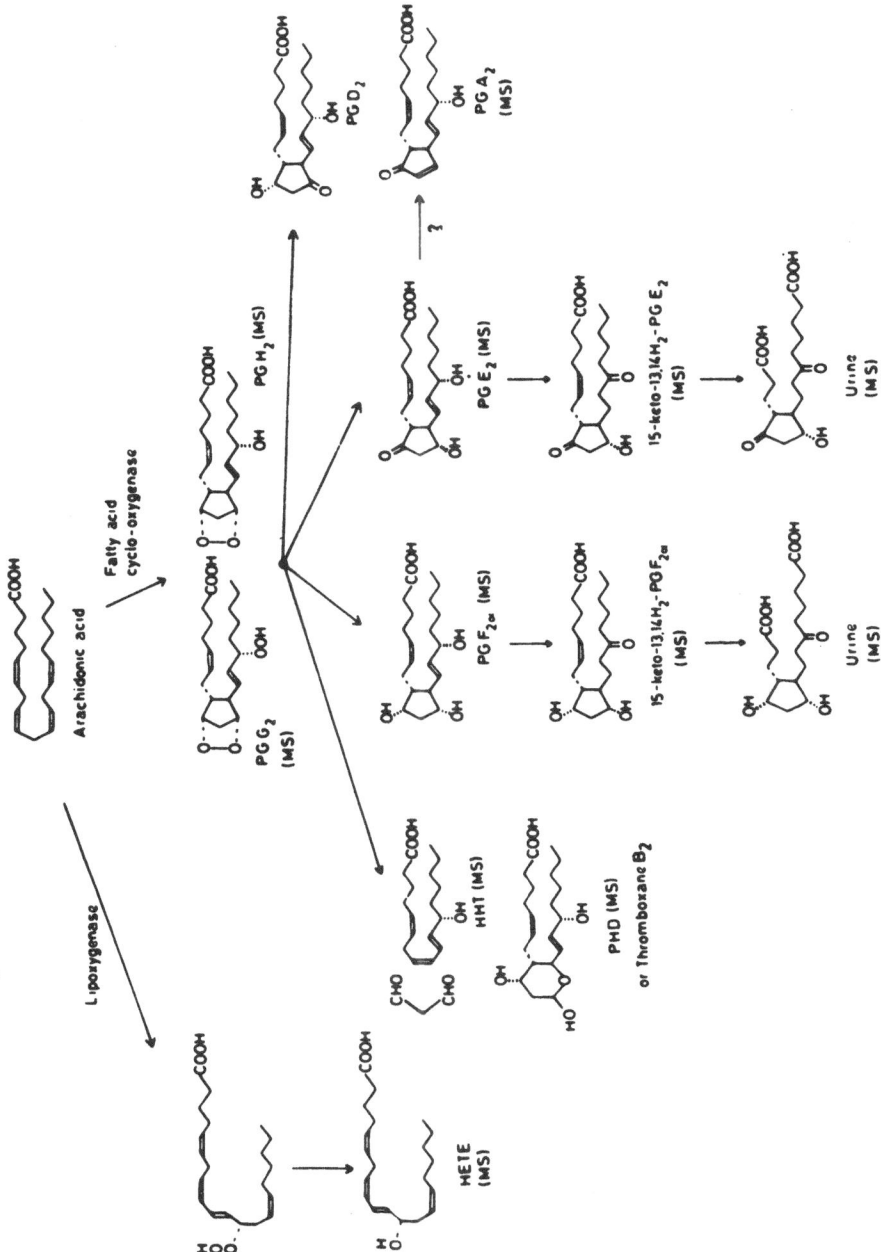

Fig. 2.   Summary of the metabolism of arachidonic acid in humans

Of great importance was the discovery that non-steroidal anti-inflammatory drugs efficiently inhibit the biosynthesis of prostaglandins (for review articles, see Refs. 21,22). This effect has been a useful tool in the study of the roles of prostaglandins in the organism. The inhibition is now known to be exerted on the fatty acid cyclo-oxygenase, viz. on the formation of the prostaglandin endoperoxides.

The metabolism of arachidonic acid in platelets has been extensively studied recently. Several findings indicated that the prostaglandin endoperoxides play a physiological role in platelet aggregation: they are released during thrombin-induced aggregation (13); they are powerful aggregating agents and induce the platelet release reaction as well (13,23); addition of arachidonic acid to platelets induces aggregation (24,25); blockade of the formation of endoperoxides by aspirin is accompanied by inhibition of the second, irreversible wave of aggregation (26); aggregating material is formed from arachidonic acid on incubation with preparations of sheep vesicular gland (27).

Arachidonic acid, when incubated with human platelets, was converted almost exclusively into compounds other than the classical prostaglandins (Fig. 2) (28). The initial step in one of the main pathways was catalyzed by a lipoxygenase, and the end product was HETE, a hydroxyeicosatetraenoic acid. The other main pathway was the formation of PGG_2, catalyzed by the cyclo-oxygenase; however, the endoperoxide was subsequently almost exclusively converted into non-prostanoate compounds. One was the earlier identified HHT, a C_{17} hydroxy acid which was formed from the endoperoxide by loss of maldondialdehyde. The other end product was a novel compound, which was first called PHD (the hemiacetal of 8-(1-hydroxy-3-oxo-propyl)-9,12L-dihydroxy-5,10-heptadecadienoic acid) but was later given the name thromboxane B_2 (TXB_2).

When platelet aggregation was induced by other agents, e.g. collagen, ADP, epinephrine, or thrombin, large amounts of end products from both pathways were released, whereas intact endoperoxides, PGE_2 and $PGF_{2\alpha}$ were found only in minor amounts (23, 29). Aspirin treatment efficiently inhibited the cyclo-oxygenase pathway; on the other hand the formation of HETE increased several-fold (29). Indomethacin treatment inhibited the endoperoxide formation and the release reaction but had no effect on PGG_2 induced release reaction (23).

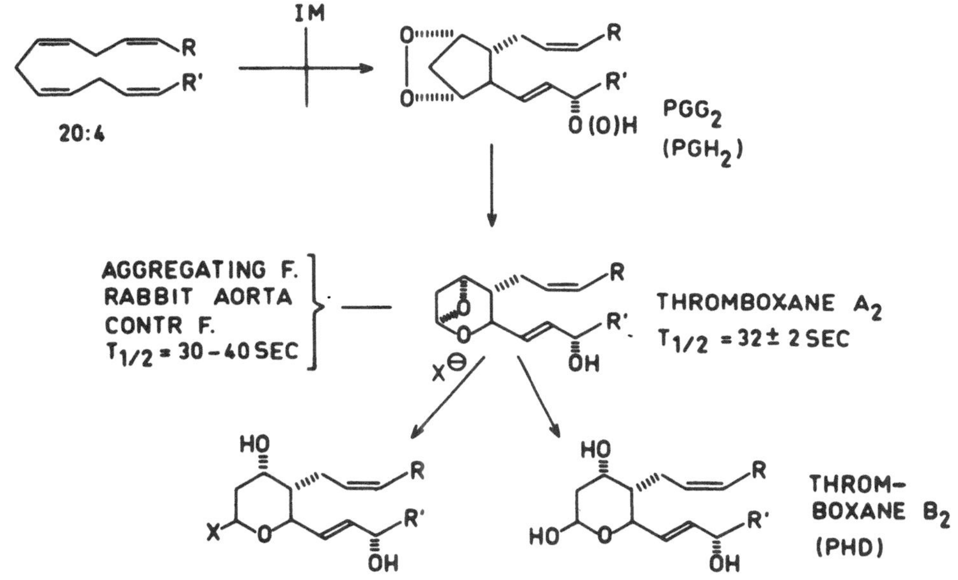

Fig. 3. Structures of aggregating factor and rabbit aorta contracting factor.

A subject with a hemostatic defect due to an abnormal release mechanism was found to have a platelet cyclo-oxygenase deficiency (23). Platelets from this subject did neither show aggregation nor release reaction on addition of collagen, arachidonic acid or epinephrine (second wave of aggregation), but responded normally in both respects to addition of PGG_2. After addition of labeled arachidonic acid, only traces of TXB_2 could be isolated; the formed amount of HETE was however normal. It was thus demonstrated that synthesis of the endo-peroxide, PGG_2, is essential in normal hemostasis.

A highly unstable and biologically very active intermediate has recently been discovered in the metabolic conversion of PGG_2 into thromboxane B_2. This intermediate had a $t_{1/2}$ of 32 sec and had been given the name thromboxane A_2 (TXA_2). Fig. 3 shows the structure of this compound, which contains a bicyclic oxane-oxetane ring system (30). TXA_2 was trapped by addition of methanol, ethanol or sodium azide, which converted the unstable intermediate into stable deriva-tives of TXB_2.

When arachidonic acid was incubated with a platelet suspension, and aliquots of this incubation were transferred to another plate-let preparation which had been preincubated with indomethacin, aggregation took place (30). This was not caused by endoperoxides, because the amounts formed were too small to induce platelet aggre-gation. Besides, the concentration of endoperoxides were highest after only about 20 sec incubation, whereas the aggregating factor appeared somewhat later (maximum about 40 to 60 sec). The half-life of this compound was about 30-40 sec. A factor with similar pro-perties could also be generated from incubations of platelets with PGG_2. The mode of formation of this aggregating factor and its pro-perties indicated that it was identical with thromboxane A_2.

Rabbit aorta contracting substance (RCS), which has been sup-posed to be an endoperoxide intermediate in prostaglandin biosynthe-sis (31-33), was recently found to contain at least two factors that contracted the rabbit aorta, i.e. PGG_2 and/or PGH_2, and an unstable factor (34). The half-life of the endoperoxides at 37ºC was about 5 min, whereas that of the unstable factor, which was responsible for the major part of the activity of RCS, was about 30 sec. The properties of this unstable component of RCS and its mode of forma-tion indicate that also this factor is identical with thromboxane A_2.

REFERENCES

1. van Dorp., D.A., Beerthuis, R.K., Nugteren, D.H. and Vonkeman, H. Biochim. Biophys. Acta 90 (1964) 204.

2. Bergström, S., Danielsson, H. and Samuelsson, B., Biochim. Biophys Acta <u>90</u> (1964) 207.

3. Bergström, S., Danielsson, H., Klenberg, D. and Samuelsson, B.,
   J. Biol. Chem. <u>239</u> (1964) PC 4006.

4. van Dorp, D.A., Beerthuis, R.K., Nugteren, D.H. and Vonkeman, H.,
   Nature, <u>203</u> (1964) 839.

5. Änggård, E. and Samuelsson, B., J. Biol. Chem. <u>240</u> (1965) 3518.

6. Samuelsson, B., Granström, E., Green, K., Hamberg, M. and
   Hammarström, S., Ann. Rev. Biochem. <u>44</u> (1975) 669.

7. Samuelsson, B., J. Am. Chem. Soc. <u>87</u> (1965) 3011.

8. Hamberg, M. and Samuelsson, B., J. Am. Chem. Soc. <u>88</u> (1966) 2349.

9. Nugteren, D.H., Beerthuis, R.K. and van Dorp, D.A., Rec. Trav.
   Chim. Pays-Bas, <u>85</u> (1966) 405.

10. Hamberg, M. and Samuelsson, B., J. Biol. Chem. <u>242</u> (1967) 5336.

11. Hamberg, M. and Samuelsson, B., Proc. Nat. Acad. Sci. USA, <u>70</u>
    (1973) 899.

12. Nugteren, D.H. and Hazelhof, E., Biochim, Biophys. Acta, <u>326</u>
    (1973) 448

13. Hamberg, M., Svensson, J., Wakabayashi, T. and Samuelsson, B.,
    Proc. Nat. Acad. Sci. USA, <u>71</u> (1974) 345.

14. Samuelsson, B. and Hamberg, M., <u>In</u>: Prostaglandin Synthetase
    Inhibitors. Eds. H.J. Robinson and J.R. Vane, Raven Press,
    New York. 1974, p.107.

15. Lee, R.E. and Lands, W.E.M., Biochim. Biophys. Acta, <u>260</u> (1972)
    203.

16. Sih, C.J., Takeguchi, C. and Foss, P., J. Am. Chem. Soc. <u>92</u>
    (1970) 6670.

17. Pace-Asciak, C. and Nashat, M. Biochim. Biophys. Acta, <u>388</u> (1975)
    243.

18. Chan, J.A., Nagsawa, M., Takeguchi, C. and Sih, C.J., Biochemistry,
    <u>14</u> (1975) 2987.

19. Miyamoto, T., Yamamoto, S., and Hayaishi, O., Proc. Nat. Acad.
    Sci. USA, <u>71</u> (1974) 3645.

20. Rome, L.H. and Lands, W.E.M., Prostaglandins, 10 (1975) 813.

21. Ferreira, S.H. and Vane, J.R., Ann. Rev. Pharmacol. 14 (1974) 57.

22. Flower, R.J. and Vane, J.R., Biochem. Pharmacol. 23 (1974) 1439.

23. Malmsten, C., Hamberg, M., Svensson, J. and Samuelsson, B., Proc. Nat. Acad. Sci. USA, 72 (1975) 1446.

24. Vargaftig, B.B. and Zirinis, P., Nature (New Biol.) 244 (1973) 114.

25. Silver, M.J., Smith, J.B., Ingerman, C. and Kocsis, J.J., Prostaglandins, 4 (1973) 863.

26. Zucker, M.B. and Petersen, J., Proc. Soc. Exp. Biol. Med. 127 (1968) 547.

27. Willis, A.L., Science, 183 (1974) 325.

28. Hamberg, M. and Samuelsson, B., Proc. Nat. Acad. Sci. USA, 71 (1974) 3400.

29. Hamberg, M., Svensson, J. and Samuelsson, B., Proc. Nat. Acad. Sci. USA, 71 (1974) 3824.

30. Hamberg, M., Svensson, J. and Samuelsson, B., Proc. Nat. Acad. Sci. USA, 72 (1975) 2994.

31. Piper, P.J. and Vane, J.R., Nature, 223 (1969) 29.

32. Vargaftig, B. and Dao, N., Pharmacology, 6 (1971) 99.

33. Vargaftig, B. and Zirinis, P., Nature (New Biol.), 244 (1973) 114.

34. Svensson, J., Hamberg, M. and Samuelsson, B., Acta Physiol. Scand. 94 (1975) 222.

METABOLISM OF PROSTAGLANDINS

Elisabeth Granström

Dept. of Chemistry, Karolinska Institutet

S-104 01 Stockholm 60, Sweden

The major initial steps in the degradation of the prostaglandins are the oxidation of the secondary alcohol group at C-15 and the reduction of the Δ^{13} double bond (for a recent review, see 1). The two enzymes that catalyze these reactions are 15-hydroxyprostanoate dehydrogenase and Δ^{13} reductase. The metabolites formed by these reactions, the 15-keto and the 15-keto-13,14-dihydro compounds, have considerably lower biologic activities than the parent prostaglandins; thus, the dehydrogenation has been considered a key step in the biological inactivation of prostaglandins. To a certain extent, reduction of the formed keto group to a hydroxyl may also take place. The biological activity of the resulting compound, the 13,14-dihydro prostaglandin, may be considerable in certain systems.

15-Hydroxyprostanoate dehydrogenase and Δ^{13} reductase have a widespread occurrence (2). In recent years, 15-hydroxyprostanoate dehydrogenases (PGDH) have been partially purified and their properties studied. The sources have been various organs from swine, cattle, monkey, chicken, human, guinea pig, rat, etc.; generally kidney, lung or placenta, but also other organs or tissues (3-12). The various dehydrogenases studied differ somewhat in physical properties, cofactor requirements and affinities for their substrates. A PGDH has recently been demonstrated in human erythrocytes (13); the meaning of this finding is not clear, since earlier reports indicate that prostaglandins are not metabolized by this pathway in blood.

The activity of the dehydrogenase is no doubt an important factor in the control of the biological inactivation of the prostaglandins. Inhibitors of this enzyme are e.g. certain diuretic drugs, xylocain, arachidonic acid, thyroid hormones, diphloretin phosphate, aspirin and indomethacin (5, 14-17); certain tricyclic antidepressant drugs on the other hand were found to stimulate the enzyme (14). The activity of PGDH has also been found to vary during certain conditions, e.g. pregnancy and nephrogenesis (18-21).

Comparatively little work has as yet been carried out with the Δ^{13} reductase. The enzyme has recently been partially purified from chicken heart and human placenta (22,23)

The formation of the 15-keto-13,14-dihydroprostaglandins is a very rapid process, and due to the widespread occurrence of the metabolizing enzymes in the body, it is highly unlikely that the prostaglandins should be circulating hormones. The half-lives of PGE_2 and $PGF_{2\alpha}$ in the circulation are very short; probably less than 30 seconds (1). In contrast, their corresponding 15-keto-13,14-dihydro metabolites have half-lives of about 8 min in the circulation and occur in 20 - 70 fold higher concentrations. From excretion data and data obtained from infusion experiments in the human, it has been possible to calculate the levels of primary prostaglandins in blood plasma: the concentration should not exceed 2 pg/ml. The levels of the 15-keto-13,14-dihydro compounds were measured in a number of human subjects (24), and they ranged between 5 and 90 pg/ml; for 15-keto-13,14-dihydro-PGE_2 the plasma levels were 28±14 pg/ml and 21* 14 pg/ml for males and females, respectively; for 15-keto-13,14-dihydro-$PGF_{2\alpha}$ the corresponding figures were 47±29 and 24±6.5 pg/ml.

The initial, rapid dehydrogenation at C-15 can be prevented by modifications of the molecule at or near this carbon atom. As a result, prolongation of the half-life and thus of the biological effects of the prostaglandin may be obtained. Investigations of the metabolism of the $PGF_{2\alpha}$ analogues, 15-methyl-, 16,16-dimethyl-, and 17-phenyl-18,19,20-trinorprostaglandin $F_2\alpha$ have been carried out in the human, and the half-lives of these biologically very potent prostaglandins were found to be considerably longer than that of the parent compound (25, Fig. 1). This was particularly the case with the two first-mentioned analogues, where the dehydrogenation was completely blocked. The increase in the half-life of the 17-phenyl analogue was somewhat less pronounced, due to the fact that this compound is to some extent metabolized by dehydrogenation at C-15 (8, 26).

It was earlier believed that prostaglandins of the E type are not converted into F derivatives in the body. However, numerous later reports have indicated that this does occur in many tissues and species. The first report demonstrated that in the guinea pig the main urinary metabolite of PGE_2 is an F_β derivative, 5β,7α-dihydroxy-11-ketotetranorprostanoic acid (1). An <u>in vitro</u> study on the

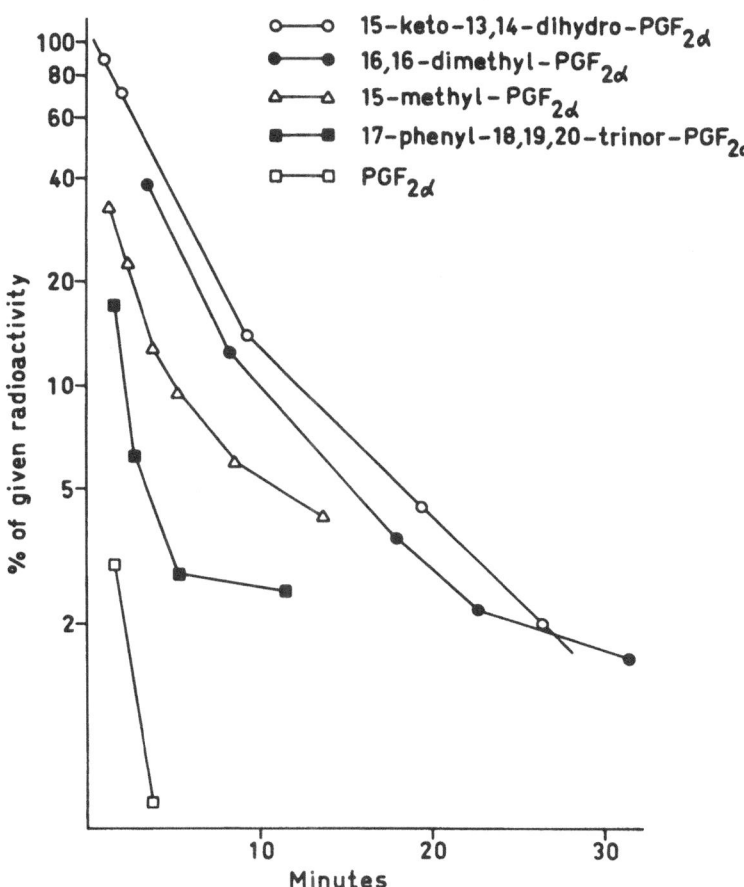

Fig. 1. Disappearance of tritium-labelled PGF$_{2\alpha}$, 15-methyl-PGF$_{2\alpha}$, 16,16-dimethyl-PGF$_{2\alpha}$, 17-phenyl-18,19,20-trinor-PGF$_{2\alpha}$ and 15-keto-13,14-dihydro-PGF$_{2\alpha}$ from plasma after intravenous injection into a human female.

metabolism of PGE_2 in guinea pig liver revealed that a considerable
part of the added PGE_2 was converted into $PGF_{2\alpha}$ and metabolites
thereof. No $PGF_{2\beta}$ derivatives were found in this study (1). Later
the presence of an enzyme designated prostaglandin E-9 ketoreductase has been detected in many tissues from several species: human,
sheep, chicken, monkey, rat, pigeon, swine, rabbit, horse (see 27
and 28, and references in these papers). The 9-ketoreductases from
these different sources seem to differ considerably in their properties, both regarding intracellular localization, cofactor requirements and substrate specificities. In addition, an enzyme that
converts 15-keto-13,14-dihydro-$PGF_{2\alpha}$ into 15-keto-13,14-dihydro-
PGE_2 was recently demonstrated in rat liver (29).

In the last few years, interest has been taken in the metabolism of prostaglandins of the A type. Earlier investigations demonstrated that the biological activity of PGA_1 and A_2 survived the
passage through the lungs in perfusion experiments(30,31),and this
was by most authors regarded a proof that PGAs are not degraded
by the lungs, but are in fact circulating hormones. Many radioimmunoassays have thus been developed for the measurement of PGA in peripheral plasma, and attempts have been made to correlate the results of these methods with various physiological and pathological
conditions. The published levels have varied between several hundred pg/ml and several ng/ml. However, a mass spectrometric method
for quantitative PGA_2 measurements demonstrated that the normal
human plasma levels are below 6 pg/ml which was the limit of detection for this method (32). Since PGAs are substrates for the PGDH,
it is not very likely that they should escape degradation; the results of the earlier bioassay studies may just as well be explained
by the formation of biologically active metabolites in the lung
(30,31). Metabolism of prostaglandins of the A type has also been
demonstrated in several recent papers. In rabbit lung the appearance of a less polar metabolite and some more polar material was demonstrated in perfusion experiments (33). In an in vitro study u-
sing rabbit kidney, Attallah et al. found several metabolites of
PGA_1 (34); two of these were tentatively identified as 13,14-dihyd-
ro-PGA_1 and 15-keto-13,14-dihydro-PGA_1. The biological activity of
13,14-dihydro-PGA_1 was similar to that of PGA_1.

The metabolism of PGAs in blood or plasma has been studied. It
was earlier demonstrated that PGA is bound to plasma proteins to a
higher degree than the more polar PGE and PGF (35, 36). Human erythrocytes convert PGA into a highly polar compound; this transformation does not occur in plasma (37,38). An isomerase that catalyzes
the conversion of prostaglandins of the A type into PGC has been
found in several species and partially purified (39-41).

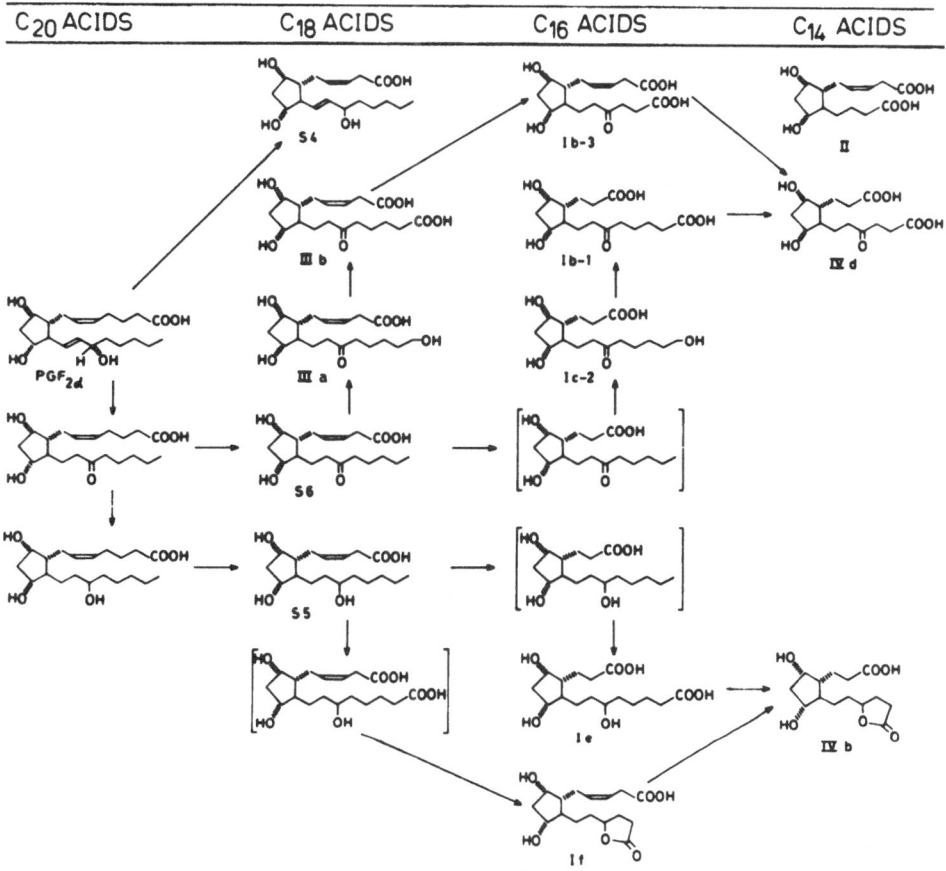

Fig. 2. Tentative pathways in the metabolism of $PGF_{2\alpha}$ in the human. Compounds within brackets have not been identified.

The prostaglandins can undergo many other metabolic fates prior to the final excretion into urine (see 1 and references therein). ω-Oxidation is a major pathway: both 19-hydroxylated and 20-hydroxylated metabolites are found. The presence of 19-hydroxylated prostaglandins of the A and B type in stored human seminal plasma was demonstrated earlier (42). Recently, it has been suggested that these compounds may be artifacts, since analysis of fresh human seminal plasma revealed mainly the presence of large quantities of 19-hydroxylated PGE_1 and PGE_2 (43, 44).

The further metabolism of prostaglandins has been extensively studied in several species: man, rat, guinea pig, rabbit (1), monkey (45). Prior to excretion in the urine, the compounds are metabolized via several pathways: one or two steps of β-oxidation from the original carboxyl end to yield dinor or tetranor compounds; ω-oxidation to afford $\omega 1$ or $\omega 2$ hydroxy compounds and eventually dicarboxylic acids, and in some species β-oxidation also from the ω end of the dioic acids with the formation of C_{14} metabolites (tetranor, ω-dinor or dinor, ω-tetranor). PGB derivatives may be found as metabolites of PGE. A deoxy-prostaglandin, $7\alpha,9\alpha$-dihydroxy-(dinor, ω-tetranor)-prost-3-en-1,14-dioic acid, was identified as one of the major metabolites of $PGF_{2\alpha}$ in the human (1); it has later also been identified as a metabolite in the rhesus monkey (45). A large number of urinary metabolites, formed by combinations of these reactions, have been identified (cf. Fig. 2). Thus, the main metabolites of PGE_2 and $PGF_{2\alpha}$ in the human are 7α-hydroxy-5,11-diketotetranorprostane-1,16-dioic acid and $5\alpha,7\alpha$-dihydroxy-11-keto-tetranorprostane-1,16-dioic acid, respectively.

Quantitative data on excretion of these metabolites have been used together with information on recovery of metabolites from their precursors to calculate the synthesis of prostaglandins of the E and F groups. For human males, a daily PGE production of 46 - 333 μg was found; for females 18 - 38 μg. The PGF production was 42 - 120 μg and 36 - 61 μg, respectively (1).

REFERENCES

1. Samuelsson, B., Granström, E., Gréen, K., Hamberg, M., and Hammarström, S., Ann. Rev. Biochem. 44 (1975) 669.

2. Änggård, E., Larsson, C., and Samuelsson, B., Acta Physiol. Scand. 81 (1971) 396.

3. Änggård, E., and Samuelsson, B., Arkiv Kemi 25 (1966) 293.

4. Oliw,E., Lundén, I. and Ånggård, E. In: Advances in Prostaglandin and Thromboxane Research. (Eds. B. Samuelsson and R. Paoletti). Raven Press, New York, 1976. Vol. 1, p. 147.

5. Nagasawa, M., Chan, J.A. and Sih, C.J., Prostaglandins 8 (1974)
   221.

6. Hansen, H.S., Prostaglandins, 8 (1974) 95.

7. Lee, S-C. and Levine, L., J. Biol. Chem. 250 (1975) 548.

8. Sun, F.F., Armour, S.B., Bockstanz, V.R. and McGuire, J.C.,
   In: Advances in Prostaglandin and Thromboxane Research. (Eds.
   B. Samuelsson and R. Paoletti). Raven Press, New York, 1976.
   Vol. 1, p. 163.

9. Braithwaite, S.S. and Jarabak, J., J. Biol. Chem. 250 (1975)2315.

10. Schlegel, W. and Greep, R.O., Eur. J. Biochem. 56 (1975) 245.

11. Yamasaki, M. and Sasaki, M., Biochem. Biophys. Res. Commun. 66
    (1975) 255.

12. Lee, S-C., Pong, S-S., Katzen, D., Wu, K-Y. and Levine, L.,
    Biochemistry 14 (1975) 142.

13. Kaplan, L., Lee, S-C. and Levine, L., Arch. Biochem. Biophys.
    167 (1975) 287.

14. Tai, H-H., and Hollander, C.S., In: Advances in Prostaglandin
    and Thromboxane Research. (Eds. B. Samuelsson and R. Paoletti).
    Raven Press, New York, 1976. Vol. 1, p. 171.

15. Thaler-Dao, H., Saintot, M., Baudin, G., Descomps, B. and Crastes
    de Paulet, A. Ibid., p. 177.

16. Pace-Asciak, C. and Cole, S. Experientia 31 (1975) 143.

17. Crutchley, D.J. and Piper, P.J., Brit. J. Pharmacol. 54 (1975)
    301.

18. Sun, F.F. and Armour, S.B., Prostaglandins 7 (1974) 327.

19. Bedwani, J.R. and Marley, P.B., Brit. J. Pharmacol. 53 (1975) 547.

20. Carminati, P., Luzzuni, F. and Lerner, L.J., Prostaglandins 8
    (1974) 205.

21. Pace-Asciak, C., J. Biol. Chem. 250 (1975) 2795.

22. Lee, S-C. and Levine, L., Biochem. Biophys. Res. Commun. 61 (1974) 14.

23. Westbrook, C. and Jarabak, J., Biochem. Biophys. Res. Commun. 66 (1975) 541.

24. Samuelsson, B. and Green, K., Biochemical Med. 11 (1974) 298.

25. Granström, E., And Hansson, G., In: Advances in Prostaglandin and Thromboxane Research. (Eds. B. Samuelsson and R. Paoletti). Raven Press, New York, 1976. Vol. 1, p. 215.

26. Granström, E., Prostaglandins 9 (1975) 19.

27. Levine, L., Wu, K-Y. and Pong, S-S., Prostaglandins 9 (1975) 531.

28. Hensby, C.N., Biochim. Biophys. Acta 409 (1975) 225.

29. Pace-Asciak, C., J. Biol. Chem. 250 (1975) 2789.

30. Piper, P.J., Vane, J.R. and Wyllie, J.H. Nature 225 (1970) 600.

31. Horton, E.W. and Jones, R.L. Brit. J. Pharmacol. 37 (1969) 705.

32. Steffenrud, S. In: Advances in Prostaglandin and Thromboxane Research. (Eds. B. Samuelsson and R. Paoletti). Raven Press, New York, 1976. Vol. 2, p. 866 (abstr.)

33. Gross, K.B. and Gillis, C.N. Biochem. Pharmacol. 24 (1975) 1441.

34. Attallah, A.A., Duchesne, M.J. and Lee, J.B. Life Sci. 16 (1975) 1743.

35. Raz, A. Biochem. J. 130 (1972) 631.

36. Attallah, A.A. and Schussler, G.C. Prostaglandins 4 (1973) 479.

37. Golub, M.S., Zia, P.K. and Horton, R. Prostaglandins 8 (1974) 13.

38. Smith, J.B., Silver, M.J., Ingerman, C.M. and Kocsis, J.J., Prostaglandins 9 (1975) 135.

39. Jones, R.L. and Cammock, S. Adv. Biosci. 9 (1973) 61.

40. Polet, H. and Levine, L. Arch. Biochem. Biophys. 168 (1975) 96.

41. Ho, P.P.K., Towner, R.D. and Sullivan, H.R. Prep. Biochem. 4 (1974) 257.

42. Hamberg, M. and Samuelsson, B.   J. Biol. Chem. <u>241</u> (1966) 257.

43. Taylor, P.L, and Kelley, R.W.   Nature (Lond) <u>250</u> (1974) 665.

44. Jonsson, H.T., Middleditch, B.S. and Desiderio, D.M.   Science <u>187</u> (1975) 1093.

45. Sun, F.F. and Stafford, J.E.   Biochim. Biophys. Acta <u>369</u> (1974) 95.

SCREENING FOR INHIBITORS OF PROSTAGLANDIN

AND THROMBOXANE BIOSYNTHESIS

Ryszard J. Gryglewski

Department of Pharmacology, Copernicus
Academy of Medicine in Cracow
31-531 Cracow, 16 Grzegórzecka, Poland

INTRODUCTION

Many in vitro methods have been employed to predict
the anti-inflammatory potency of newly synthetized compounds. These methods are essentially based on the interaction of drugs with enzymic and non-enzymic proteins.
Anti-inflammatory drugs are supposed to protect albumin
against heat denaturation (Mizushima, 1964), to displace
marker compounds from binding sites of albumin (Skidmore
and Whitehouse, 1965), to induce fibrinolysis in plasma
clots (Gryglewski, 1966), to accelerate disulfide inter-
change reaction between serum protein and sulfhydryl
reagents (Gerber et al., 1967), to stabilize erythrocyte
(Brown et al., 1967) and lysosomal (Miller and Smith,
1966) membranes as well as to inhibit a wide range of
enzymes including uncoupling of oxydative phosphorylation (Whitehouse and Haslam, 1962) and inhibition of
cyclic-AMP phosphodiesterase (Weinryb et al., 1972;
Moffat et al., 1972).
    None of these methods seems to be satisfactory
for a predictive assessment of anti-inflammatory drugs
(Glenn et al., 1973). These methods lack of specificity,
though that occasionally a parallelism between in vitro
and in vivo data might occur. In most instances mili-
molar concentrations of drugs have to be used to induce
an effect in vitro, whereas only micromolar concentrations are expected to be reached in body fluids in vivo.
Neither there is a convincing evidence that any of the
above in vitro reactions is essentially vital for the

mechanism of anti-inflammatory activity of drugs. In 1971
Vane and his colleagues (Vane, 1971; Smith and Willis
1971; Ferreira et al., 1971) have discovered that the
target biomolecule for non-steroidal anti-inflammatory
drugs is arachidonic acid cyclo-oxygenase, a component
of prostaglandin synthetase system.

Vane (1972 a,b) has postulated that aspirin-like
drugs exert their pharmacological action through inhibition of prostaglandin biosynthesis in vivo, and his
concept has been supported by numerous evidence from
different laboratories, as reviewed by Flower (1974).
Inhibition of prostaglandin biosynthesis in vitro is
therefore the most promising rational approach to predict
anti-inflammatory activity for newly synthetized compounds
Unfortunately, it has become soon evident that there
hardly exists a direct interrelationship between anti-
prostaglandin synthetase potency in vitro and anti-inflammatory efficiency in vivo. Several factors dissociate
the concordance of the in vitro and in vivo data, the
most important being pharmacokinetic properties of drugs
which cannot be foreseen in a single in vitro test. For
instance acidic non-steroidal anti-inflammatory drugs
are unequally bound to albumin (table I). It can be
expected that their anti-prostaglandin synthetase potency
in vivo is an exponent of their "plasma unbound" fraction.
Consequently only this fraction will contribute to an
acute anti-inflammatory effect. Therefore it is not enough
in vitro to measure the anti-enzymic potency of a drug.
This has to be supplemented by measuring its avidity to
bind to albumin. Combining of these two measurements
allows to obtain an approximate in vitro index for pre-
diction of anti-inflammatory potencies within a series
of prostaglandin synthetase inhibitors (Gryglewski, 1976a)

We have recently found (Dembińska-Kieć et al.
1976) that anti-enzymic activity of any prostaglandin
synthetase inhibitor is the same in various microsomal
preparations, providing that concentrations of the substrate and cofactors are the same. No matter how much
is different the basic activity of prostaglandin synthetases from brain, seminal vesicles or kidney medulla
microsomal preparations derived from various species -
all microsomal enzymes are equally susceptible to the
inhibitory action of aspirin (fig. 1). Most of acidic
anti-inflammatory drugs listed in table I behave like
aspirin, except for dipyrone. Dipyrone is a selective
inhibitor of brain synthetase sharing this property with
paracetamol (Flower and Vane, 1972).

Uniform susceptibility of microsomal prostaglandin
synthetases to most of anti-inflammatory drugs enables

Table I
Inhibition of prostaglandin biosynthesis in bovine seminal vesicle microsomes and displacement of 8-anilino-1-naphthalene sulfonate from bovine serum albumin by anti-inflammatory drugs.

Drugs	Inhibition of PG synthetase IC_{50} (μM)	Binding to albumin IC_{50} (μM)
Acidic non-steroidal anti-inflammatory drugs		
Indomethacin	0.10	140
Mefenamic acid	0.25	160
Naproxen	13	10000
Niflumic acid	33	160
Ibuprofen	50	630
Fenoprofen	56	500
Phenylbutazone	148	1000
Aspirin	189	5600
Dipyrone	500	10000
Non-acidic non-steroidal anti-inflammatory drugs		
Flumizole	0.10	inactive at 10^3
L 8027[+]	5.90	650
Benzydamine	820	inactive at 10^4
Amidopyrine	1000	inactive at 10^4
Anti-inflammatory steroids		
Hydrocortisone	inactive at 300	not tested
Dexamethasone	inactive at 300	not tested
Fluocinolone	inactive at 300	not tested

[+]L 8027 is 1´(isopropyl-2-indolyl)-3-pyridyl-3-ketone (Deby et al., 1971).

to use a single microsomal preparation (e.g. bovine seminal vesicle microsomes, see table I) and to extrapolate the calculated anti-enzymic potencies of drugs on other tissues. There is a biological system that does not fit this scheme. In blood platelets arachidonic acid is converted via prostaglandin endoperoxides to thromboxane A_2 and those labile substances induce release reaction and aggregation of platelets (Hamberg et al., 1975). Non-steroidal anti-inflammatory drugs inhibit platelet aggregation, however, their anti-aggregatory and anti-prostaglandin synthetase potencies may differ from each other (compare tables I and III). Aspirin, dipyrone (table III), ditizole (Caprino et al., 1973) and L 8027 (Deby et al., 1971; Gryglewski et al., to be published) are the most representative examples.

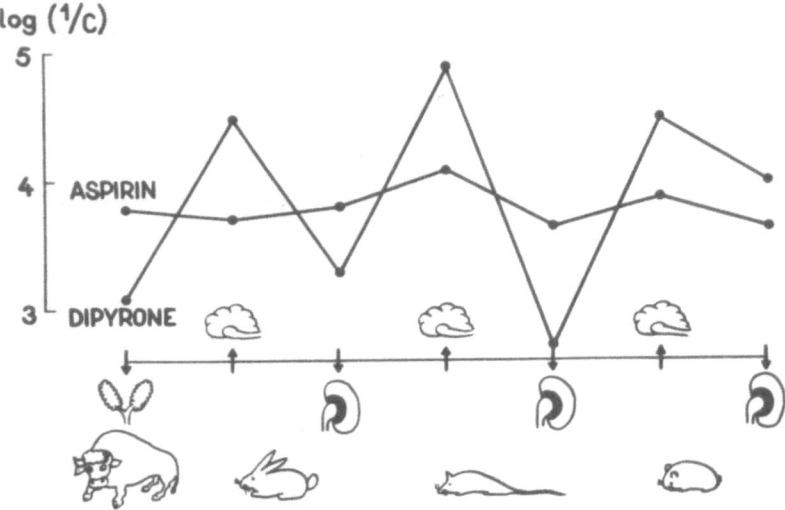

Figure 1. Schematic representation of uniform inhibition
by aspirin and selective inhibition by dipyrone of
microsomal prostaglandin synthetases derived from bovine
seminal vesicle microsomes and from rabbit, rat and
guinea pig brains and kidney medullas (Dembińska-Kieć
et al., 1976). Ordinate shows log (1/C), where C is
molar concentration of either aspirin or dipyrone
which results in a 50% inhibition of enzymic activity.

One possibility is that certain prostaglandin synthetase
inhibitors are also inhibitors of thromboxane A$_2$ isomerase. Another possibility is that platelet cyclo-
oxygenase has a preferential susceptibility to certain
prostaglandin synthetase inhibitors. In any case it is
strongly recommended to include the third test for
in vitro activity of prostaglandin synthetase inhibitors,
namely to investigate their anti-aggregatory properties
and their efficiacy to inhibit formation of thromboxane
A$_2$ by platelets.

INHIBITION OF PROSTAGLANDIN SYNTHETASE

Ram (Samuelsson et al., 1967) and bovine (Takeguchi et al., 1971) seminal vesicle microsomes are the
richest source of arachidonic acid cyclo-oxygenase,
a component of prostaglandin synthetase, which is sensitive to inhibitory action of non-steroidal anti-
inflammatory drugs (see Flower, 1974). The incubation

mixture contains microsomal enzyme, arachidonic acid
and usually glutathione and a phenolic compound, e.g.
adrenaline or hydroquinone, in a 0.05 - 0.1 M buffer
of pH 8.0 - 8.3. Ku and Wasvary (1975) have recommended
for studies with inhibitors to use a low concentration
of arachidonic acid (0.2 - 2.0 μM), low temperature
(25°C) and short time (10 min) of incubation. The
activity of the enzyme is estimated by measuring the
generated products (PGE$_2$, PGF$_{2\alpha}$, PGD$_2$ or malonyldialdehyde). The formed prostaglandins can be quantified
by bioassay (Vane, 1971) by radiochemical assay (Tomlinson et al., 1972), by radioimmunoassay (Levine,1972)
or by gas liquid chromatography - mass spectrometry
(Hamberg, 1972). The indirect methods of measuring of
enzymic activity comprise polarographic measuring of
oxygen uptake (Lands et al., 1971) and spectrophotometric measuring of transformation of adrenaline to
adrenochrome (Takeguchi and Sih, 1972). Other methods
for assay of prostaglandin synthetase activity have
been reviewed by Sih and Takeguchi (1973).

The concentration of arachidonic acid in the
incubation mixture dramatically influences the rate
of reaction, the ratio of products formed and the
anti-enzymic potencies of certain inhibitors (Flower
et al., 1973; Gryglewski, 1974; Ho and Esterman, 1974;
Robak et al., 1975). Unfortunately, in various laboratories arachidonic acid was used at a wide range of
concentrations varying from 0.1 - 1000 μM and reported
K$_m$ values also varied from 1 - 100 μM (see Robak et al.
1975). This may explain that anti-prostaglandin synthetase potencies of anti-inflammatory drugs differ considerably from one laboratory to another (see Gryglewski,
1974).

In our laboratory prostaglandin synthetase activity is assayed as follows. Homogenate of bovine seminal
vesicle glands in 0.25 M sucrose (1 : 4 w/v) is centrifuged at 10,000 x g for 10 min and the resulted supernatant is centrifuged at 100,000 x g for 60 min. The
precipitate is resuspended in distilled water and lyo-
filized. The microsomal powder (BSVM) usually contains
60 - 65% of protein. Incubation mixture comprises
3 - 6 mg of BSVM in 2 ml of 66 mM phosphate buffer pH
8.0, glutathione (166 μM), hydroquinone (45 μM) and
a test compound (0.01 - 1000 μM). After 5 min of pre-
incubation at 37°C sodium arachidonate (10 μM) is added
and incubation carried on for 20 min. The reaction is
stopped by boiling for 30 sec. The formed prostaglandins
are bioassayed in terms of PGE$_2$ equivalents directly in
the boiled and centrifuged (1000 x g) incubation mixture.

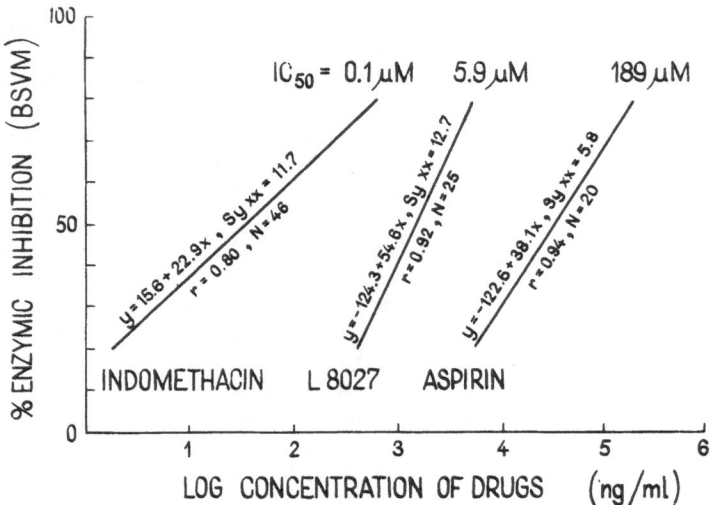

Figure 2. Regression lines and equations of regression lines for inhibition of prostaglandin synthetase in bovine seminal vesicle microsomes by indomethacin, L 8027 and aspirin. $S_{y.xx}$, standard deviation of regression line. r, Regression coefficient. N, Number of pairs. IC_{50}, The concentration of a drug needed for 50% inhibition of enzymic activity (μM).

Prostaglandins are assayed using a rat stomach strip (Vane, 1957) which is superfused with Krebs bicarbonate (3 ml/min, 37°C) containing a mixture of combined antagonists (Gilmore et al., 1968) and indomethacin (1 μg/ml) An average enzymic activity is 1360 ± 84 ng PGE_2 equivalents per 1 mg of BSVM protein (mean ± S.E., n = 100). K_m for arachidonic acid is 8 μM. Our radiochromatographic studies (Darska, 1976) revealed that in the above experimental conditions 68% of arachidonic acid is converted to radioactive products comprising in 44% of PGE_2, in 13% of PGD_2, in 2 % of $PGF_{2\alpha}$ and in 9% of unidentified products. On the other hand 97.7% of biological activity that can be extracted from from chromatographic plate is located at the narrow region of PGE_2 standard. No trace of malonyldialdehyde can be detected by tiobarbituric method in the incubation mixture (Robak et al., 1975). We have found that the potency of indomethacin to inhibit prostaglandin synthetase in BSVM is practically the same either calculated by direct bioassay of prostaglandin E_2-like activity in the crude incubation mixture or by bioassay of PGE_2 separated chromatographically or by radiochemical assay of PGE_2 formed from $^{14}C-1$-arachidonic acid or by measuring of oxygen uptake by Clark electrode. Similar concordance of the results for these four assay techniques has been found in case of aspirin and mefenamic acid (table I). Therefore we felt free to use larga manu our simple and fast technique for bioassay of anti-prostaglandin synthetase activity of anti-inflammatory drugs and newly synthetized compounds. Figure 2 shows the method for calculation of anti-prostaglandin synthetase potencies.

Non-steroidal anti-inflammatory drugs inhibit prostaglandin biosynthesis at a very early stage of the cyclo-oxygenase activity (Smith and Lands, 1971; Lands et al., 1975). Because of that aspirin-like drugs also supress the formation of cyclic endoperoxides (PGG_2 and PGH_2) and thromboxanes (Hamberg et al., 1975; Samuelsson et al., 1975). This last effect was observed in biological experiments as the supression of generation of a "rabbit aorta contracting substance" (RCS) by aspirin-like drugs (Piper and Vane, 1969; Vargaftig and Dao Hai, 1971; Gryglewski and Vane, 1972a,b; Willis, 1974) before the identity of RCS with prostaglandin endoperoxides and/or thromboxane A_2 was established (Samuelsson et al., 1975; Hamberg et al., 1975; Svensson et al., 1975; Needleman et al., 1976a). Most of acidic anti-inflammatory drugs are equipotent inhibitors of the release of RCS and the release of prostaglandins from chopped guinea pig lungs (Fjalland, 1974). Since RCS from lungs seems to be comprised mainly of thromboxane A_2 (Hamberg et al., 1975; Svensson et al., 1975; Gryglewski et al. 1976b) the results obtained by Fjalland (1974) are in agreement with the observation made by the Vane's group (Bunting et al., 1976a) that indomethacin and other acidic anti-inflammatory drugs do not inhibit thromboxane A_2 isomerase. The same group has reported that a basic anti-inflammatory drug - benzydamine is a weak (IC_{50} = 320 µM) but selective inhibitor of thromboxane A_2 isomerase in horse platelet microsomes. This finding has stimulated us to look for potential thromboxane A_2 isomerase inhibitors within a group of basic anti-inflammatory drugs. Indeed, a potent inhibitor of thromboxane A_2 biosynthesis has been found and its properties will be discussed later.

Mechanism of inhibition of prostaglandin synthetase by anti-inflammatory drugs is complex and might differ for various drugs (Flower et al., 1973; Flower, 1974; Horodniak et al., 1974; Ku and Wasvary, 1975). The most obvious differences exist between aspirin and

the rest of acidic polycyclic aspirin-like drugs, as
well as between acidic and non-acidic anti-inflammatory
drugs. The anti-enzymic effect of indomethacin is time-
dependent, substrate-dependent (competetive) but amazing-
ly enough it is also irreversible. In other words indomethacin is a competetive inhibitor of prostaglandin
synthetase in BSVM provided that the substrate and the
inhibitor are added at the same time into the enzymic
preparation. However, when indomethacin is preincubated
for 5 min with microsomes then inhibition of the enzymic
activity is no longer dependent on the amount of arachidonic acid used (Robak et al., 1975). Most acidic anti-
inflammatory drugs behave like indomethacin (Ku and
Wasvary, 1975). It well might be that this type of inhibition is similar to the "active-site-directed irreversible inhibition" which has been described for inhibition
of adenosine desaminase by adenine derivatives (Schaeffer, 1971).

Inhibitory action on prostaglandin synthetase is
a common feature for all acidic non-steroidal anti-inflammatory drugs.Certain acidic drugs have an asymmetric
carbon atom in their molecules; in each case the in vivo
active enantiomer has been found to be a more potent
prostaglandin synthetase inhibitor than its partner (Ham
et al., 1972; Takeguchi and Sih, 1972; Tomlinson et al.,
1972; Shen et al., 1974; Gaut et al., 1975). This stereospecific effect could not be detected in other in vitro
tests (Mizushima et al., 1975).

Non-acidic anti-inflammatory drugs may or may not
inhibit the enzyme (table I). The best known non-acidic
inhibitors of prostaglandin synthetase are: 1'(isopropyl-
-2-indolyl)-3-pyridyl-3-ketone (L 8027) (Deby et al.,
1971), indoxole (Ham et al., 1972), benzydamine (Flower
et al., 1973) and flumizole (Wiseman et al., 1975). The
most potentis flumizole. It is at least as potent (table
I) or even more potent (Wiseman et al., 1975) than indomethacin. A comparative kinetic study of anti-enzymic
activities of these two potent prostaglandin synthetase
inhibitors should reveal the differences between the
mechanism of action of acidic and non-acidic non-steroidal anti-inflammatory drugs.

Steroidal anti-inflammatory drugs are not inhibitors of prostaglandin synthetase (Flower et al., 1972;
table I)though that they are capable to inhibit prostaglandin release from tissues (Gryglewski, 1976). This
type of activity implies the possibility of a direct
interaction of corticosteroids with biomembranes, which
might either impair the supply of endogenous substrates
for prostaglandin biosynthesis or inhibit transmembrane
transport of prostaglandins (Lewis and Piper, 1975).

BINDING TO ALBUMIN

Several reviews have been published on the character of drug-protein interaction (Brodie and Hogben,
1957; Meyer and Guttman, 1969; Settle et al., 1971) and
on the methods for studing of this interaction (Chignell,
1971). The binding of non-steroidal anti-inflammatory
drugs to protein is usually assessed by methods of equilibrium dialysis (Zaroslinski et al., 1974), circular
dichroism (Chignell, 1969) and displacement of a probe
from the probe-protein complex (Whitehouse et al., 1971).
We have chosen the last procedure. 1-Anilino-8-naphtha-
lene sulfonate (ANS) was used as a probe and bovine serum
albumin Cohn fraction V) (BSA) as an acceptor protein.
Displacement of ANS from BSA by prostaglandin synthetase
inhibitors was determined by measuring the intensity of
fluorescence due to the albumin-bound ANS (Daniel and
Weber, 1966). The excitation wavelenght was 380 nm and
the fluorescence maximum 485 nm. BSA (m.w. 66,000) at
a concentration of 10 μM in 0.1 M phosphate buffer of
pH 7.0 was incubated at 22°C for 30 min with a prostaglandin synthetase inhibitor at three to six concentrations (0.01 - 40 mM). Then ANS was added to yield the
final concentration of 40 μM. Quenching of ANS-BSA
fluorescence by prostaglandin synthetase inhibitors was
measured against fluorescence of a solution containing
BSA plus ANS only. A concentration of prostaglandin
synthetase inhibitor that quenched the fluorescence by
50% (IC_{50}) was graphically calculated. Each compound in
the highest concentration used was tested for intrinsic
fluorescence excited at 380 nm and for quenching ANS
fluorescence in methanol.
For special purposes the above experiments were
performed at various concentrations of BSA (1 - 50 μM)
in the presence of ANS at a concentration of 40 μM, as
well as at various concentrations of ANS (4 - 80 μM) in
the presence of BSA at a concentration of 10 μM. These
experiments allowed to calculate number of binding sites
for ANS in BSA (n) and an apparent association constant
(K_{app}) using the method of Scatchard (Sctachard, 1949;
Flangan and Ainworth, 1969; Wiethold et al., 1973).
In table II there is shown the influence of six anti-
inflammatory drugs on the number of binding sites in
BSA for ANS and on the affinity of ANS to BSA. Indomethacin, phenylbutazone and mefenamic acid decrease
number of binding sites and the affinity of ANS to BSA.
Aspirin decreases only the affinity, whereas non-acidic
chloroquine and amidopyrine are ineffective in both

Table II

Apparent association constants (K_{app}) and number of binding sites (n) of 1-anilino-8-naphthalene sulfonate to bovine serum albumin in the absence and in the presence of anti-inflammatory drugs at a concentration of 1 mM.

Drugs	$K_{app} \times 10^5 M^{-1}$	n
None	0.82	1.71
Indomethacin	0.52	0.32
Mefenamic acid	0.12	0.70
Phenylbutazone	0.51	0.93
Aspirin	0.45	1.51
Amidopyrine	0.90	1.66
Chloroquine	1.03	1.50

respects. This finding is in the favour of the concept that polycyclic acidic prostaglandin synthetase inhibitors are stronger ligands to enzymic and non-enzymic proteins than aspirin. Their anionic radicals interact with the surface polar groups of proteins (Skidmore and Whitehouse, 1966), whereas their lipophylic moieties are anchored into hydrophobic clefts of protein molecules (Chignell, 1971). Both sites of binding are essential for a profound distortion of the tertiary structure of proteins, and this effect may be quite helpful to kill cyclo-oxygenase (Gryglewski, 1974). However, an excessive potency to bind to proteins is not a desired property for prostaglandin synthetase inhibitors. Then the avidity to interact with any protein has to be compensated by a highly stereospecific and electronic arrangement of the molecule of an inhibitor which would fit to the code of an active site or to an allosteric area of the cyclo-oxygenase molecule. It is the case of indomethacin. On the other hand aspirin (a weak ligand to protein) has a unique property to act as a selective active-site acetylating agent for cyclo-oxygenase (Roth et al., 1975). This finding can explain a relatively strong (table III) and persistent anti-aggregatory action of aspirin.

For screening purposes it is sufficient to find IC_{50} for binding to albumin of a prostaglandin synthetase inhibitor (table I). The ratio of anti-enzymic and albumin-binding potencies roughly correlates with anti-inflammatory potencies in a series of compounds.

ANTI-AGGREGATORY ACTIVITY AND INHIBITION
OF THROMBOXANE A_2 GENERATION IN AGGREGATING PLATELETS

Screening for anti-aggregatory activity of prostaglandin synthetase inhibitors is recommended not only
in aim to confirm their anti-cyclo-oxygenase potency but
also there might be a chance to find a selective thromboxane A_2 isomerase inhibitor in platelets. Benzydamine is
an example for this possibility (Bunting et al., 1976 a).
Another possibility is that a selective inhibitor of
platelet cyclo-oxygenase will be found. In either case
one can stop thinking about anti-inflammatory drugs and
try to develop an antiplatelet drug with potential activity in thromboembolic diseases.

Aspirin (Weiss et al., 1968), phenylbutazone
(O'Brien, 1968), Indomethacin (Glenn et al., 1972),
sudoxicam (Constantine and Purcell, 1973), suprofen
(De Clerck et al., 1975) and a number of other prostaglandin synthetase inhibitors (Mustard and Packham,1975)
release ADP, serotonin and acid hydrolases from aggregating platelets. Anti-inflammatory drugs supress also
the generation of prostaglandins (Smith and Willis,1971;
Glenn et al., 1972; Patrono et al., 1976), prostaglandin
endoperoxides and thromboxanes (Hamberg et al., 1974,
1975; Samuelsson et al., 1975). These last two (or one
of them) are triggers of the release reaction in platelets, which is followed by platelet aggregation (Malmsten
et al., 1975). Therefore a great number of non-steroidal
anti-inflammatory drugs inhibits platelet aggregation
in vitro and in vivo in various species including human
beings (Mustard and Packham, 1975). In these experiments
aggregation has been usually induced by collagen that
probabely activates phospholipase A_2 in platelets thus
liberating endogenous arachidonic acid that is transformed to prostaglandin endoperoxides and thromboxane A_2
(Flower et al., 1975). We prefer to use for aggregation
exogenous arachidonic acid (Silver et al., 1973; Vargaftig and Zirinis, 1973; Willis and Kuhn, 1973) and to
measure in parallel platelet aggregation and thromboxane A_2 formation.

Our procedure is as follows. Blood from healthy
donors, who have not taken any drugs for at least 10 days
is collected from the anti-cubital vein (50 - 100 ml)
with 0.12 vol. of 0.1 M trisodium citrate. Alternatively
rabbit blood can be used. Platelet rich plasma (PRP) is
prepared by centrifugation at 400 x g for 10 min at room
temperature. Aggregation of platelets in 1 ml PRP is
monitored in a Chrono-log aggregometer at 37° C.

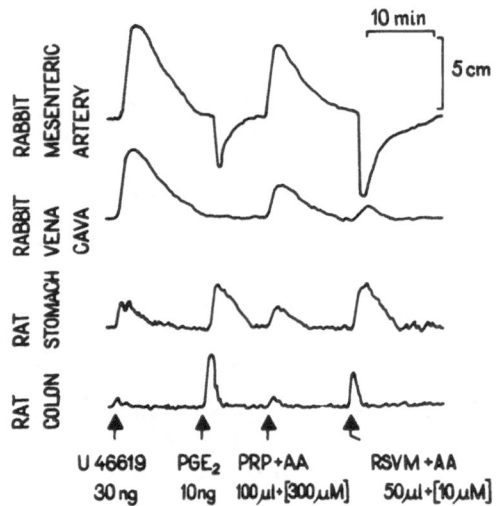

Figure 3. Bioassay of thromboxane A$_2$ which was generated
by human platelet rich plasma from arachidonic acid
(PRP + AA) after sixty seconds of incubation. The assay
organs were: rabbit mesenteric artery, rabbit vena cava,
rat stomach and rat colon superfused in cascade. The
organs were calibrated with 30 ng of a prostaglandin endoperoxide analog - (15S)-hydroxy-11α,9α-(epoxymethano)
prosta-5Z,13E-dienoic acid (U 46619), with 10 ng of PGE$_2$
and with the crude extract of prostaglandin endoperoxides
that were formed during 3 min incubation of ram seminal
vesicle microsomes with arachidonic acid (RSVM + AA).
Note that the profile of biological action of thromboxane
A$_2$ is similar to that of U 46619 but not to those of
PGE$_2$ and prostaglandin endoperoxides.

Drugs and the corresponding solvents (20 - 50 µl) are
added to PRP at zero time and six minutes later kalium
arachidonate at concentrations of 150 - 1500 µM for
human PRP and 30 - 300 µM for rabbit PRP is instilled.
Thirty to sixty seconds later a 100 ul aliquot of PRP
is withdrawn and immediately bioassayed for thromboxane
A$_2$, prostaglandin endoperoxides and prostaglandins.
          Differential bioassay of the above substances has
been based on the recent discovery of the Vane's group
(Bunting et al., 1976b) that strips of rabbit coeliac
and mesenteric arteries are relaxed by prostaglandin
endoperoxides PGG$_2$ and PGH$_2$) and contracted by thromboxane A$_2$ (fig. 3). A bank of assay organs consisting of

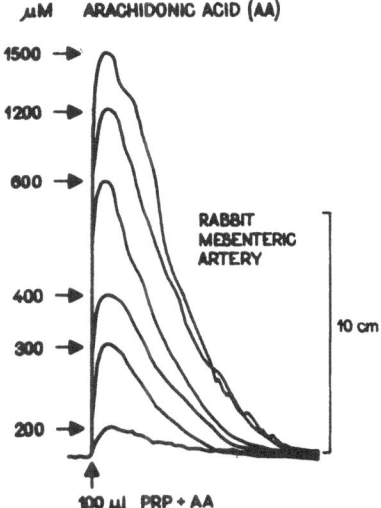

Figure 4. Dependance of the amount of thromboxane A$_2$
generated by human platelet rich plasma (PRP) on the
concentration of arachidonic acid (AA) which was used
for aggregation. Thromboxane A$_2$ was bioassayed by
contractions of rabbit mesenteric artery.

a rabbit mesenteric artery, a rabbit aorta or vena
cava, a rat stomach and a rat colon is superfused in
cascade (Vane, 1964) with Krebs bicarbonate (37°C,
3 ml/min) which contains a mixture of antagonists
(Gilmore et al., 1968) and indomethacin (1 μg/ml). The
contractile potency of thromboxane A$_2$ on rabbit aorta
(and vena cava) is about 50 times higher than that of
prostaglandin endoperoxides (Needleman et al., 1976 a),
whereas mesenteric artery is contracted by thromboxane
A$_2$ and relaxed by PGG$_2$ and PGH$_2$ (Bunting et al., 1976a).
All three vascular strips are insensitive to PGF$_{2\alpha}$ while
PGE$_2$ relaxes only a strip of mesenteric artery. On the
other hand rat colon is insensitive to thromboxane A$_2$
and prostaglandin endoperoxides, being conracted by
PGF$_{2\alpha}$ and PGE$_2$. Rat stomach is contracted by PGE$_2$, PGF$_{2\alpha}$
prostaglandin endoperoxides and thromboxane A$_2$ in a
decreasing order of potency (Bunting et al. 1976 a).
This natural and excellent differentiation between thromboxane A$_2$ and other biologically active products of arachidonic acid metabolism (fig. 3) may be reinforced by
checking the instability of a substance that behaves
like thromboxane A$_2$. A half life time of thromboxane A$_2$
is about 30 sec at 37°C (Hamberg et al., 1975).

Table III

Threshold anti-aggregatory concentrations of acidic non-steroidal anti-inflammatory drugs. Aggregation of rabbit platelet rich plasma was induced by arachidonic acid at a concentration of 150 μM. The listed concentrations of anti-inflammatory drugs also inhibited formation of thromboxane A_2 by platelet rich plasma, as measured by contractions of mesenteric artery (fig. 3 and 4). Each drug was tested at 4 - 6 concentrations in 3 - 10 separate experiments.

Drugs	Range of the threshold anti-aggregatory concentrations (μM)
Indomethacin	0.3 - 1
Dipyrone	3 - 30
Mefenamic acid	4 - 12
Aspirin	5 - 55
Niflumic acid	10 - 106
Phenylbutazone	32 - 81
Fenprofen	38 - 227
Ibuprofen	73 - 121
Naproxen	130 - 260

The aggregated PRP is sucked into a syringe through a millipore filter (0.8 μm) and the filtrate is kept for 2 min at 37°C. Then its contractile activity on vascular strips disappears or is greatly supressed.

Thromboxane A_2 is generated by aggregating human and rabbit PRP and the amount of formed thromboxane A_2 is dependant on the concentration of arachidonic acid which is used for aggregation (fig. 4). Minute amounts of thromboxane A_2 are also generated by low concentrations of arachidonic acid which are unable to induce platelet aggregation (e.g. 200 uM in fig. 4). Acidic anti-inflammatory drugs that inhibit platelet aggregation also inhibit formation of thromboxane A_2 (table III) at the very early stage of cyclo-oxygenation of arachidonic acid. Dipyrone and aspirin are much more active and naproxen is less active as anti-aggregatory agents as compared to their anti-cyclo-oxygenase activity in BSVM (tables I and III). However, all acidic anti-inflammatory drugs have something in common : their anti-aggregatory potencies hardly depend on the pro-aggregatory concentrations of arachidonic acid used (fig. 5). At a wide range of pro-aggregatory concentrations of

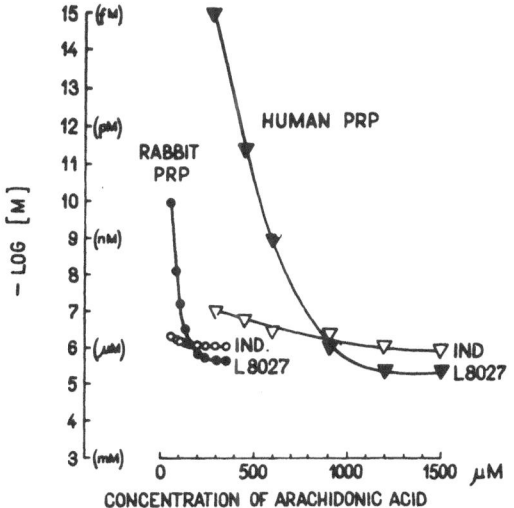

Figure 5. Threshold anti-aggregatory potencies of
1´(isopropyl-2-indolyl)-3-pyridyl-3-ketone (L 8027)
and indomethacin (IND) in human and rabbit platelet
rich plasmas (PRP). Ordinate: - log of threshold
anti-aggregatory concentration . Abscissa: micromolar
concentrations of arachidonic acid which were used to
induce platelet aggregation.

arachidonic acid, the anti-aggregatory concentrations
of these drugs may vary not more than in order of 10
in magnitude. The same is true for potencies of acidic
anti-inflammatory drugs to inhibit formation of thromboxane A_2 in PRP.
        There is a non-acidic anti-inflammatory agent,
L 8027 (Deby et al., 1971, fig. 5) that deserves a special attention. L 8027 is an inhibitor of cyclo-oxygenase in BSVM with IC_{50}= 5.9 uM (Fig. 1, table I). This
anti-cyclo-oxygenase potency is too low to explain its
potent anti-aggregatory action that is observed at a
concentration of 1 nM in rabbit PRP and at a concentration of 1 fM (!) in human PRP (fig. 5). This powerful
antiplatelet action of L 8027 occurs at the threshold
pro-aggregatory concentrations of arachidonic acid.When
concentrations of arachidonic acid are rised then the
anti-aggregatory potency of L 8027 declines to reach its
plateau at a region of 0.4 - 4 µM (fig. 5). A completely
different picture is seen for indomethacin. Its anti-
aggregatory potency hardly depends on the concentration
of arachidonic acid used (fig. 5). Indometacin is an

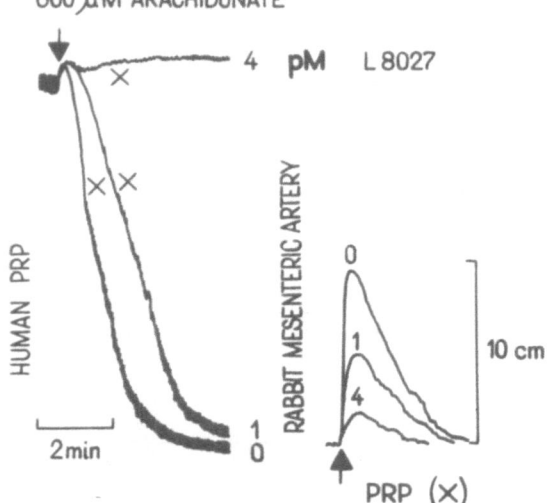

Figure 6. Simultaneous assay of anti-aggregating and
anti-thromboxane A_2 synthetase activities of L 8027
at concentrations of 1 and 4 pM. Human platelet rich
plasma (PRP) was aggregated with arachidonic acid at
a concentration of 600 μM in the absence (0) or in
the presence (1 and 4 pM) of L 8027. Crosses denote
the withdrawal of   100 μl aliquots of PRP for bioassay
of thromboxane A_2 using a strip of rabbit mesenteric
artery (fig. 3 and 4). Note that depression of thromboxane A_2 formation by a half with L 8027 at a concentration of 1 pM was not sufficient to inhibit platelet
aggregation. Inhibition of platelet aggregation occured
with L 8027 at concentration of 4 pM and then only minute
amounts of thromboxane A_2 were formed.

irreversible inhibitor of cyclo-oxygenase (Ku and Wasvary
1975; Robak et al., 1975). This fact can explain the
independence of anti-aggregatory potency of indomethacin
from the amount of arachidonic acid used for aggregation.
One cannot explain the mode of anti-aggregatory action
of L 8027 in terms of its anti-cyclo-oxygenase activity.
Indeed, L 8027 has been found to inhibit thromboxane
A_2 isomerase in platelet microsomes when crude extract
of prostaglandin endoperoxides is used as the substrate
(Gryglewski et al. to be published). Thus a simple
technique for platlet aggregation can help to differentiate between cyclo-oxygenase and thromboxane A_2 isomerase inhibitors. Fig. 5 shows that the inhibition of
arachidonate-induced platelet aggregation is parallel to

inhibition of thromboxane A_2 formation by platelets. These results strongly indicate that thromboxane A_2 is an essential link in process of platelet aggregation induced by arachidonic acid, what seems to contrast the opinion of Needleman et al. (1976b).

Summing up, additional tests for antiplatelet and anti-thromboxane A_2 synthetase activities of prostaglandin synthetase inhibitors are valuable complementary procedures that should be incorporated into the scheme of the in vitro screening.

CONCLUSIONS

Oxydative metabolism of arachidonic acid in microsomes gives rise to many biologically active products. Pharmacological interference in the enzymic conversion of arachidonic acid is or might be useful in the treatment of inflammation, pain, pyresis and arterial thrombosis. All acidic non-steroidal anti-inflammatory drugs inhibit prostaglandin biosynthesis by inactivation of arachidonic acid cyclo-oxygenase. A single in vitro test for inactivation of microsomal cyclo-oxygenase is not sufficient to predict anti-inflammatory potency of an inhibitor. This enzymic test should be supplemented by measuring the potency of an inhibitor to bind to albumin. The ratio of anti-enzymic and albumin-binding potencies gives a rough approximation of an index for anti-inflammatory potency in vivo within a series of chemical analogs. Some of non-acidic anti-inflammatory drugs also inhibit cyclo-oxygenase. One of them, 1´(isopropyl-2-indolyl)-3-pyridyl-3-ketone (L 8027) with a moderate anti-cyclo-oxygenase activity has been found to be a powerful inhibitor of thromboxane A_2 isomerase in platelets. This biochemical activity of L 8027 can explain its antiplatelet action. Therefore it is strongly recommended to test any prostaglandin synthetase inhibitor for its anti-aggregating and anti-thromboxane A_2 activites. Hydrocortisone and synthetic anti-inflammatory steroids inhibit the release of prostaglandins and thromboxane A_2 from stimulated intact cells, tissues and organs, but do not inhibit microsomal cyclo-oxygenase.

ACKNOWLEDGEMENTS

I gratefully acknowledge the generous grant of equipment from the Trustees of The Wellcome Trust, London, Great Britain.

REFERENCES

Brodie, B.B., and Hogben, C.A.M., 1957, Some physico-
    chemical factors in drug action, J.Pharm. Pharmacol.
    $\underline{9}$: 345 - 380.
Brown, J.H., Mackey, H.K., and Rigillo, D.A., 1967,
    A novel in vitro assay for anti-inflammatory agents
    based on stabilization of erythrocytes, Proc.Soc.
    exp. Biol. Med. $\underline{125}$: 837 - 843.
Bunting, S., Moncada, S., Needleman, P., and Vane, J.R.,
    1976a, Prostaglandin endoperoxides and thromboxane
    generating systems and their selective inhibition,
    Br. J. Pharmac. $\underline{56}$: 334P - 345P.
Bunting, S., Moncada, S., and Vane, J.R., The effects
    of prostaglandin endoperoxides and thromboxane A_2
    on strips of rabbit coeliac artery and certain
    other smooth muscle preparations, Proc. Br. Pharma-
    col. Soc. 1-2 April 1976b, p.48.
Caprino, L., Borelli, F., and Falchetti, R., 1973b,
    Effect of 4,5-diphenyl-2-bis-(2-hydroxyethyl)-ami-
    noxazol (Ditazol) on platelet aggregation, adhesiv-
    ness and bleeding time, Arzneimittel Forsch. $\underline{23}$;
    1277 - 1283.
Chignell, C.F., 1969, Optical studies of drug-protein
    complexesm III. Interaction of flufenamic acid and
    other N-aryl-anthranilates with serum albumin.
    Molec. Pharmac. $\underline{5}$: 455 - 462.
Chignell, C.F., 1971, Physical methods for studing drug-
    protein binding, in: Handb. Exp. Pharm. 28/1,Con-
    cepts in Biochemical Pharmacology, (B. Brodie and
    J.R. Gilette, eds.), pp. 187-212, Springer-Verlag,
    Berlin, Heidelberg, New York.
Constantine, J.W., and Purcell, I.M., 1973, Inhibition
    of platelet aggregation and of experimental thrombo-
    sis by sudoxicam. J. Pharmacol. Exper. Ther. $\underline{187}$:
    653-665.
Daniel, E., Weber, C., 1966, Cooperative effects in
    binding by bovine serum albumin. The binding of 1-
    anilino-8-naphthalene sulphonate. Fluorimetric
    titrations. Biochemistry $\underline{5}$: 1893-1907.
Darska, J., 1976, Comparison of radiochemical and bio-
    logical assays of prostaglandin synthetase activity
    in bovine seminal vesicle microsomes. PhD thesis.
    Copernicus Academy of Medicine in Cracow.
Deby, C., Descamps, M., Binon, F., and Bacq, Z.M., 1971,
    Inhibition de la biosynthése in vitro de la prosta-
    glandine E_2 par des substances anti-inflammatoires.
    C.R. Soc. Biol. $\underline{165}$: 2465-2468.

De Clerck, F., Vermylen, J., and Reneman, R., 1975,
    Effects of suprofen, an inhibitor of prostaglandin
    biosynthesis, on platelet function, plasma coagula-
    tion and fibrynolysis. I. In vitro experiments.
    Arch. int. Pharmacodyn. 216 : 263-279.
Dembińska-Kieć, A., Żmuda, A., and Krupińska, J., 1976,
    Inhibition of prostaglandin synthetase by aspirin-
    -like drugs in different microsomal preparation.in:
    Advances in Prostaglandin and Thromboxane Research,
    Vol.1 (B. Samuelsson and R. Paoletti, eds.), p.99,
    Raven Press New York.
Ferreira, S.H., Moncada, S., and Vane, J.R., 1971,
    Indomethacin and aspirin abolish prostaglandin re-
    lease from the spleen. Nature New Biol. 231 : 237-239.
Fjalland, B., 1974, Inhibition by non-steroidal anti-
    inflammatory agents of the release of rabbit aorta
    contracting substance and prostaglandins from
    chopped guinea pig lungs. J.Pharm. Pharmacol. 26 :
    448-451.
Flanagan, M.T., and Ainworth, S., 1968, The binding of
    aromatic sulphonic acids to bovine serum albumin.
    Biochim. Biophys. Acta 168 : 16-26.
Flower, R.J., 1974, Drugs which inhibit prostaglandin
    biosynthesis. Pharmac. Rev. 26 : 33-67.
Flower, R.J., Blackwell, G.J., and Parsons, M.F., 1975,
    Mechanism of collagen induced platelet aggregation.
    Abstr. VIth Internat. Congress Pharmac. p.292, Hel-
    sinki.
Flower, R.J., Cheung, H.S., and Cushman, D.W., 1973,
    Quantitative determination of prostaglandins and
    malonyldialdehyde formed by arachidonate oxygenase
    system in bovine seminal vesicles. Prostaglandins
    4 : 325-341.
Flower, R.J., Gryglewski, R., Herbaczyńska-Cedro, K.,
    and Vane, J.R., 1972, The effect of anti-inflamma-
    tory drugs on prostaglandin biosynthesis. Nature,
    New Biol. 238 : 104-106.
Flower, R.J., and Vane, J.R., 1972, Inhibition of
    prostaglandin synthetase in brain explains the anti-
    pyretic activity of paracetamol (4-acetamidophenol).
    Nature 240 : 410-411.
Gaut, Z.N., Baruth, H., Randall, L.O., Ashley, C., and
    Paulsrud, J.R., 1975, Stereometric relationships
    among anti-inflammatory activity, inhibition of
    platelet aggregation, and inhibition of prostaglan-
    din synthetase. Prostaglandins 10 : 59-66.

Gerber, D.A., Cohen, Nm., and Giustra, R., 1967, The
    ability of nonsteroidal anti-inflammatory compounds
    to accelerate a disulfide interchange reaction of
    serum sulfhydryl groups and 5,5´-dithio-bis-(2-
    nitrobenzoic) acid. Biochem. Pharmac. 16: 115-120.
Gilmore, N., Vane, J.R., and Wyllie, J.H., 1968, Prosta-
    glandins released by the spleen. Nature 218 : 1135-
    1137.
Glenn, E.M., Rohloff, N., Bowman, B.J., and Lyster, S.
    C., 1973, The pharmacology of 2-(2-fluoro-4-biphe-
    nylyl)propionic acid (flurbiprofen) a potent non-
    steroidal anti-inflammatory drug. Agents and Actions
    3/4 : 210-216.
Glenn, E.M., Wilks, J., and Bowman, B.J., 1972, Plate-
    lets, prostaglandins, red cells, sedimentation rates,
    serum and tissues proteins and non-steroidal anti-
    inflammatory drugs. Proc.Soc.Exp.Med. 141:879-886.
Gryglewski, R., 1966, The fibrinolytic activity of anti-
    inflammatory drugs. J.Pharm. Pharmac. 18: 474.
Gryglewski, R., 1974, Structure-activity relationships
    of some prostaglandin synthetase inhibitors, in:
    Prostaglandin Synthetase Inhibitors (H.J. Robinson
    and J.R. Vane, eds.), pp. 33-52, Raven Press, New
    York.
Gryglewski, R.J., 1976, Steroid hormones, anti-inflam-
    matory steroids and prostaglandins. Pharmacol. Res.
    Commun. 8: 337-348.
Gryglewski, R.J., Ryznerski, Z., Gorczyca, M., and
    Krupińska, J., 1976a, Design of new prostaglandin
    synthetase inhibitors in a group of N-(2-carboxy-
    phenyl)phenoxyacetamides and their anti-inflamma-
    tory activity, in: Advances in Prostaglandin and
    Thromboxane Research, Vol. 1 (B. Samuelsson and R.
    Paoletti, eds.), pp. 117-120. Raven Press, New York.
Gryglewski, R.J., Dembińska-Kieć, A., Grodzińska, L.,
    and Panczenko, B., 1976b, Differential generation
    of substances with prostaglandin-like and thrombo-
    xane-like activities by guinea pig trachea and by
    lung strips, in:Lung cells in disease (A. Bouhuys,
    ed.), in press. Elseview Pub. Amsterdam.
Gryglewski, R., and Vane, J.R., 1972a, The release of
    prostaglandins and rabbit aorta contracting sub-
    stance (RCS) from rabbit spleen and its antagonism
    by anti-inflammatory drugs. Br.J.Pharmac. 45: 37-47.
Gryglewski, R., and Vane, J.R., 1972b, The generation
    from arachidonic acid of rabbit aorta contracting
    substance (RCS) by microsomal enzyme preparation
    which also generates prostaglandins. Br.J.Pharmac.
    46: 449-457.

Ham, E.A.,Cirillo, K.J., Zanetti, M., Shen, T.Y., and
    Kuehl, F.A., 1972, Studies on the mode of action
    of non-steroidal anti-inflammatory agents, in:
    Prostaglandins in cellular biology (P.W. Ramwell
    and B.B. Pharriss, eds.), pp. 345-352. Plenum Press,
    New York.
Hamberg, M., 1972, Inhibition of prostaglandin synthesis
    in man. Biochem. Biophys. Res. Commun. $\underline{49}$: 720-726.
Hamberg, M., Svensson, J., and Samuelsson, $\overline{B}.$, 1975,
    Thromboxanes: A new group of biologically active
    compounds derived from prostaglandin endoperoxides.
    Proc. Nat. Acad. Sci. USA, $\underline{72}$: 2994-2998.
Hamberg, M., Svensson,J., Wakabayashi, T., and Samuels-
    son, B., 1974, Isolation and structure of two prosta-
    glandin endoperoxides that cause platelet aggrega-
    tion. Proc.Nat. Acad. Sci. USA, $\underline{71}$: 345-349.
Ho, P.P., and Esterman, M.A., 1974, Fenoprofen: inhibi-
    tion of prostaglandin synthesis. Prostaglandins, $\underline{6}$:
    107-113.
Horodniak, J.W., Julius, M., Zarembo, J.E., and Bender,
    D., 1974, Inhibitory effects of aspirin and indo-
    methacin on the biosynthesis of PGE_2 and PGF_2
    Biochem. Biophys. Res. Commun. $\underline{57}$: 539-545.
Ku, E.C., and Wasvary, J.M., 1975, Inhibition of prosta-
    glandin synthetase by pirprofen. Studies with sheep
    seminal vesicle enzyme. Biochim. Biophys..Acta $\underline{384}$:
    360-368.
Lands, W.E.M., Cook, H.W., and Rome, L.H., 1975, Prosta-
    glandin biosynthesis: consequences of oxygenase me-
    chanism upon in vitro assays of drug effectiveness.
    Abstr. Int. Conference on Prostaglandins, p.3,
    Florence.
Lands, W., Lee, R., and Smith, W., 1971, Factors regula-
    ting the biosynthesis of various prostaglandins.
    Ann. N.Y. Acad. Sci. $\underline{180}$: 107-122.
Levine, L., 1972, Prostaglandin production by mouse
    fibrosarcoma cells in culture: Inhibition by indo-
    methacin and aspirin. Biochem. Biophys. Re. Commun.
    $\underline{47}$: 888-896.
Lewis, G.P., and Piper, P.J., 1975, Inhibition of release
    of prostaglandins as an explanation of some of the
    actions of anti-inflammatory corticosteroids.
    Nature $\underline{254}$: 308-311.
Malmsten, C., Hamberg, M., Svensson, J., and Samuelsson,
    B., 1975, Physiological role of an endoperoxide in
    human platelets: hemostatic defect due to platelet
    cyclo-oxygenase deficiency. Proc. Nat. Acad. Sci.
    USA, $\underline{72}$:1446-1450.

Meyer, M.C., and Guttman, D.E., 1968, The binding of
    drugs by plasma proteins. J. Pharm. Sci. 57: 895-917.
Miller, W.S., and Smith, J.G., 1966, Effect of acetyl-
    salicylic acid on lysosomes. Proc. Soc. Exp. Biol.
    Med. 122: 634-636.
Mizushima, Y., 1964, Inhibition of protein denaturation
    by anti-rheumatic or antiphlogistic agents. Arch.
    Int. Pharmacodyn. 149: 1-7.
Mizushima, Y., Ishi, Y., and Masumoto, S., 1975,
    Physico-chemical properties of potent non-steroidal
    anti-inflammatory drugs. Biochem. Pharmacol. 24:
    1589-1592.
Moffat, A.C., Patterson, D.A., Curry, A.S., and Gwen,
    P., 1972, Inhibition in vitro of cyclic 3´,5´-
    -nucleotide phosphodiesterase activity by drugs.
    Eur. J. Toxicol. 5: 160-162.
Mustard, J.F., and Packham, M.A., 1975, Platelets,
    thrombosis and drugs. Drugs 9: 19-76.
Needleman, P., Moncada, S., Bunting, S., Vane, J.R.,
    Hamberg, M., and Samuelsson, B., 1976a, Identifica-
    tion of an enzyme in platelet microsomes which
    generates thromboxane A_2 from prostaglandin endo-
    peroxides. Nature 261: 558-560.
Needleman, P., Minkes, M., and Raz, A., 1976b,
    Thromboxanes: Selective biosynthesis and distinct
    biological properties. Science 193: 163-165.
O´Brien, J.R., 1968, Effect of anti-inflammatory agents
    on platelets. Lancet 1: 894-895.
Patrono, C., Ciabattoni, G., Greco, F., and Grossi-
    Belloni, D., 1976, Comparative evaluation of the
    inhibitory effects of aspirin-like drugs on prosta-
    glandin production by human platelets and synovial
    tissue, in: Advances in Prostaglandin and Thrombo-
    xane Research, Vol. 1 (B. Samuelsson and R. Paoletti,
    eds.), pp. 125-131, Raven Press, New York.
Piper, P.J., and Vane, J.R., 1969, Release of additional
    factors in anaphylaxis and its antagonism by anti-
    inflammatory drugs. Nature 233: 29-35.
Robak, J., Dembińska-Kieć, A., and Gryglewski, R.,
    1975 , The influence of saturated fatty acids on
    prostaglandin synthetase activity. Biochem. Pharmac.
    24 : 2057-2060.
Roth, G.J., Stanford, N., and Majerus, P.W., 1975,
    Acetylation of prostaglandin synthetase by aspirin.
    Proc.Nat. Acad. Sci. USA, 72 : 3073-3076
Samuelsson, B., Granström,E., and Hamberg, M., 1967,
    On the mechanism of biosynthesis of prostaglandins.
    in: Nobel Symposium 2 Prostaglandins (S.Bergstrom
    and B. Samuelsson, eds.), pp.31-44, Almqvist and
    Wiksell, Stockholm.

Samuelsson, B., Hamberg, M., Svensson, J., and Malmsten,
    C., 1975, The role of prostaglandin endoperoxides
    in human platelets. Abstr. VIth Congress Pharmac.
    p. 480, Helsinki.
Scatchard, G., 1949, The attractions of proteins for
    small molecules and ions. Ann. N.Y. Acad. Sci. 51:
    660-672.
Schaeffer, H.J., 1971, Factors in the design of rever-
    sible and irreversible enzyme inhibitors, in: Drug
    Design, Vol. II (E.J. Ariëns, ed.), pp. 129-160,
    Academic Press, New York, London.
Settle, W., Hegeman, S., and Featherstone, R.M., 1971,
    The nature of drug-protein interaction, in: Handb.
    Exp. Pharm. 28/1, Concepts in Biochemical Pharmaco-
    logy, (B. Brodie and J.R. Gilette, eds.), Springer-
    Verlag, Berlin, Heidelberg, New York, pp. 175-186.
Shen, T.Y., Ham, E.A., Cirillo, V.J., and Zanetti, M.,
    1974, Structure-activity relationship of certain
    prostaglandin synthetase inhibitors, in: Prosta-
    glandin Synthetase Inhibitors (H.J. Robinson and
    J.R. Vane, eds.), pp. 19-31. Raven Press, New York.
Sih, C.J., and Takeguchi, C.A., 1973, Biosynthesis, in:
    The Prostaglandins, Vol. I. (P.W. Ramwell, ed.),
    pp. 83-100, Plenum Press, New York, London.
Silver, M.J., Smith, J.B., Ingerman, C., and Kocsis, J.
    J., 1973, Arachidonic acid-induced human platelet
    aggregation and prostaglandin formation. Prosta-
    glandins 4: 863-875.
Skidmore, I.F., Whitehouse, M.W., 1965, Effect of non-
    steroidal anti-inflammatory drugs on aldehyde binding
    to plasma albumen: a novel in vitro assay for poten-
    tial anti-inflammatory activity. J. Pharm. Pharmac.
    17: 671-673.
Skidmore, I.F., and Whitehouse, M.W., 1966, Concerning
    the regulation of some diverse biochemical reactions
    underlying the inflammatory response by salicylic
    acid, phenylbutazone and other acidic antirheumatic
    drugs. J. Pharm. Pharmac. 18: 558-560.
Smith, W.L., and Lands, W.E.M., 1971, Stimulation and
    blockade of prostaglandin biosynthesis. J. Biol.
    Chem. 246: 6700-6702.
Smith, J.B., and Willis, A.L., 1971, Aspirin selectively
    inhibits prostaglandin production in human platelets.
    Nature New Biol. 231: 235-237.
Svensson, J., Hamberg, M., and Samuelsson, B., 1975,
    Prostaglandin endoperoxides IX. Characterization of
    rabbit aorta contracting substance (RCS) from guinea
    pig lung and human platelets. Acta Physiol. Scand.
    94: 222-228.

Takeguchi, C., Kohono, E., Sih, C.J., 1971, Mechanism of prostaglandin biosynthesis. I. Characterization and assay of bovine prostaglandin synthetase. Biochemistry 10: 2372-2376.

Takeguchi, G., and Sih, C.J., 1972, A rapid spectrophotometric assay for prostaglandin synthetase: Application to the study of non-steroidal antiinflammatory agents. Prostaglandins 2: 169-184.

Tomlinson, R.V., Ringold, H.J., Quershi, M.C., and Forchielli, E., 1972, Relationship between inhibition of prostaglandin synthesis and drug efficacy: Support for the current theory on mode of action of aspirinlike drugs. Biochem. Biophys. Res. Commun. 46: 552-559.

Vane, J.R., 1957, A sensitive method for the assay of 5-hydroxytryptamine. Br. J. Pharmac. Chemother. 12: 344-349.

Vane, J.R., 1964, The use of isolated organs for detecting active substances in the circulating blood. Br. J. Pharmac. Chemother. 23: 360-373.

Vane, J.R., 1971, Inhibition of prostaglandin synthesis as a mechanism of action for aspirin-like drugs. Nature New Biol. 231: 232-235.

Vane, J.R., 1972a, Prostaglandins in inflammation, in: Inflammation, mechanisms and control (I.H. Lepow, and P.A. Ward, eds.), pp. 261-279, Academic Press, New York, London.

Vane, J.R., 1972b, Prostaglandins and aspirin-like drugs. Hosp. Pract. 7: 61-71.

Vargaftig, B.B., and Dao Hai, N., 1971 , Release of vasoactive substances from guinea-pig lungs by slow reacting substance C and arachidonic acid. Pharmacology 6: 99-108.

Vargaftig, B.B., and Zirinis, P., 1973, Arachidonic acid induced platelet aggregation is accompanied by release of potential inflammatory mediators distinct from PGE_2 and PGF_2. Nature New Biol. 244: 114-116.

Weinryb, I., Chasin, M., Free, C.A., Harris, D.M., Goldenberg, H., Michel, I.M., Raik, V.S., Phillips, M., Samamiego, S., and Hess, S., 1972, Effects of therapeutic agents on cyclic AMP metabolism in vitro: J. Pharm. Sci. 61: 1556-1567.

Weiss, H.J., Aledort, L.M., and Kochwa, S., 1968, The effect of salicylates on the hemostatic properties of platelets in man. J. Clin. Invest. 47: 2169-2180.

Whitehouse, M.W., and Haslam, J.M., 1962, Ability of some antirheumatic drugs to uncouple oxydative phosphorylation. Nature 196: 1323-1324.

Whitehouse, M.W., Kippen, I., and Klinenberg, J.R., 1971, Biochemical properties of anti-inflammatory drugs - XII. Inhibition of urate binding to human albumin by salicylate and phenylbutazone analogues and some novel anti-inflammatory drugs. Biochem. Pharmac. 20: 3309-3320.

Wiethold, G., Hellenbrecht, D., Lemmer, B., and Palm, D., 1973, Membrane effects of beta-adrenorgic blocking agents: investigations with the fluorescence probe 1-anilino-8-naphthalene sulphonate (ANS) and antihemolytic activities. Biochem. Pharmacol. 22: 1437-1449.

Willis, A.L., 1974, An enzymic mechanism for the anti-thrombotic and antihemostatic actions of aspirin. Science 183: 325-327.

Willis, A.L., and Kuhn, D.C., 1973, A new potential mediator of arterial thrombosis whose biosynthesis is inhibited by aspirin. Prostaglandins 4: 127-129.

Wiseman, E.H., McIlhenny, H.M., and Bettis, J.W., 1975, Flumizole, a new nonsteroidal anti-inflammatory agent. J. Pharm. Sci. 64: 1469-1475.

Zaroślinski, J.F., Keresztes-Nagy, S., Mass, R.F., and Oester, Y.T., 1974, Effect of temperature on the binding of salicylate by human serum albumin. Biochem. Pharmac. 23: 1767-1776.

PROSTAGLANDIN SYNTHETASE INHIBITORS

T. Y. Shen

Merck Sharp & Dohme Research Laboratories
Rahway, New Jersey 07065, U.S.A.

1. Introduction

Since the discovery of aspirin and indomethacin as prostaglandin synthesis inhibitors in 1971 (1), within a short span of
five years, literally hundreds of chemical structures have been reported as inhibitors of the prostaglandin synthetase system. These
compounds, either as therapeutic agents or as research tools, have
become a new class of pharmacological agents. Standard drugs such
as indomethacin and aspirin, have been used widely and effectively
in delineating the involvement of prostaglandins in various biological phenomena. The general correlation of PG synthetase inhibitory activity and antiinflammatory property in many cases has stimulated an active search for new synthetase inhibitors as potential
antiarthritic agents. Recently, notable advances were made in
several aspects, which not only facilitated the search for new inhibitors, but also suggested possible directions for further improvements. In this communication some notable chemical features
of PG synthetase inhibitors will be reviewed briefly. Some of our
findings with two new antiarthritic drugs, sulindac and diflunisal,
will be highlighted to illustrate the influence of chemical and
pharmacokinetic properties on the safety and selectivity of PG
synthetase inhibitors. Finally, some future directions in this area
will be considered.

Details of the "Prostaglandin Synthetase" pathway are still
being elucidated. Our current understanding is outlined as follows:

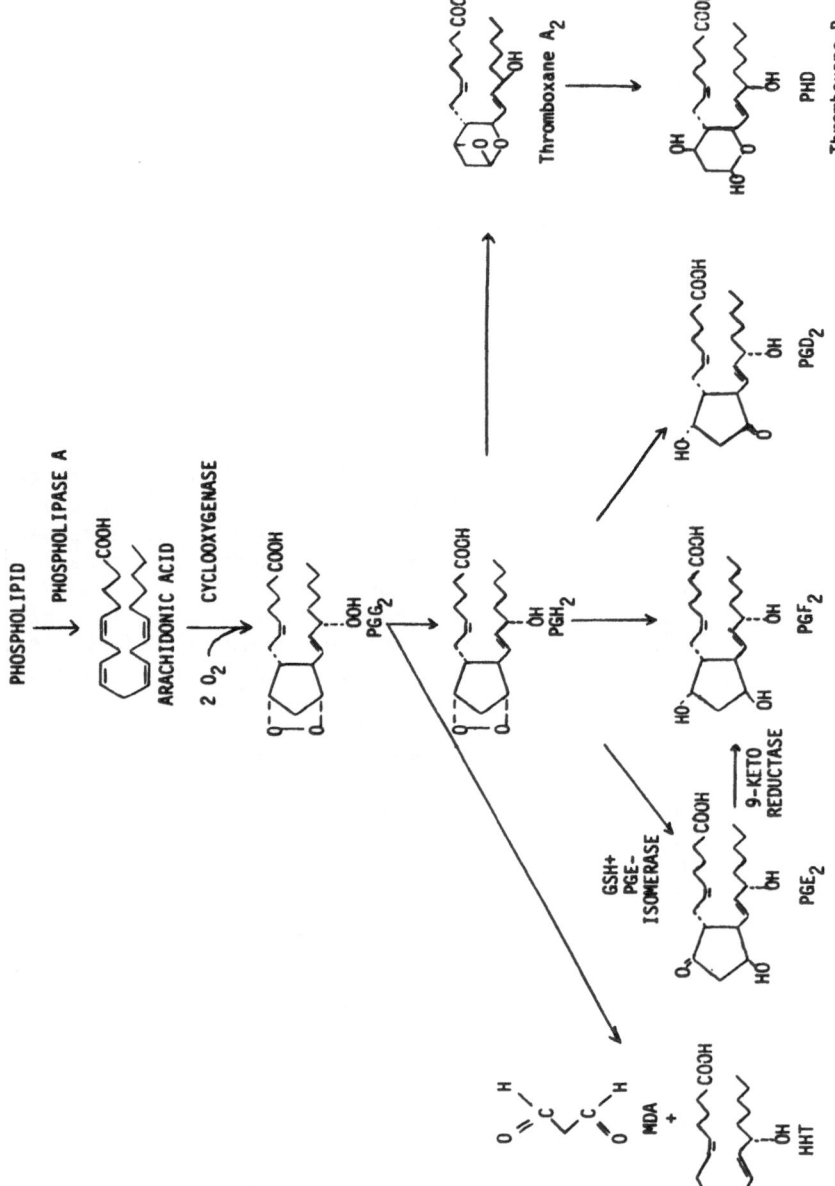

URINARY METABOLITE

In the past two years, an important new branch, the thromboxane
pathway, has been established (2). A key enzyme, cyclooxygenase,
has been fractionated and purified (3). The characterization of
this enzyme should clarify some confusions in the literature regarding the relative potency, reversibility and interaction of cyclo-
oxygenase inhibitors. It would also lay the foundation for quantitative enzymology, essential for the design and search for better
inhibitors. The acetylation of cyclooxygenase by aspirin was another interesting finding (4). The conversion of PGE to PGF by 9-
keto reductase illustrates the interrelationship of different
prostaglandins. Obviously the inhibition of this enzyme would
spare PGE_2 and selectively suppress the formation of PGF, a process
of special interest to renal physiology.

A large number of PG synthetase inhibitors have appeared in the
literature. Among these, some were derived from chemical or biochemical reasoning, but many were discovered by efficient in vitro
screening assays. They can be conveniently grouped into the following categories:

Table 1
"PG SYNTHETASE" INHIBITORS

1. Analogs of substrates and intermediate metabolites
   1.1 Unsaturated fatty acids
   1.2 Bicyclic analogs of PGG and PGH
   1.3 Stable thromboxane analogs

2. Acidic antiinflammatory agents
   2.1 Salicylates
   2.2 Indomethacin, sulindac and congeners
   2.3 Aryl propionic acids
   2.4 Fenamates and analogs
   2.5 Acidic enols and others

3. Non-acidic antiinflammatory agents

4. Other pharmacological agents
   4.1 Antiarthritic drugs
   4.2 Psychotropic drugs
   4.3 Sulfhydryl reagents and derivatives
   4.4 Hormones and mediators

As several comprehensive reviews are available (5,6,7), only
a few salient remarks and some current findings about these inhibitors are given below.

2. Analogs of Substrates and Intermediate Metabolites

As shown in the diagram of the PG synthetase pathway, the substrates for the first synthetic enzyme, cyclooxygenase, are arach-

idonic acid (for PG_2) and di-homo-α-linolenic acid (for PG_1), released by the action of lipases upon activation. Interestingly,
the blockade of this release process by corticosteroids has been
suggested as part of their antiinflammatory action (8,9,10). As
expected, many unsaturated fatty acids as shown below in Table 2
are competitive inhibitors.

Table 2

FATTY ACID ANALOGS

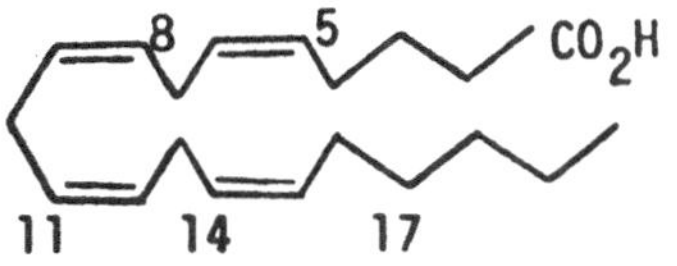

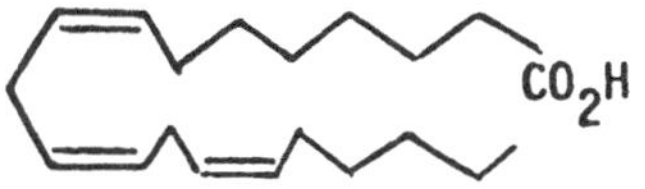

Fatty Acid		Unsaturation						Ki µM
Arachidonic	20:4		5c	8c	11c	14c		1.7
	20:6	2c	"	"	"	"	17c	2.5
	20:5		"	"	"	"	"	
	20:4		"	"	12t	"		
				"		"		
Eicosatrienoic	20:3				12t			.12
Linolenic	18:3			9c	12c	15c		15
Linoleic	18:2			9c	12c			
Oleic	18.1			9c				22
	18:1			10a				10

c = *cis* olefin;   t = *trans* olefin;   a = acetylene

It is of interest to note that acetylenic acids are particularly
effective. The irreversible inhibition by the acetylenic analog
of arachidonic acid, eicosa-5,8,11,14-tetraynoic acid, at 2 µm (11)
may be attributed to a Kcat, or enzyme suicide, mechanism. Analogous to other acetylenic enzyme inhibitors (12), the extraction of
an α-proton may lead to the formation of a highly reactive allenic
intermediate, which may form a covalent bond with a nucleophilic
group at the active site. This postulated mechanism could readily
be verified with a purified enzyme preparation and labelled acetylenic acid.

ACETYLENIC ANALOGS

$$C_5H_9C\overset{14}{\equiv}C\text{-}CH_2C\overset{11}{\equiv}C\text{-}CH_2\text{-}C\overset{8}{\equiv}C\text{-}CH_2\text{-}C\overset{5}{\equiv}C\text{-}(CH_2)_3CO_2H$$

$$C_5H_9C\equiv C\text{-}CH_2\text{-}C\equiv C\text{-}CH_2\text{-}C\equiv C\text{-}(CH_2)_6CO_2H$$

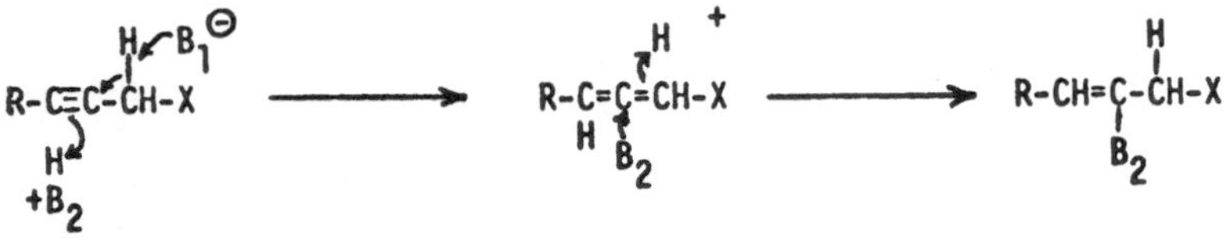

Kcat Inhibitor

Bicylic analogs of PGH, with the peroxide linkage replaced by isosteres, such as CH_2-O, N=N, CH=CH and CH_2-CH_2, have received some renewed attention. The first two variations gave potent and more stable agonists (13,14).

BICYCLIC ANALOGS

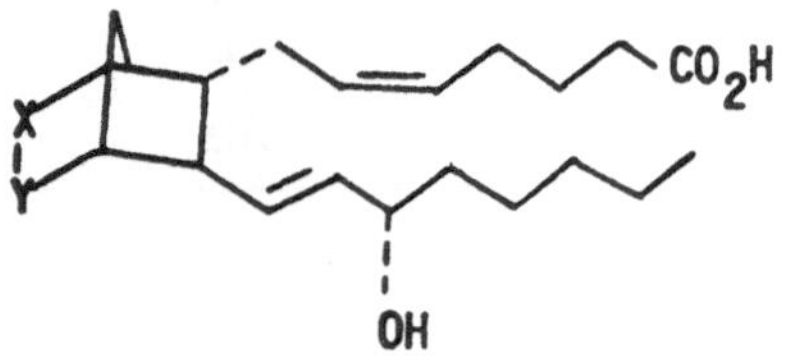

X - Y

O - O PGH_2
O - CH_2
CH_2 - O
N = N
CH = CH
CH_2 - CH_2

As an extension, many other bicylic analogs of PGH can readily be conceived. For instance, several hydrophobic ring systems, e.g. bicyclo[2,2,1], [2,2,2], [3,2,1] alkanes or a cyclooctyl group, have been found to be comparably effective in other biological systems. Possibly PGH analogs derived from these may also show differential PG agonistic or inhibitory activities.

BICYCLIC EQUIVALENTS

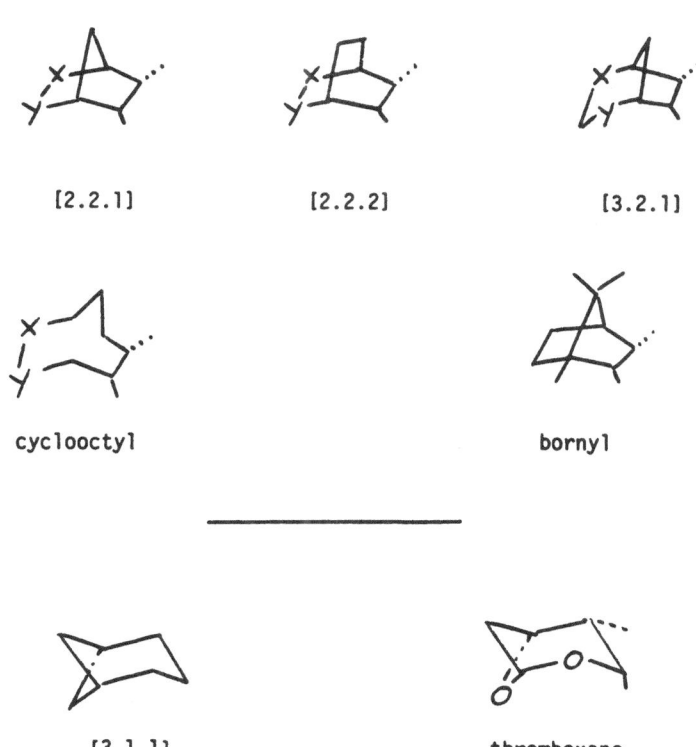

[2.2.1]   [2.2.2]   [3.2.1]

cyclooctyl   bornyl

[3.1.1]   thromboxane

The highly unstable thromboxane A (2) has a bicyclo [3.1.1] ring system. The synthesis of more stable analogs of thromboxanes as potential inhibitors is another attractive medicinal chemical goal. A stable thromboxane mimic may also be useful in the affinity chromatography of thromboxane receptor. Variations of this ring system and the two aliphatic side-chains are only limited by the chemical ingenuity of several laboratories actively engaged in this approach.

3. Aspirin and Diflunisal

The O-acetyl group in aspirin is known to be chemically reactive and capable of acetylating proteins such as albumin, and platelets. In fact, some of the potential side-effects, such as allergenicity and platelet dysfunction, of aspirin have been attributed to this chemical reaction in a general manner (16). Interestingly, the acetylation of cyclooxygenase by 14C-labelled aspirin was found to be a quantitative and selective process (4,17). The rate of acetylation of cyclooxygenase at its functional stage is much higher than that of nonspecific acetylation of serum protein. The acetylation is promoted by the presence of a hemeprotein activator and is partially blocked by other cyclooxygenase inhibitors, e.g. indomethacin. This indicates that the position of acetylation is probably near or within the active site of the enzyme.

INACTIVATION OF CYCLOOXYGENASE BY ASPIRIN

The tissue specificity of aspirin was demonstrated by its inhibitory potencies on the release of $PGF_{2\alpha}$ from human platelets and synovium (18). A six-fold difference in ID_{50} (29 vs. 179 µm respectively) was observed. As both systems were equally sensitive to indomethacin, this may explain the profound effect of aspirin on platelet function at its antiarthritic doses.

Salicylic acid, not having an O-acetyl group, is a weaker antiinflammatory agent and much poorer PG synthetase inhibitor. However, the O-acetyl group, and the consequent ability to acetylate cyclooxygenase, are not necessary for antiinflammatory-analgesic actions of substituted salicylates. For example, the 5-difluorophenyl derivative of salicylic acid, diflunisal, is a more potent and longer acting analgesic antiinflammatory agent (19). It inhibits the PG synthetase system prepared from sheep seminal vesicle at 2 µM (20). It also decreased the excretion of the major urinary metabolite of PGE, 7-α-hydroxy-5,11-diketotetranorprostane-1,6-dioic acid by 70% after five daily dosing of 750 mg in volunteers

(21). Naturally, without the O-acetyl group, it can not acetylate cyclooxygenase. In the platelet system its inhibition of platelet aggregration is reversible and has a shorter duration than aspirin. Indeed, at the therapeutic dose of 1/2-1 g/day no apparent inhibition of platelet function, in terms of bleeding time, platelet aggregation, etc. was observed. Diflunisal is also less irritating to the G.I. tract than aspirin in chronic animal models as well as in patients. It seems that the changing nature of PG synthetase inhibition by diflunisal may contribute to its improved tolerance in man. Further clarification of the mechanism of diflunisal and related compounds in the synthetase pathway is in progress.

DIFLUENISAL
(MK-647)

ANTIINFLAMMATORY	8 X	ASPIRIN
ANALGESIC	15 X	
ANTIPYRETIC	2 X	
PG SYNTHETASE INHIBITION	$ID_{50}= 2$ μM	

4. Indomethacin and Sulindac

With indomethacin analogs, a close parallelism between in vitro PG synthetase inhibitory activity and in vivo antiinflammatory potency, as measured by the carregeenan foot-edema assay in rats, was observed (20).

Table 3

INDOMETHACIN AND SULINDAC ANALOGS

ANTIINFLAMMATORY ACTIVITY

INDOLE SERIES	MAN DAILY DOSE	RAT CARRAGEENAN EDEMA ED_{50} MG/KG	PG SYNTHETASE INHIBITION (SSV)* ID_{50} μM
INDOMETHACIN	75-100 mg	2.4	0.4
MK-825	200 mg	4	0.6
p-F ANALOG		ca. 1.1	0.6
5-OH METABOLITE		ca. 50	1.5
MK-555	2 - 3 g	25	10
MK-410	1.5 g	15	2.2
INDENE SERIES			
MK-715	200-400 mg	4	2
SULINDAC	300-400 mg	4.9	-
SULFIDE METABOLITE		2	2.2

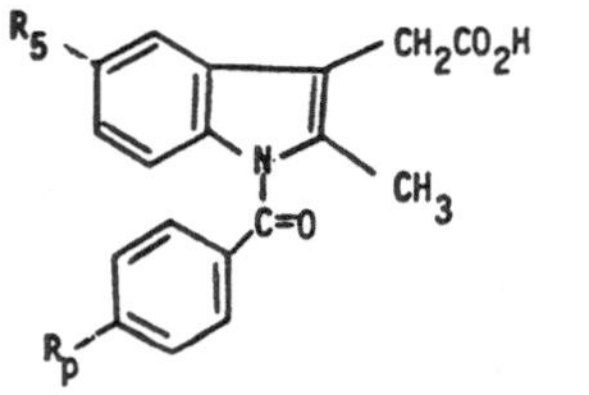

	R_5	R_p
INDOMETHACIN	CH_3O	Cl
MK-825	$(CH_3)_2N$	Cl
p-F ANALOG	CH_3O	F
5-OH METABOLITE	OH	Cl

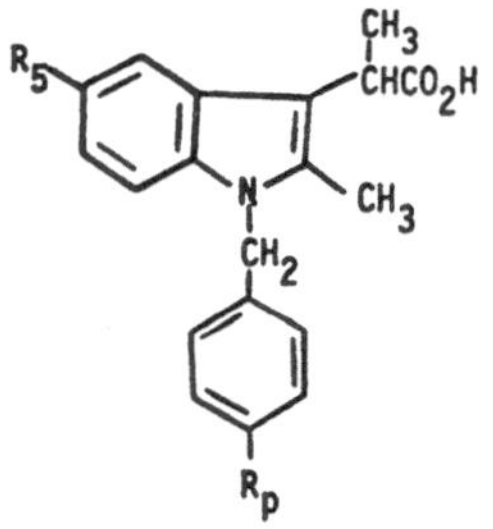

	R_5	R_p
MK-555	CH_3O	Cl
MK-410	CH_3O	SCH_3

Similar structure activity relationships in terms of activity-enhancing substituents, e.g. the 5-methoxy and p-chloro groups, and
the requirement for an S absolute configuration of the chiral center
in the α-methyl acetic acid side-chain, were observed in vitro and
in vivo. The indene isosteres of indomethacin are also active
compounds. Interestingly, the cis geometrical isomer is more than
5X more active than the trans isomer in vitro as well as the in vivo.
The high degree of stereo specificity for an active molecule is

consistent with the data from X-ray crystallographic studies (22,23).
The configuration of indomethacin in a crystalline state is comparable to that of the cis geometrical isomer of the indene analogs.
Clearly with indomethacin and its indole and indene analogs, the
dominent factors affecting their in vitro PG synthetase inhibition
and in vivo antiinflammatory properties are the aromatic ring-
system, optimal substituents, and some well defined geometrical and
absolute configurations. Presumably other metabolic parameters,
e.g. absorption, serum binding, distribution and metabolism, remain
relatively constant for these analogs without distorting the correlation of their in vitro and in vivo activities.

SULINDAC AND METABOLITES

	R_5	R_p	
MK-715	CH_3O	Cl	(5 X trans isomer)
SULFIDE METABOLITE	F	CH_3S	
SULINDAC	F	$CH_3S \to 0$	
SULFONE METABOLITE	F	$CH_3S \lessgtr^0_0$	

In our studies, a conspicuous exception to the above conclusion was found with the new antiarthritic drug, sulindac. Sulindac
is a 5-F-p-methylsulfinyl derivative in the indene series (24). It
is fully active in in vivo antiinflammatory assays, ca. 1/2 X indomethacin, but inactive in in vitro PG synthetase and only weakly
active in platelet aggregation assays (25). This apparent discrepancy is readily explained by its pharmacokinetic properties.

PHARMACODYNAMICS OF SULINDAC

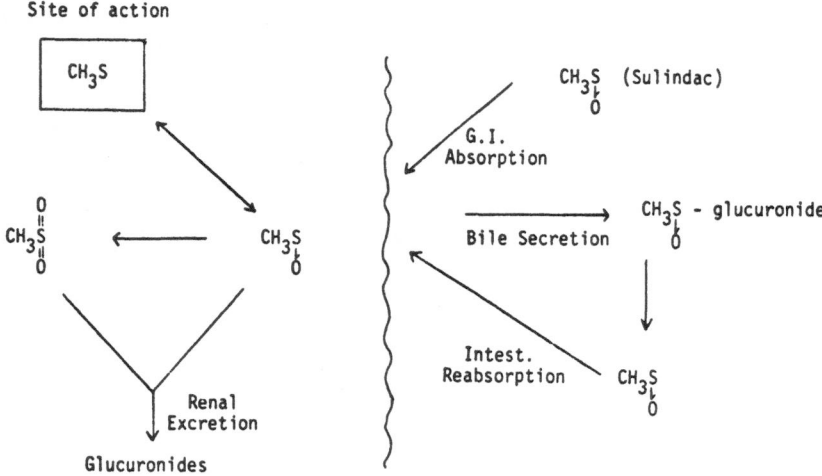

After oral absorption sulindac, a sulfoxide, is in equilibrium with its sulfide metabolite in the serum. The sulfide is fully active both in vivo and in vitro. It is also several times more active than the sulfoxide when applied locally into the dog knee joint or topically on the mouse ear. These observations suggested that the sulfide may be primarily responsible for the in vivo activity of sulindac. In other words, sulindac is being administered as a prodrug. The lack of PG synthetase inhibition of sulindac per se may help to minimize the local G.I. irritation following oral administration. Furthermore, most antiinflammatory aryl acids undergo an enterohepatic recirculation in the form of their acyl glucuronides. As the free drug is liberated from the conjugate in the gut and reabsorbed, the intestine is continually exposed to the irritating action of the active drug. Fortuitously in this case, the major biliary metabolite turns out to be the glucuronide of sulindac and not its active sulfide metabolite. That means, the intestinal tract is also exposed to the inactive prodrug only. The reabsorption of sulindac continues to sustain the serum level of its sulfide metabolite. The prolonged serum half life of the sulfide, being greater than 20 hrs., forms the basis of its b.i.d. regimen in man. Finally the inactive and more soluble sulfoxide, its sulfone metabolite and their glucuronides are excreted in the urine. This sequence of metabolic conversions of a prodrug resulted the optimization of both the duration of action and G.I. tolerance of a systemic PG synthetase inhibitor.

5. Aryl Aliphatic Acids

Another example of metabolic conversion of a weak PG synthetase inhibitor to a more active metabolite is fenbufen (26), but detailed pharmacokinetic data are yet available.

FENBUFEN

The active metabolite of fenbufen, 4-biphenylyl acetic acid, was an early prototype of many aryl propionic acids of interest currently. Some prominent examples of this broad class of PG synthetase inhibitors are shown below in Table 4.

Table 4

SUBSTITUTED PHENYL PROPIONIC ACIDS

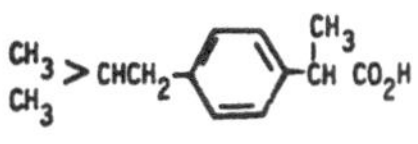

Ibuprofen

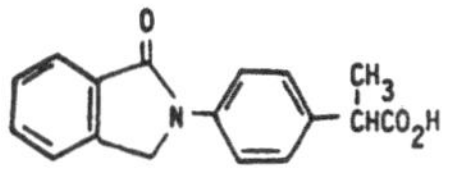

Indoprofen

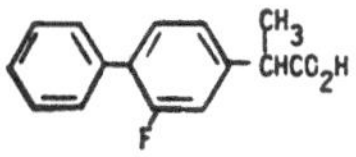

Flurbiprofen

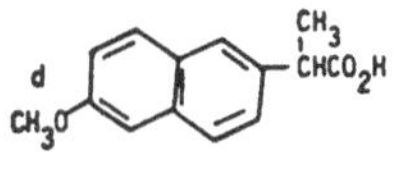

Naproxen

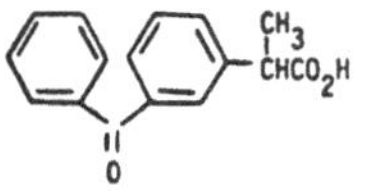

Ketoprofen

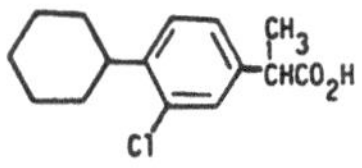

MK-830

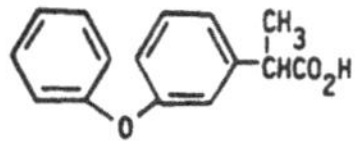

Fenoprofen

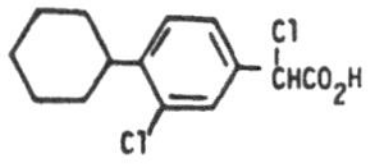

Fenclorac

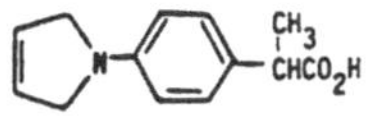

Piroprofen

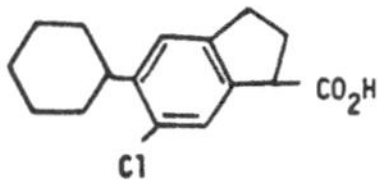

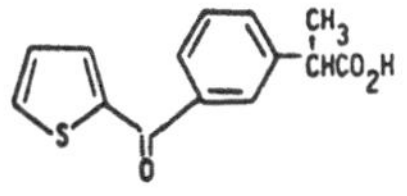

Suprofen

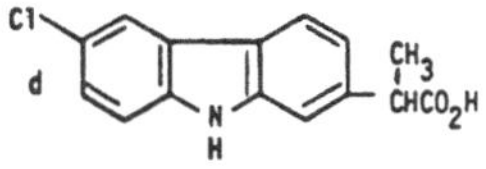

C - 8012

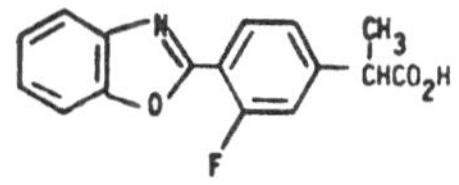

OTHER ARYL ALIPHATIC ACIDS

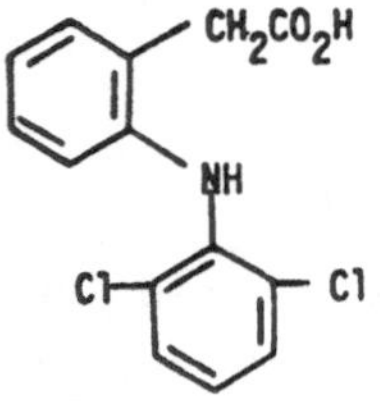

Dichlofenac

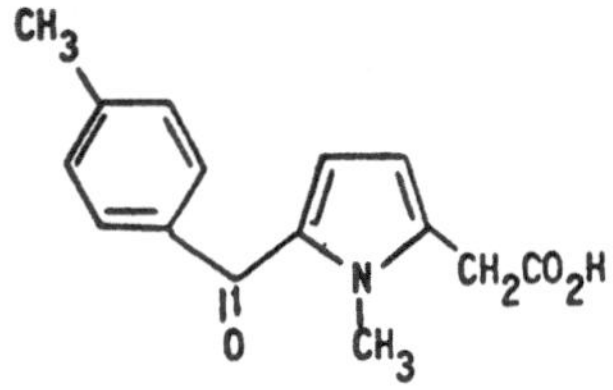

Tolmetin

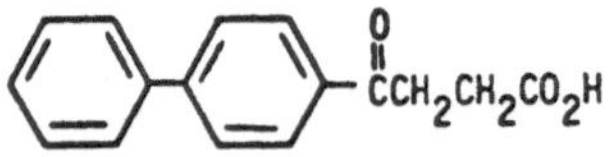

Fenbufen

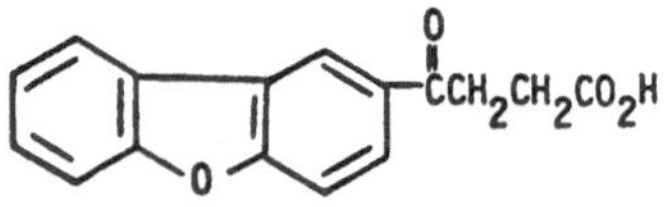

Furoprofen

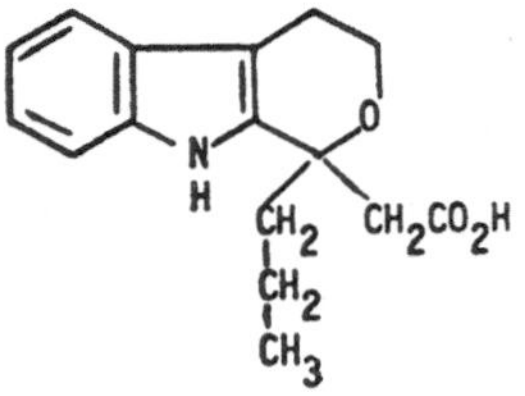

Prodolic Acid

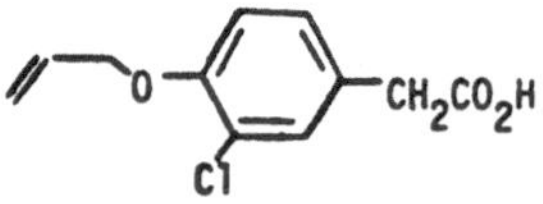

Alclofenac

Previously we have noted the specific requirement of a sinister (S) absolute configuration of the chiral center in the indole and substituted phenyl-α-propionic acid series for in vitro and in vivo activities (27). This generalization remains largely true for many new inhibitors such as the d-3-fluoro-4-(2-benzoxazolyl)phenyl-α-propionic acid, which is more potent than indomethacin and flurbiprofen in our in vitro and in vivo assays and with a more favorable therapeutic index. In a few cases, possibly due to metabolic or other factors, no significant difference between the S and R optical isomers have been noted. Most aryl propionic acids, e.g. pirprofen (28) and fenoprofen (29) are reversible inhibitors. Several more potent compounds, often prossessing the activity-enhancing halogen substituents, appeared to be irreversible inhibitors presumably due to higher affinity in their binding to the active site (30).

6. The Fenamates and Phenolic Compounds

The fenamates represent another family of aryl acidic PG synthetase inhibitors. A notable feature is their multi action potential, being PG antagonists and inhibitors of both mouse pancreas lipase and PG synthetase (31). Interestingly, the inhibition of cyclooxygenase by fenamates is very much enhanced by the addition of phenols and related cofactors, which stimulate the enzyme. In contrast, the sensitivity of cyclooxygenase to inhibition by other aryl acids is not altered by the presence or absense of phenolic cofactors (32).

FENAMATES AND ANALOGS

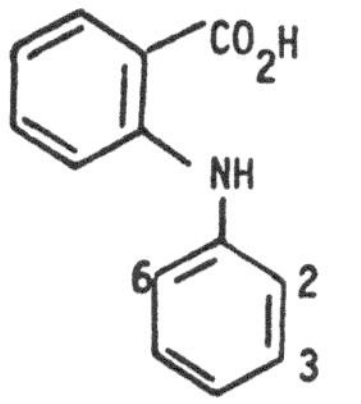

2,3 di Me   Mefanamic Acid

3 - CF_3 Flufenamic Acid

2,6-diCl-3-Me   Meclofenamic Acid

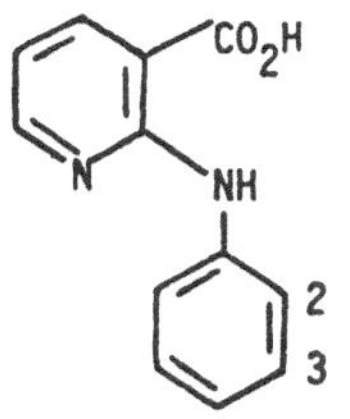

3 - CF_3 Niflumic Acid

3-Cl-2-Me   Clonixin

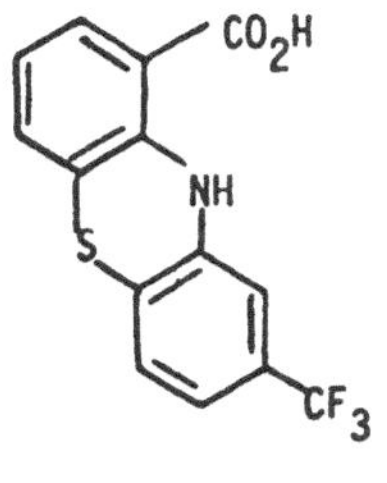

SKF 22908

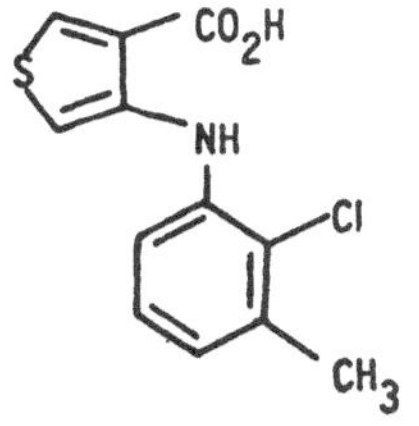

HOE 895

Phenolic compounds such as eugenol and 5-HIAA were known to be stimulators. Others like α-tocophenol and the 5-OH indomethacin metabolite are inhibitors of phenol-activated enzyme preparations. Still other, like the flavonoid rution and derivatives, their inhibitory or stimulatory effects vary according to minor structural modifications (33). Further clarification of these observations would be of much interest.

PHENOLIC COMPOUNDS

1. STIMULATORS

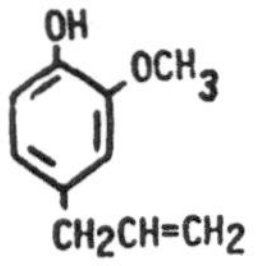

EUGENOL 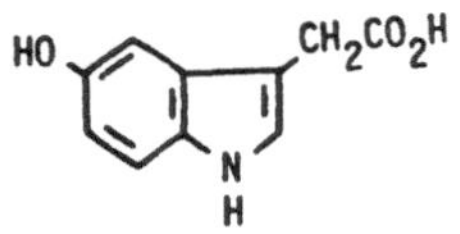   5-HIAAA

2. INHIBITORS

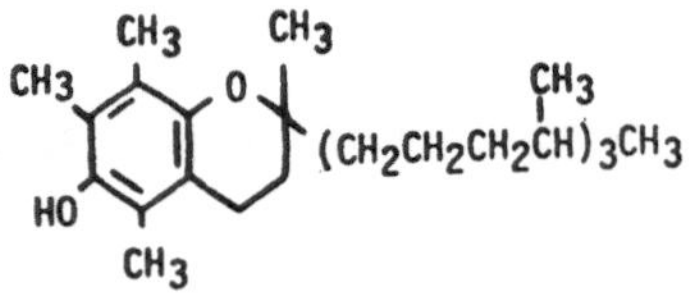

α-TOCOPHEROL 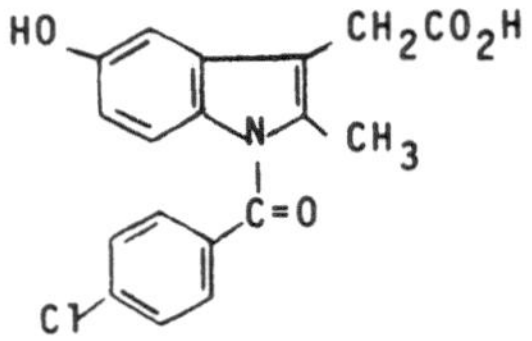   INDOMETHACIN METABOLITE

3. FLAVONIDS

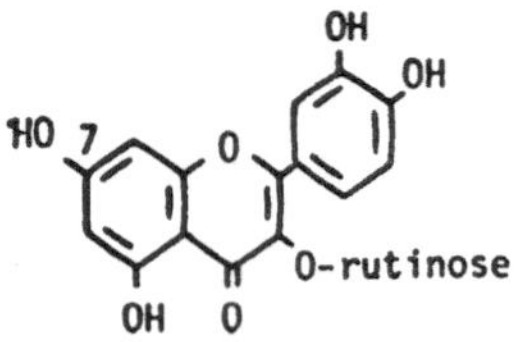

RUTIN AND
DERIVATIVES

   Phenols are feebly acidic. A class of more acidic enols (pKa
∿ 4), typified by phenylbutazone and analogs, are also active inhibitors. Phenylbutazone and oxyphenbutazone are comparable in
their antiinflammatory properties but the later, with an extra
phenolic group in the molecule, is only 1/10 as active vs. PG synthetase _in vitro_. Interestingly, oxyphenbutazone is a substrate of
cyclooxygenase. The enzymic product, a 4-hydroxy derivative, is
inactive. The uricosuric analog in the phenylbutazone family,
sulfinpyrazone, is a platelet aggregation inhibitor, currently being evaluated as a potential antithrombotic agent. Like sulindac,
sulfinpyrazone has a sulfoxide group. In our laboratory (D. H.
Minsker, private communication), the corresponding sulfide analog
was found to be a more potent platelet aggregation inhibitor than
sulfinpyrazone. One would speculate that the _in vivo_ activity of
sulfinpyrazone may also be attributable to its sulfide metabolite.
Several examples of antiinflammatory compounds containing acidic
enolic groups are shown below.

ACIDIC ENOLS

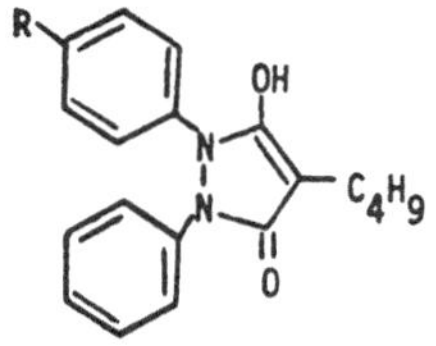

R = H   Phenylbutazone

OH   Oxyphenbutazone

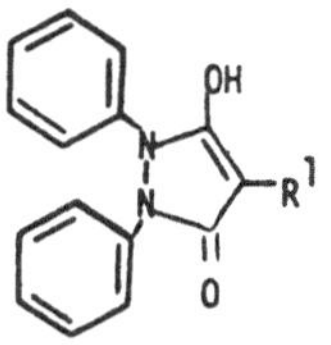

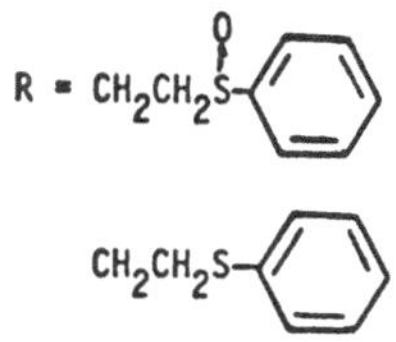

$R = CH_2CH_2\overset{O}{\underset{}{S}}$—⟨phenyl⟩   Sulfinpyrazone

CH_2CH_2S—⟨phenyl⟩

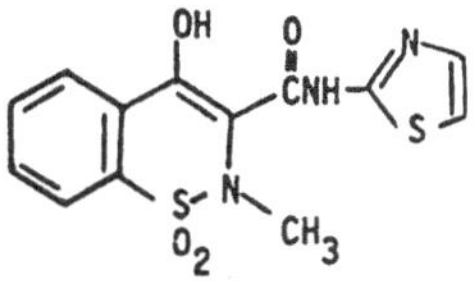

Sudoxicam

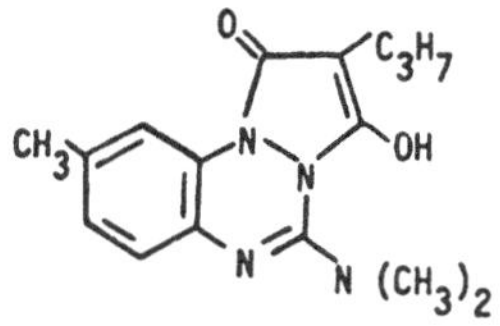

Azapropazone

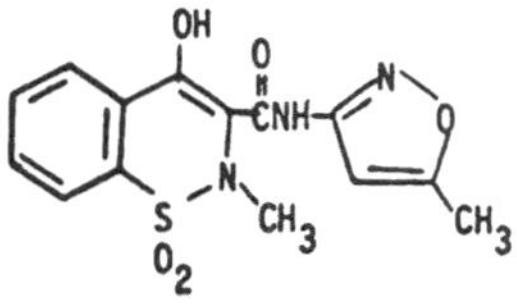

W 8495

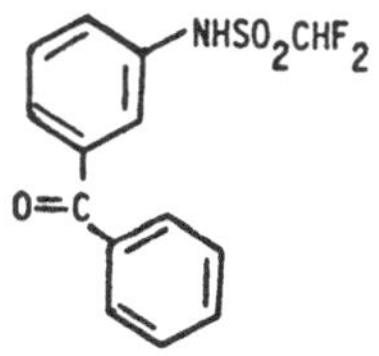

R 807

7. Nonacidic Inhibitors

Several nonacidic antiinflammatory agents have also displayed PG synthetase inhibitory activities. A common structural feature seems to be a five-membered heteroaryl group substituted with 2 or 3 phenyl groups either fused or in an angular fashion.

AROMATIC MOIETIES OF NONACIDIC INHIBITORS

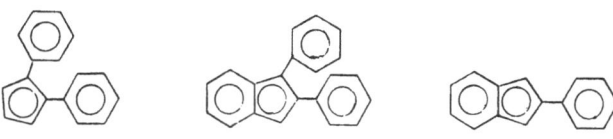

Typical examples are indoxole, dimetopyrrol and the most potent
compound is flumisole, in accordance with their reported antiinflammatory activities (34).

NON-ACIDIC ANTIINFLAMMATORY AGENTS

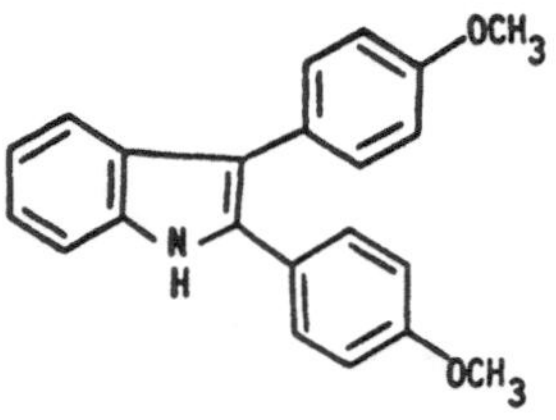

Indoxole

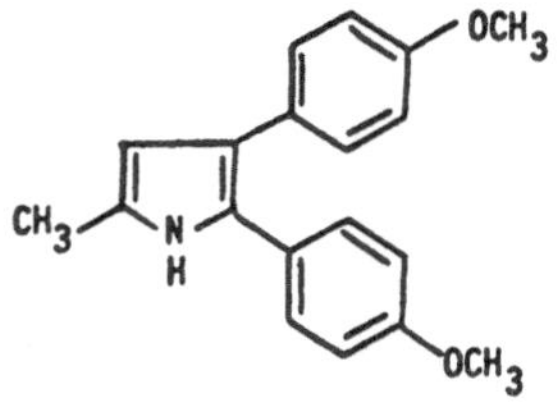

Bimetopyrol

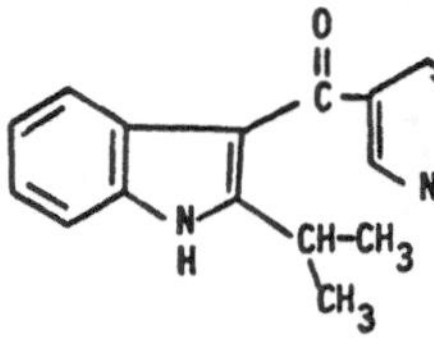

L 8027

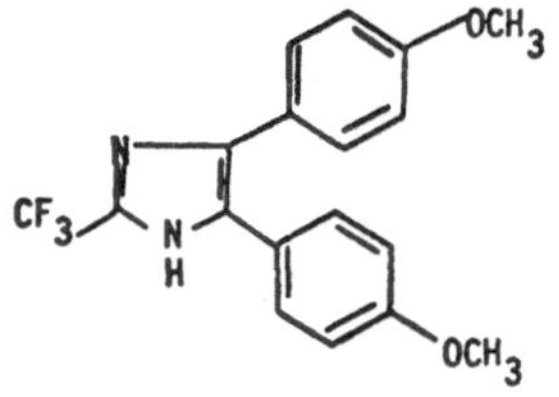

Flumizole

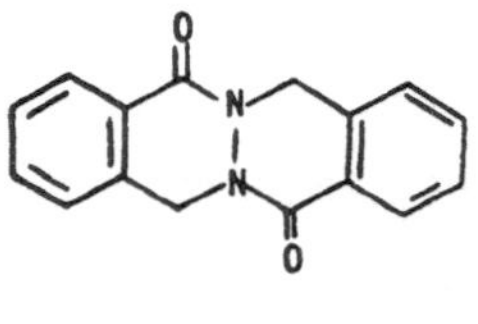

Diftalone

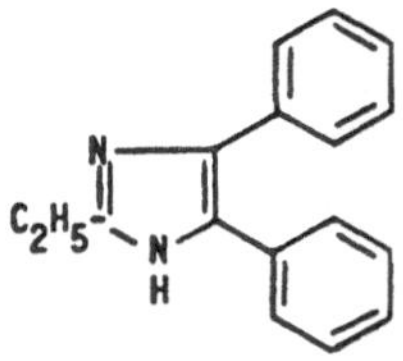

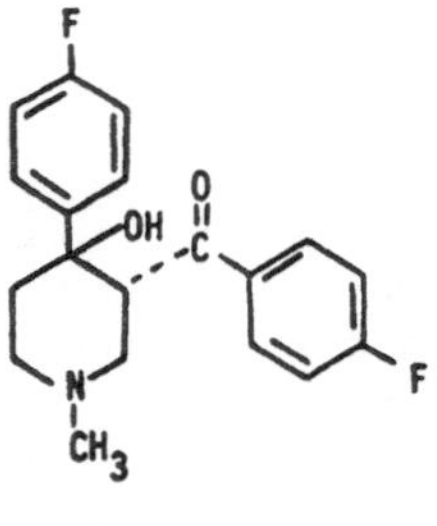

Flazalone

In our laboratory, the modest activity of 2-phenyl benzoxazole was
greatly enhanced by the synthesis of its aza analogs, the [4,5-b]
and [5,4-b] oxazolo pyridines (20,35). The _in vitro_ activity of
some members are in the same range of potent aryl aliphatic acids.

However, without a carboxyl group, the metabolic inactivation of
these compounds are highly·structure-dependent. The systemic anti-
inflammatory activity and the relative toxicity of several potent synthetase inhibitors in this series vary widely with·their rate of
oxidative metabolism.

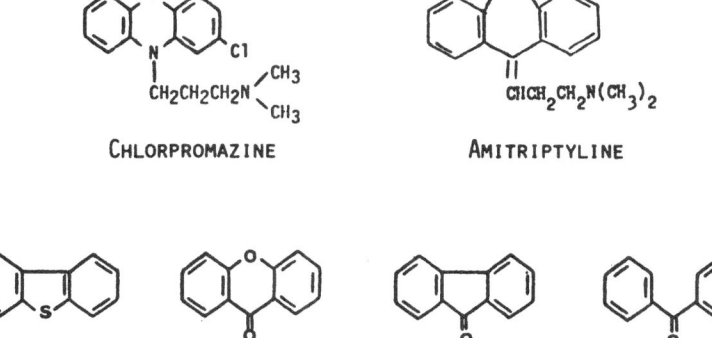

ID$_{50}$ (µM)   50   5   2.5

(SSV)*

THIABENDAZOLE

50

Other examples of nonacidic inhibitors are tricylic ring systems such as dibenzothiophene, xanthone, fluorenone and an uncy-
clized benzophenone. Introduction of an amino alkyl side-chain or
a carboxyl alkyl side-chain to these nuclei would give structures
similar to the tricyclic psychotropic agents, such as chloroproma-
zine or amitriptyline, on one hand and the familiar aryl propionic
acids on the other. In spite of their partial structural resem-
blance to these basic and acidic synthetase inhibitors, it is not
yet clear whether their sites of action are overlapping or totally
apart.

TRICYCLIC INHIBITORS

CHLORPROMAZINE   AMITRIPTYLINE

A group of hydrophobic aryl substituted α-β-unsaturated ketones,
with high sulfhydryl reactivity, have also been found to be potent
inhibitors in vitro. Unfortunately, their therapeutic potential is
severely limited by their nonspecific toxicities.

SULFHYDRYL REAGENTS

ETHACRYNIC ACID   VULPINIC ACID

8.   The Search for New PG Synthetase Regulators

     With a better understanding of the biosynthesis and physiolo-
gical roles of prostaglandins and related mediators, with a variety
of chemical structures showing synthetase inhibitory, or even stimulatory, activities, the emergence of a new generation of chemical
agents with improved efficacy as well as selectivity are to be expected. In addition to arthritis, the application of new synthetase regulators to other inflammatory conditions, e.g. dermatolog-
ical, ocular and periodontal diseases would require compounds with
different physicochemical and metabolic characteristics. The complex PG synthetase pathway has offered multiple inhibitory sites
for more selective regulation of various active metabolites, e.g.
PGG, PGH, thrombaxane, etc. The feasibility of finding multiaction
regulators, operating at the levels of biosynthesis, receptor binding and metabolic inactivation, on the basis of some common chemical
features of these metabolites is almost evident.

     Traditionally model building and retrospect analysis have often
provided hypothetical "receptor contour" to guide the synthesis of
new enzyme inhibitors. The methodology of this approach has been
refined with the aid of X-ray coordinates and computer modeling.
Such an analysis of cyclooxygenase substrates and inhibitors, along
with some mechanistic considerations, has suggested a new receptor
model in our laboratory (36).

POSSIBLE CONVERSION OF ARACHIDONIC ACID

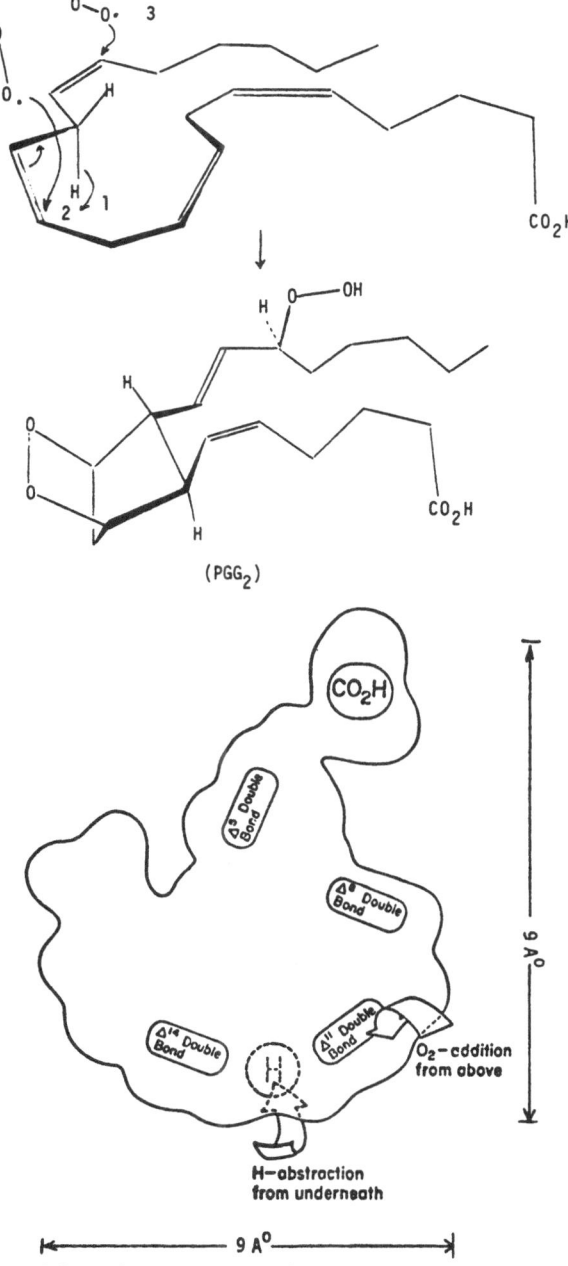

The Binding Site of Fatty Acid Substrate and Inhibitors of Prostaglandin Synthetase.

It is hoped that the recent purification of cyclooxygenase may
eventually lead to the X-ray crystallographic and other spectro-
scopic studies of the real enzyme and the description of a real
receptor in the near future.

The tissue specificity of PG synthetase inhibitors has now
been demonstrated in several cases with brain, platelets, synovial,
ocular and other tissues. In addition to differential distribution
and metabolism at the local tissue level, we have also seen the
effectiveness of a prodrug approach through the manipulation of
pharmacokinetic characteristics. Finally, other selective drug
delivery schemes, chemical conjugates, liposomes or simple topical
formulations may also contribute to the therapeutic value of this
newly defined major class of pharmacological agents.

REFERENCES

(1)   J. R. Vane, Inhibition of Prostaglandin Synthesis as a Mech-
       anism of Action for Aspirin-Like Drugs, Nature New Biol. <u>231</u>,
       232 (1971).

(2)   M. Hamberg, J. Svensson and B. Samuelsson, Thromboxanes: A
       New Group of Biologically Active Compounds Derived from
       Prostaglandin Endoperoxides, Proc. Nat. Acad. Sci. USA, <u>72</u>,
       2994-2998 (1975).

(3)   T. Miyamoto, N. Ogino, S. Yamamoto and O. Hayaishi, Purifi-
       cation of Prostaglandin Endoperoxide Synthetase from Bovine
       Vesicular Gland Microsomes, J. Biol. Chem. <u>251</u>, 2629-2636
       (1976).

(4)   G. J. Roth, N. Stanford and P. W. Majerus, Acetylation of
       Prostaglandin Synthase by Aspirin, Proc. Nat. Acad. Sci.,
       <u>72</u>, 3073-3076 (1975).

(5)   H. J. Robinson and J. R. Vane, Editors, Prostaglandin Synthe-
       tase Inhibitors, Raven Press, New York (1974).

(6)   R. J. Flower, Drugs which Inhibit Prostaglandin Biosynthesis,
       Pharmacol. Rev. <u>26</u>, 33-67 (1974).

(7)   T. Y. Shen, Prostaglandin Synthetase Inhibitors I, Handbook
       of Exper. Pharmacol. (in press).

(8)   F. Kantrowitz, D. R. Robinson, M. B. McGuire and L. Levine,
       Corticosteroids Inhibit Prostaglandin Production by Rheumatoid
       Synovia, Nature, <u>258</u>, 737-739 (1975).

(9)   A. H. Tashjian, Jr., E. F. Voelkel, J. McDonough and L. Levine,
       Hydrocortisone Inhibits Prostaglandin Production by Mouse
       Fibrosarcoma Cells, Nature (London), <u>258</u>, 739-741 (1975).

(10)   R. J. Gryglewski, Steroid Hormones, Antiinflammatory Steroids
       and Prostaglandins, Pharmacol. Res. Commun., <u>8</u>, 337-348 (1976).

(11)   J. Y. Vanderhoek and W. E. M. Lands, Acetylenic Inhibitors of
       Sheep Vesicular Gland Oxygenase, Biochim. Biophys, Acta, <u>296</u>,
       374-381 (1973).

(12)   R. H. Abeles and A. L. Maycock, Suicide Enzyme Inactivators,
       Accounts of Chem. Res., <u>9</u>, 313-319 (1976).

(13)   C. Malmsten, Some Biological Effects of Prostaglandin Endo-
       peroxide Analogs, Life Sciences <u>18</u>, 169-176 (1976).

(14) E. J. Corey, M. Shibasaki, K. C. Nicolaou, C. L. Malmsten and
      B. Samuelsson, Simple Stereocontrolled Total Synthesis of a
      Biologically Active Analog of the Prostaglandin Endoperoxides
      (PGH_2, PGG_2), Tetrahedron Lett. 737-740 (1976).

(15) T. J. Leeney, P. R. Marsham, G. A. F. Ritchie and M. W.
      Senior, Inhibitors of Prostaglandin Biosynthesis: A Bicyclo -
      [2,2,1] - Peptene Analogue of "2" Series Prostaglandins and
      Related Derivatives, Prostaglandins, 11, 953-960 (1976).

(16) R. S. Farr, Editorial, J. Allergy, 45, 321 (1970).

(17) L. H. Rome, W. E. M. Lands, G. J. Roth and P. W. Majerus,
      Aspirin as a Quantitative Acetylating Reagent for the Fatty
      Acid Oxygenase that Forms Prostaglandins, Prostaglandins 11,
      23-29 (1976).

(18) C. Patrono, G. Ciabattoni, F. Greco and D. Grossi-Belloni,
      Comparative Evaluation of the Effects of Aspirin-Like Drugs
      on Prostaglandin Production by Human Platelets and Synovial
      Tissue, Advances in Prostaglandins and Thromboxane Research,
      Editors B. Samuelsson and R. Paoletti, Raven Press, New York,
      1976, Vol. 1, pp. 125-131.

(19) J. Hannah, W. V. Ruyle, A. Matzuk, H. Jones, K. Kelley, W. J.
      Holtz, B. E. Witzel, R. Houser, L. H. Sarett and T. Y. Shen,
      Diflunisal, A Novel Long-Acting and Potent Salicylate, In Prep.

(20) T. Y. Shen, E. A. Ham, V. J. Cirillo and M. Zanetti, Structure-
      Activity Relationship of Certain Prostaglandin Synthetase In-
      hibitors, Prostaglandin Synthetase Inhibitors, Ed. J. R. Vane
      and H. R. Robinson, Raven Press, New York, pp. 19-31 (1974).

(21) S. L. Steelman, C. T. S. Sibinga, P. Schulz, W. J. H. Vanden
      Heuvel and K. F. Tempero, The Effect of Diflunisal on Prosta-
      glandin Excretion and Blook Platelet Function in Normal Sub-
      jects, In Press.

(22) K. Hoogsteen and N. R. Trenner, Structure and Conformation of
      cis and trans Isomers of 1(p-chlorobenzylindene)-2-methyl-5-
      methoxy-3-indenyl acetic acid, J. Org. Chem., 35, 521 (1970).

(23) T. J. Kistenmacher and R. E. Marsh, Crystal and Molecular
      Structure of an Antiinflammatory Agent, Indomethacin, 1-(p-
      chlorobenzoyl)-5-methoxy-2-methyl-indole-3-acetic acid, J.
      Am. Chem. Soc., 94, 1340-1345 (1972).

(24) T. Y. Shen, B. E. Witzel, H. Jones, B. O. Linn, J. McPherson,
      R. Greenwald, M. Fordice and A. Jacobs, Synthesis of a New

Antiinflammatory Agent, Cis-5-Fluoro-2-Methyl-1-[P-(Methylsulf-inyl) Benzylidenyl] Indene-3-Acetic Acid.

(25) C. G. Van Arman, E. A. Risley, G. W. Nuss, H. B. Hucker and D. E. Duggan, Pharmacology of Sulindac in Clinoril, The Treatment of Rheumatic Disorders, Edited by E. C. Huskisson and P. Franchimont, Raven Press, New York, 1976, pp. 9-36.

(26) E. L. Tolman, J. E. Birnbaum, F. S. Chiccarelli, J. Panagides and A. E. Sloboda, Inhibition of Prostaglandin Activity and Synthesis by Fenbufen (A New Nonsteroidal Antiinflammatory Agent) and One of its Metabolites, Advances in Prostaglandin and Thromboxane Research, Ed. B. Samuelsson and R. Paoletti, Vol. 1, 133-138, Raven Press, New York (1976).

(27) T. Y. Shen, Perspectives in Nonsteroidal Antiinflammatory Agents, Angew Chemic (int. ed.) 11, 460-472 (1972).

(28) E. C. Ku and J. M. Wasvary, Inhibition of Prostaglandin Synthetase by Pirprofen Studies with Sheep Seminal Vesicle Enzyme, Biochim. Biophy. Acta 384, 360-368 (1975).

(29) C. Patrono, G. Ciabattoni and D. Grossi-Belloni, In Vitro and In Vivo Inhibition of Prostaglandin Synthesis by Fenoprofen, A. Nonsteroid Antiinflammatory Drug, Pharmacol. Res. Comm. 6, 509-518 (1974).

(30) L. H. Rome and W. E. M. Lands, Structural Requirements for Time-Dependent Inhibition of Prostaglandin Biosynthesis by Antiinflammatory Drugs, Proc. Natl. Acad. Sci. 72, 4863-4865 (1975).

(31) H. Zwarenstein, N. Sapeika and J. H. Holmes, Effect of Anti-Inflammatory Drugs on Lipase In Vitro, Res. Comm. Chem. Pathol. Pharmacol., 13, 563 (1976).

(32) R. W. Egan, J. L. Humes, C. A. Eckert, M. Galavage and F. A. Kuehl, Jr., The Influence of Phenols and Related Compounds on Inhibition of Prostaglandin Biosynthesis by Nonsteroidal Antiinflammatory Agents, Fed. Proc. 35, 1652 (1976).

(33) G. Arthurson and C. E. Jonsson, Stimulation and Inhibition of Biosynthesis of Prostaglandins in Human Skin by Some Hydroxy-ethylated Rutosides, Prostaglandins, 10, 941-948 (1975).

(34) T. Y. Shen, Non-acidic Antiinflammatory Agents, R. A. Scherrer and M. W. Whitehouse Editors, Antiinflammatory Agents, Acad. Press, New York (1974) Vol. II, pp. 179-207.

(35) R. L. Clark, N. Jensen, A. A. Penolano, T. Lanza and T. Y.

Shen (Unpublished).

(36) P. Gund and T. Y. Shen, A Model of the Antiinflammatory Re-
      ceptor Mechanism of Prostaglandin Biosynthesis (1976).

RESPIRATORY SYSTEM (GENERAL INTRODUCTION)

L. Puglisi and F. Maggi

Institute of Pharmacology and Pharmacognosy

University of Milano, 20129 Milano, Italy

OCCURRENCE, BIOSYNTHESIS, RELEASE AND METABOLISM IN THE LUNG

Prostaglandins occur naturally in lungs and in bronchial tissues. The presence of a lipid extract from normal lung, possessing smooth muscle stimulating activity, had been detected (1,2) some years before Bergstrøm reported the structural formulas of the primary prostaglandins, isolated from sheep vesicular glands (3,4).

The lung parenchima is a rich source of prostaglandins as it allowed Bergstrøm and coworkers (5) to obtain from normal lungs of sheep and swine the first member of a new series of prostaglandins: $PGF_{2\alpha}$. Prostaglandins were found later in the lungs of other species, practically in all the chosen animals: cattle (6), guinea pig, monkey (7), human (7,8), cat, rat, rabbit and chicken (9) and swine (10). No prostaglandins were detected in the lungs of the dog (9). The distribution of prostaglandins in the lung is particularly interesting since, unlike many other tissues, they contain more prostaglandins of the F series ($PGF_{2\alpha}$), with the exception of the cat (Table 1). $PGF_{1\alpha}$ and $PGF_{3\alpha}$ were tentatively identified in sheep lung by Anggard (7); $PGF_{3\alpha}$ was previously isolated in low yield from bovine lung by Samuelsson (6). Neither PGE_1 nor PGE_3 could be detected from lung tissues

of any of the examined species, neither any prostaglandins of other series (7). The amounts of $PGF_{2\alpha}$ in human
lung parechima, reported by Karim et al. (8), as it
is shown in Table 1, are very high and are comparable
with the concentrations found in tissues of other human
organs, known to contain a lot of this material, such
as thyroid and kidney medulla. Instead the data related
to human and monkey, obtained by Anggard (7), appear to
be too lower in comparison with that of other examined
species, so he believes his results could be an artifact
due to post-mortem degradation. In any case it appears
from more recent works, that, except that in seminal
vesicles and in seminal fluid, normal prostaglandin
content, reported in tissues and tissue fluids, is
probably incorrect, as the prostaglandin concentration
in the cell are the result of an increased(or sometimes
decreased) synthesis during experimental preparation (II).

The lungs as many other tissues, if not all, normally do not store prostaglandins but they are capable of
synthetizing them. The biosynthetic ability of this organ
to form prostaglandins, as measured by their conversion
from labeled precursors or by their release in media
during incubation or perfusion exceeds that of most of
the other organ systems (12) with the exception of that
of seminal vesicles. The biosynthesis of prostaglandins
is catalized by a prostaglandin-synthetase system acting
on precursor unsaturated "essential" fatty acids. The
precursor is generated through the action of phospholipase enzyme system, acting on phospholipids, which are
widely present in cell membranes (13). The enzyme system
consists of a fatty acid cyclo-oxygenase (5) and an endoperoxide isomerase (15). PGE and PGF_{α} compounds originate
from common endoperoxide intermediates (14, 15, 16). The
PG synthetase complex involves several steps during which
have been recently identified time precursors (cycloendoperoxides) (17) and non prostaglandin products (Thromboxanes) (18). Prostaglandin synthetase activity is high
in the lung and it was demonstrated that it is able to
operate on tritiated di-homo-γ-linolenic acid (19) as well
as on labeled arachidonic acid (20). Since from Table 1
the only two prostaglandins extracted prevalently from
lung tissues, no question on the genuine concentrations,
are PGE_2 and $PGF_{2\alpha}$, it seems reasonable to argue that
arachidonic acid too is the physiological precursor in
lung cell membrane phospholipids.

TABLE 1   OCCURRENCE OF PROSTAGLANDINS IN THE LUNG

Species	$F_{2\alpha}$ (ng/g tissue)	E_2 (ng/g tissue)	Reference
Chicken	30,4	7,7	Karim, Hillier and Devlin,1968 (9)
Rat	30-375,0	2,5-6,6	Karim, Sandler, Williams, 1967 (8)
Rat	90,4	16,6	Karim, Hillier and Devlin, 1968 (9)
Guinea Pig	500,0	2,5	Anggard,1965 (7)
Guinea Pig	375,0	2,5	Karim, Hillier and Devlin, 1968 (9)
Rabbit	8,0	5,4	Karim, Hillier and Devlin, 1968 (9)
Cat	15,5	65,0	Karim, Hillier and Devlin, 1968 (9)
Cat	1,5	15,5	Karim, Sandler, Williams, 1967 (8)
Dog	n.d.	n.d.	Karim, Sandler, Williams, 1967 (8)
Sheep	500,0	40,0	Anggard,1965 (7)
Pig	+ +	n.d.	Anggard, and Bergstrom 1963 (10)
Monkey	200,0		Anggard,1965 (7)
Man	20,0		Anggard,1965 (7)
Man	12,4-50,0	1,3-2,4	Karim, Sandler, Williams, 1967 (8)

n.d. = not detected

TABLE 2   RELEASE OF PROSTAGLANDINS FROM LUNG TISSUES BY VARIOUS STIMULI

Species	Stimulus	PGE	PGF$_\alpha$	RCS	Reference
	Anaphylaxis	+	++		Piper and Vane, 1969 (21)
	Anaphylaxis	+	+		Piper and Vane, 1969 (22)
	Anaphylaxis			+	Piper and Walker, 1973 (23)
	Phospholipase A				Vogt, Meyer, Kunze, Lufft and Babilli, 1969 (24)
Guinea Pig	Embolism	+			Piper and Vane, 1969 (21-22)
	Embolism	+		+	Palmer, Piper and Vane, 1970 (25)
	Embolism	+			Lindsey and Willie, 1970 (26)
	Embolism				Piper and Vane, 1971 (27)
	Hyperinflation	+			Piper and Vane, 1971 (27)
	Hyperinflation				Berry, Edmonds and Willie, 1971 (28)
Rat	Anaphylaxis				Piper and Walker, 1973 (23)
	Embolism				Palmer, Piper and Vane, 1970 (25)
	Embolism				Lindsey and Willie, 1970 (26)
	Hypoxia				Weir, McMurtry, Tucker, Reeves and Grover, 1976 (29)
Cat	Hypoxia	+			Said, Hova, Yoshida, 1975 (30)
Dog	Hyperinflation				Said, Kitamura and Vreim, 1973 (31)
	Hypoxia				Weir, McMurtry, Tucker, Reeves and Grover, 1976 (29)
Calf	Anaphylaxis	+			Burka and Eyre, 1974 (32)
	Hypoxia	+	+		Weir, McMurtry, Tucker, Reeves and Grover, 1976 (29)
Human	Anaphylaxis	+	+		Piper and Walker, 1973 (23)
	Hyperventilation	++	+		Said, Kitamura and Vreim, 1972 (31)
	Hypoxia		++		Said, Yoshida, Kitamura and Vreim, 1974 (33)

The fact that prostaglandins can be extracted and
biosynthetized from lung preparation does not necessarily mean that they have a physiological role in this organ.
There is no clear-cut evidence that they are involved
in normal or abnormal pulmonary function. This may be
still unsettled because, in spite of the very early
stage of knowledge of the lungs as a potent source of
prostaglandins, mainly $PGF_{2\alpha}$, until recently their
effects on respiratory system have received little
attention. In order to demonstrate such a role for prostaglandins in this apparatus, first of all it is
important to find out and to detect prostaglandin release
upon some physiological or pathological stimulation.
Among the circumstances which have been employed to
stimulate the prostaglandin formation and release from
the lung there are those summarized in Table 2. In almost all the conditions reported in Table 2, mixtures
of PGE and PGF_{α} compounds are usually released and measured by a dynamic bioassay in which more than one isolated
assay organ is superfused with an artificial salt solution containing many antagonists to eliminate responses
to histamine, acetylcholine, catecholamine and serotonin
(21 - 34).Nevertheless, it appears very difficult to
ascertain which of the two series of prostaglandins was
preminently synthetized relating to a specific experimental stimulation. During guinea pig lung anaphylaxis (21)
it seems to be $PGF_{2\alpha}$ and it was shown more recently by
Said and coworkers that prostaglandin of the E type pre-
vails in hyperinflation and hyperventilation (35) whereas PGF_{α} in hypoxia (33) in human but not in other species
(29, 30). It seems worthly to remark that when in 1969
Piper and Vane (21 , 22) demonstrated for the first time
the release of prostaglandins during anaphylaxis in guinea
pig they found also an unstable substance, other than
acid lipid, discharged from sensitized guinea pig lung
challenged with antigen. This material was named RCS
(Rabbit aorta Contracting Substances). RCS was recently
characterized (36) and it was found to contain at least
two factors that induce contraction of the rabbit aorta:
a cycloendoperoxide (PGG_2 and/or PGH_2) which is a time
precursor of prostaglandins (E or F_{α}) and a very unstable
factor which is responsible for the major part of this
contraction i.e. thromboxane A_2. Therefore we can assert
that the three main groups of mediators of the actions
of polyunsaturated fatty acids, transformed via the cycloxygenase pathway, i.e.: stable prostaglandins, endoperoxides and thromboxanes, may be formed in the lung. Prostaglandin formation and release are also known to happen
during perfusion of the lung with some naturally occurring
substances, such as: kinins and slow-reacting substances
of anaphylaxis (SRS-A) (22), serotonin (37) and histamine
(38). All these are wellknown mediators of anaphylaxis.

Furthermore also mechanical stimuli, such as gentle
stretching, squeezing, stroking the perfused organ (27)
and stirring the chopped tissue (23, 27), provoke the
release of prostaglandins from the lung. Since less
prostaglandins can be extracted than it is released (39,
40, 41), all the procedure which induce prostaglandin
liberation must also stimulate their synthesis. The only
preformed smooth-muscle stimulating substance released
by the lung is indeed hystamine, also for the others is
involved fresh synthesis. Piper and Vane (27) point out
that the only feature common to those chemical and physical reported stimuli seems to be a disturbance of
cell membrane. Therefore any slight provocation at the
cell membrane level leads to degeneration and release
of prostaglandins, this release occurring as a result
of stimulation of their biosynthesis. The endogenous
synthesis of prostaglandins in intact animal could be
determined by monitoring specific urinary metabolites
(42-47). The half life of the major metabolites of
PGE_2 and $PGE_{2\alpha}$ in the circulation is very short (42, 48).
Thus, the quantitative determination of these metabolites
could be an index of an acute altered synthesis (increased) of prostaglandins. In fact, using the method suggested
(42-47), it was possible to determine that in the guinea pig,
anaphylaxis in vivo increases the formation of $PGF_{1\alpha}$ and
$PGF_{2\alpha}$ (47). More recently (49) it has been confirmed that
endoperoxides and thromboxanes are also involved in anaphylactic reactions together with prostaglandins. The endoperoxides (50) and probably also thromboxanes are more
potent on tracheal smooth-muscles in vitro as well as in
vivo in inducing contraction of these muscles. Furthermore
these unstable compounds are present in the lung or, better,
they may be synthetized and released in large amounts in
comparison with the bronchodilator PGE_2. Since their discovery in 1935 (51, 52) as a lipid soluble smooth-muscle
stimulating principle extracted from human semen and sheep
vesicular glands, prostaglandins were also known to lower
systemic arterial blood pressure. The prostaglandins of
the E type (PGE_1, PGE_2 and PGE_3) are uniformally potent
vasodepressors in all the laboratory animals studied and in
human too (53, 54, 55). The F_α prostaglandin effects on
systemic circulation are complicated by qualitative species
variations; when they are depressor (cat and rabbit) this
action is quite moderate, if they exert pressor action (rat,
dog, monkey and human) this action is less potent than of
angiotensin or noradrenaline (54, 55). This particular ex-
isting state suggests, first of all, that prostaglandins
may not be circulating substances and that they must be
very rapidly removed from venous blood.

Prostaglandins which originate in the intestine or spleen are largely (80%) removed by the liver (56) but the portal circulation receives only part of the cardiac output. In addition to synthesis and release of prostaglandins, the lung is an important organ for their metabolism. The lung is perhaps ideally located and structured to change circulating vasoactive substances and its metabolic role could be vital for survival. This organ may be considered as a machinery designed to rapidly modify the concentrations of compounds, elsewhere born in the body, as the entire cardiac output flows through it and comes in contact with a large surface area. This fact ensures the removal or the change of material entering from the venous blood and the release of inactive or active humoral agents into the arterial circulation. In recent years the lung has been shown to modify many circulating vasoactive substances: some become biologically active, such as angiotensin II; some become inactive, such as serotonin, bradykinin and prostaglandins (57, 58). Although prostaglandins are stable in blood (56) their half life in circulation is short because of rapid removal by pulmonary circulation. Vane and coworkers (56, 57, 59) have shown, using blood-bathed organ technique, that the lung is able to remove from the blood by one passage through it more than 90-95% of prostaglandin E_1 and E_2 and about 80% of $PGF_{2\alpha}$ whereas only 35-70% of the A series. Other investigators have found that PGA_1 and PGA_2 appear unchanged by passage through dog lungs (60). Following intravenous injection of tritium labeled PGE_1 in healthy man subjects most of the radioactivity disappears from the blood within 10 minutes (61) because the pulmonary circulation removes 90% of injected prostaglandin (62). The inactivation of prostaglandins in the lung has been attributed to the enzyme 15-hydroxy prostaglandin dehydrogenase (PGDH=prostaglandin dehydrogenase). This enzyme catalyzes oxidation of PGs at C-15 and was first isolated from homogenates of swine lung (63). Furthermore lung tissue contains among the highest levels of this enzyme in the body (64). In contrast with "angiotensin-converting enzyme" which seems to be located on the luminal surface of pulmonary endothelial cells (65), the PGDH is mainly in the particle free fraction (63, 64). Prostaglandin metabolism involves another enzyme which catalyzes the reduction of $\Delta 13$ double bond: prostaglandin-$\Delta 13$- reductase (PGR). High concentration of this enzyme occurs also in the lung (66). Since prostaglandin PGDH abides inside the cell prostaglandin must go intracellularly for the enzyme system to be effective. According to Ryan et al.(67) as the transit time of $PGF_{2\alpha}$ through the lung is longer than that of blue dextran, an intravascular not diffusable marker (68), this delay could be due to the transport of this prostaglandin into the cell. Same results were obtained recently also using PGE_I (69). Since prostaglandins of the A series , which in vitro experiments are destroyed by the lung metabolizing enzymes (39) and are a good substrate for these enzymes (70, 7I), are not readily inactivated on passage through the dog lung, it appears likely that at least in this species they do not enter the cells. Another explanation for the delay of the transit time of PGs could be related to a rather specific uptake mechanism for PGE and PGF_α combined, as suggested by Bito (72),with a specific carrier mediator. The time inactivation of PGE_2 in human is very rapid, as it was observed that after 90 seconds by the i.v. injection of labeled PGE_2 into one arm the metabolite (II-α - hydroxy-9,I5 diketo prostanoic acid) appears in the venous blood of the other arm (Hamberg and Samuelsson, unpublished, quoted by P.J. Piper (73). The metabolic function of the lung may be a very important one, although to date most reports have dealt with the normal lung and no diseases have been reported to be related to a failure of this function. It proceeds normally also in the presence of minor abnormalities in the lung due to mitral valve disease (74). It seems important to underline that premature birth is commonly associated with an immature lung which could be unable to meet its demanding metabolic requirements.

An interesting observation of more extensive special attention has been made by Pace-Asciak (75): there is a little change in prostaglandin biosynthetic capability during animal development, whereas prostaglandin catabolism varied showing higher activity. This suggests a protective mechanism for the immature organs against the potentially harmful of vasoactive properties of prostaglandins during certain critical stages of organogenesis. In immature rat and sheep the most active catabolyzing enzymes are PGDH and PGR and they are both organ and species-specificity. Also the stages of the development of the two PGs catabolyzing enzymes are different. PGDH in the prenatal lungs of the rabbit reaches adult levels at least several days before birth whereas the levels of PGR keep increasing slowly with the age (76).

The major prostaglandin metabolites coming from the lung into the arterial circulation are the I3:I4 dihydro-I5-keto derivatives. These metabolites of PGE_2 and $PGF_{2\alpha}$ have been reported to have minimal spasmogenic activity when compared with their parent compounds (77, 78). An exception is the metabolite of PGA_2 which mantains the biological potency of its parent (39). But as a result of lung metabolism either the I3:I4 dihydro derivatives or the I5-keto derivatives can possibly be formed by the enzymatic breakdown, and they may have or not biological activity. PGE_2 is usually bronchorelaxant, however its I5-keto derivative,

which is IOO times less active, after an initial relaxation exerts a potent contraction (79). The I5-keto $PGF_{2\alpha}$
has been found to be more potent than its parent on isolated human bronchial smooth muscle (80). These data suggest
that even a light modification of PGs metabolism may be an
important factor as well as a PGs biosynthesis change in
the respiratory system in the pathogenesis of pulmonary
edema, pulmonary hypertension, bronchial asthma and anaphylaxis.

The prostaglandin synthesis may be prevented by non
steroidal antiinflammatory compounds (NOSAC), such as indomethacin and aspirin (8I, 82) and by TYA (5,8, II,I4 eicosatetraynoic acid) (83). NOSAC and TYA block also the
generation of thromboxane A_2 as they act on the prostaglandin cycloxygenase enzyme. Very recently (84) it was found
in perfusates from guinea pig lung during anaphylaxis a substance more stable than RCS, which was called RCS-releasing
factor (RCS-RF). This partially purified factor, injected
into pulmonary artery of perfused lungs from unsensitized
guinea pig, induces a release of RCS. RCS-RF stimulates
synthesis of prostaglandins, endoperoxides and thromboxane
A_2 probably by releasing the arachidonic acid from lung
cell membrane phospholipids via activation of phospholipase
enzyme activity. This release seems to be blocked by steroid
antiinflammatory drugs, such as dexametasone. Inhibition of
prostaglandin release by antiinflammatory steroids was
previously suggested also by Lewis and Piper (85, 86). In
perfused lungs of sensitized guinea-pig hydrocortisone
completely blocks the release of prostaglandins and RCS
induced by anaphylactic shock. This block is reversed by
arachidonic acid (87). More detailed informations about this
argument are reported in another chapter of this volume.

The inhibition of pulmonary prostaglandin metabolism,
from the academical point of view, nowadays may be useful,
as it gives a pharmacological tool in order to provide more
information on the mechanism of the PGs incactivation. In
isolated perfused guinea pig lungs the most active compounds
able to block PG metabolism were polyphloretin phosphate
(PPP) and di-4-phloretin phosphate (DPP) (88). PPP selectively antagonizes the exogenous prostaglandin activity in vitro
(89, 90) and in vivo (9I, 92). But the enzymatic breakdown
inhibition is achieved at much lower concentration. Therefore it is possible to use PPP to block metabolism without
antagonizing prostaglandin action. This blocking effect of
PPP seems to be specific because bradykinin or serotonin
were normally removed by guinea pig isolated lungs. According to the authors the mechanism of action of PPP and DPP in
lung metabolism may be due to the block of PGDH, the enzyme
which is known to be the crucial step of PGs breakdown, how-

ever it is also possible that these compounds inhibit PGs
transport into the cell. Other drugs can repress pulmonary
PGs metabolism inhibiting their biotransport in the lung
cells (93). The PGs transport process could take place,
as it happens for other tissues, against a concentration
gradient (94) and may be "carried-mediated". One PGs transport inhibitor is probenecid which was shown to block the
metabolism of $PGF_{2\alpha}$ by the isolated perfused rat lung. In
the future could be possible to use such drugs (PPP and
Probenecid) to overcome the deficiency of the prostaglandin synthetizing system as it is coming to do in the
opposite condition (hyperproduction of PGs) with the biosynthesis inhibitors (NOSAC).

PHARMACOLOGICAL EFFECTS

It is known that a number of factors can cause pulmonary vasoconstriction preminent enough to raise pulmonary
arterial pressure under conditions in which cardiac output
remains constant. Usually the same factors that constrict
the systemic circulation will also cause pulmonary vasoconstriction. These vasopressor stimuli include adrenaline
and noradrenaline, angiotensin II and an increased sympathetic nervous activity. In some pathological conditions
several vasoactive substances may induce pulmonary vasoconstriction. Serotonin, histamine and various peptides
may be released in the general circulation or locally
within a discrete part of the pulmonary circulation and all
these bring about vasocostriction in the lung. According to
Vane (57) the naturally occurring vasoactive substances
can be divided into two types: local or circulating. The
circulating autocoids are those which pass through the
lungs unchanged; the local ones, including prostaglandins,
are those which effectively are removed by lungs and, if
they do have a physiological function, this will be probably localized at/or near the site of their release. Therefore endogenous prostaglandins, released from the lung it-
selt, could affect pulmonary vessels.

Isolated pulmonary vessels

The pharmacological effects of prostaglandins on
isolated pulmonary vascular smooth muscles have been the
object of few studies although many extrapulmonary vessels
have been studied extensively (95). Calf pulmonary arteries
and veins are relaxed by PGE_I and PGE_2 when both vessels
are submaximally contracted by some agonist, such as histamine, serotonin or acetylcholine, whereas $PGF_{2\alpha}$ elicited a
dose-related contraction (96). In higher doses PGE_I and PGE_2

TABLE 3   EFFECT OF PROSTAGLANDINS ON INTRAPULMONARY VESSELS

Species	E Vein	E Artery	F$_{2\alpha}$ Vein	F$_{2\alpha}$ Artery	Reference
Guinea Pig	0	R[++]	0	C	Okpako, 1972 (97)
Dog	R[+]	R[+]	C	0	Kadowitz et al., 1973 (98)
Dog	C[++]	0	C	0	Kadowitz et al., 1975 (101)
Sheep	R[+]	R[+]	C	0	Kadowitz et al., 1975 (101)
Calf	R	R	C	C	Lewis and Eyre, 1972 (96) Burka and Eyre, 1974 (99)
Swine	0	0	0	0	Kadowitz et al., 1975 (101)
Baboon	0 or slC	0 or slC	C	C	Kadowitz et al., 1975 (101)
Chimpanzee	R[+]	R[+]	n.d.	C	Kadowitz et al., 1975 (101)
Monkey	n.d.	Sl C[+]	Sl C	C	Kadowitz et al., 1975 (101)
Man	Sl R[+]	Sl C[+]	C	C	Kadowitz et al., 1975 (101)

C = Contraction
R = Relaxation
0 = no effect
n.d. = not detected
sl = slight
+ PGE$_1$
++ PGE$_2$

are able to relaxe also uncontracted preparations. Prostaglandin $F_{2\alpha}$ induces a dose-related concentration also on
the guinea pig pulmonary arterial strip, which is not
antagonized by phentolamine or mepyramine (97). An antagonistic effect (dose-dependent) was observed with PGE_2
against contracted smooth muscle by histamine or noradrenaline; but relaxation of the arterial strip preparations could be demonstrated at high doses, only when the
tone of the muscle was raised by prior addition to the
organ bath of one agonist (97). On isolated helically-cut
strip of lobar veins and arteries of dog, PGE_I causes a
relaxation whereas $PGF_{2\alpha}$ induces a marked contraction on
the veins and a very slight one on the arteries (98).
Bovine pulmonary vein in vitro is contracted by $PGF_{2\alpha}$ and
relaxed by PGE_I (99). These data are similar to those
previously obtained by Lewis and Eyre (96). PGE_2 was
shown to have a dual effect inducing either relaxation
or contraction depending on the concentrations used (99).
In 1974 the effects of PGE_I, PGE_2 and $PGF_{2\alpha}$ (which have
been found in maternal and fetal circulation during labour)
were analyzed in fetal smaller pulmonary arteries of calves
(100). In this case exogenous prostaglandin $F_{2\alpha}$ and E_I
relaxed pulmonary arteries whereas PGE_2 caused contraction.
It appears particularly interesting to note that later in
life the pulmonary artery responses to PGE_2 and $PGF_{2\alpha}$ are
reversed (96, 99). Kadowitz and coworkers have reviewed
experiments they performed on isolated intrapulmonary vessels from several animal species (101). These authors demonstrated that $PGF_{2\alpha}$ usually contracts veins and arteries
of most species, whereas PGE_2 exibits a species dependent
activity which could be relaxation, slight contraction or
no detectable action on both vessels. These experiments,
however, are lacking of quantitative details. According to
Strong and Bohr (102) the PGE_I and PGE_2 compounds yielded
a biphasic response, lower concentration causing relaxation
and the higher inducing contraction on isolated strips of
various extrapulmonary arteries. The data available from
the studies conducted on pulmonary vascular smooth muscle
are summarized in Table 3. Generally the effects of prostaglandins E and F_α on pulmonary vessels parallel their
effects on tracheal and bronchial smooth muscle (see in
the next pages).

Pulmonary circulation

Hauge et al. demonstrated in the rabbit that PGE_I decreases pulmonary vascular resistance (PVR) in vitro, an
effect comparable with that of adrenaline but not abolished by α or β adrenergic inhibitors (103). In guinea pig
isolated perfused whole lung $PGF_{2\alpha}$ causes an increase in
PVR whereas PGE_2 induces a decrease (97). The physiological role of prostaglandins on pulmonary vascular tone

is unknown. In fact in the isolated perfused-organ, the
interpretation of the obtained results is difficult as
the experimental conditions are far from the normal physiological ones. The in vivo studies of prostaglandin
effects on pulmonary circulation have been quite extensive (IO , IO4 -III). The results obtained are often
contradictory. The pulmonary circulation has been described as a passive low-pressure system capable of little
regulation. Thus pulmonary blood flow distribution can
easily be profoundly altered. In the early experiments
it seems, according to Kadowitz (II2), that pulmonary
blood flow was not always kept constant.In fact the
circulation in the lungs depends also on the cardiac
output and on the myocardial activity. Therefore to study
the direct effects of prostaglandins on pulmonary vascular
bed,Hyman (II3, II4) developed a particular right heart
and transseptal catheterization techniques which allowed
controlled blood flow in lightly anaestetized spontaneous
breathing animals. With this technique it was possible
to use isolated left lower lung lobe with its intact
autonomic innervation. With this preparation it was demonstrated (IOI, II2) that $PGF_{2\alpha}$ induces vasoconstriction
in both lobar veins and arteries in intact dog whereas in
intact swine and lamb this effect was observed only at
lobar artieries. Furthermore the canine pulmonary vessels
responded at concentration of $PGF_{2\alpha}$ as low as IO-I2 M, being more sensible than that of the others species in which
had been reported large amounts of PGs in the lung or released by the lung. PGE_I has an opposite effect in both dog,
lamb and swine,being a potent pulmonary vasodilating agent.
On the other hand PGE_2 increases PVR by constricting lobar
veins and too lesser extent arteries too, in all the three
species studied. In most extrapulmonary vascular beds PGE_2
induces a potent vasodilatation (II5, II6). The reason for
this opposite effect elicited by PGE_2 on different vascular beds is still unknown. The possibility that in the
pulmonary district this action of PGE_2 was mechanic, via
platelet aggregation, was excluded (II3). Another explanation could be the convertion of PGE_2 to $PGF_{2\alpha}$ in the lung
through the activity of the enzyme PGE_2-ketoreductase (II7).
In any case several other naturally occurence substances,
such as histamine, bradykinin and acetylcholine cause systemic vasodilatation together with pulmonary vasoconstric-
tion.Furthermore the opposite effects between the two so
closely related PGE_I and PGE_2 are not so surprising, as also
for platelets aggregation PGE_2 is an enhanced agent whereas PGE_I is a potent inhibitor (II8).

Respiratory changes

Respiration, in the broadest sense, applies both to
the processes whereby O_2 and CO_2 are exchanged within the
environment, i.e. between air and blood, and to utilization of O_2 and production of CO_2 by individual cells.
This special and vital work is critically dependent on
the activity of smooth muscle both in the pulmonary vessels which regulate the blood flow through the lung
capillaries and in the bronchial tree which regulates
the flow of air into the alveoli. Respiration can be
stimulated by certain prostaglandins, including PGE_2
and $PGF_{2\alpha}$ in various animal species: guinea pig (95),
rat and cat (119), rabbit (120), dog (106, 121) and
Human (122, 123). Recently McQueen demonstrated in
anaesthetized cat that PGE_1, E_2, $F_{2\alpha}$, A_1 and A_2 increase
respiratory frequency which was abolished by bilateral
vagotomy (124). The tidal volume can be increased by
PGE and PGA but not by $PGF_{2\alpha}$ and it seems to be preceded
by a fall in mean blood pressure. However, if cats were
anaesthetized with chloralose, instead that with pentobarbital, no modification is observed in tidal volume although blood pressure falls. In spontaneous breathing
anaesthetized dog $PGF_{2\alpha}$ was found to be the most active
agent on pulmonary airways resistance (increasing) and
dynamic lung compliance (decreasing) in comparison with
histamine and acetylcholine (125, 126). The effect of
autonomic nerves block, such as atropine, reduces the
bronchopulmonary responses evoked by $PGF_{2\alpha}$ whereas propranolol does not. The authors suggest therefore that sympathetic innervation of pulmonary airway, in the dog at
least, may be minimal and may be less important in the
regulation of their caliber. Some analogues of naturally
occurring endoperoxides have been reported (127) to be
more potent than $PGF_{2\alpha}$ in producing pharmacological
modification in canine pulmonary airways resistance, dynamic lung compliance, expiratory air flow rate and
respiratory frequency.

A respiratory pathophysiological situation in which
prostaglandins may be implicated is bronchial asthma (79,
95, 128, 129), a desease characterized by increased responsiveness of trachea and bronchi to various stimuli associated with narrowing of the airways.

Airway smooth muscle

Tracheal and bronchial smooth muscle are controlled
by the autonomic nervous system and can be influenced by
many chemical agents including naturally occurence substances as well as drugs, which modify neurohumoral physiological activity.

β-Adrenergic stimulating agents normally produce tracheo-
bronchodilatation whereas β-blocking and cholinergic stimuli cause
bronchoconstriction. The smooth muscle relaxing action of drug
stimulating β-adrenoceptors is mediated by an increase of the intracellular level of cyclic AMP (130, 131, 132). α-Adrenergic stimulation (adrenaline + propranolol) decreases lung levels of cAMP (133)
and causes bronchoconstriction in guinea pig and cat preparation
in vivo (Puglisi, Berti and Maggi unpublished). The effects of
vagal stimulation appear to be mediated by an increase of cyclic
GMP (131). On bovine tracheal smooth muscle the cyclic GMP level
also increases rapidly at the beginning of its contraction elicited
by carbachol (135). It was also shown that on isolated bronchial
human smooth muscle PGE_1 increases cAMP (136). Recently Kadowitz
et al. performed a study to assess the effects of PGE_1 and $PGF_{2\alpha}$ on
cyclic nucleotide levels in canine isolated pulmonary veins and
arteries (137). PGE_1 induces a significant increase in intracellular
cAMP concentration, both in veins and in arteries, in agreement with
its effect on smooth muscle i.e. relaxation. On the other hand $PGF_{2\alpha}$
does increase the cGMP levels only in the vein but not in the artery.
Again this result parallels with the action of $PGF_{2\alpha}$ on smooth muscle
mechanical response. Thus prostaglandins may be related to cyclic
nucleotides but the exact relationship between prostaglandins and
cAMP and cGMP in controlling contraction-relaxation cycle of
respiratory smooth muscle at cellular level is not yet established.
The only data concerning the occurrence of prostaglandins in
bronchial smooth muscle are those of Karim et al. (8) in human. In
this tissue PGE_2 is present in higher amounts than $PGF_{2\alpha}$, i.e. there
is an opposite situation in regard to the lung.

Generation of prostaglandin E-like material by guinea pig
trachea was found (138) during contraction induced by histamine. The
amount of PGE_2 and $PGF_{2\alpha}$, released from normal untreated guinea pig
trachea in the bath medium, also indicates a prevalent biosynthesis
of PGE_2 compared with $PGF_{2\alpha}$ (139). The trachea, isolated from healthy
and sensitized guinea pig, released more PGE_2 than $PGF_{2\alpha}$ (140).

The pharmacological effects of prostaglandins on the respiratory
smooth muscle in several mammalian species were first described by
Main (141) and Horton and Main (142). Prostaglandins of the E type
induce in vitro a relaxation of the tracheal smooth muscle, which is
more evident in a preparation with inherent tone or when the tone is
pharmacologically increased; whereas PGF_α compounds elicit contraction at the same dose range. The same effects were originally found
by Sweatman and Collier (143) on isolated human bronchial smooth
muscle. Since the prostaglandins action in vitro as well as in vivo
on respiratory smooth muscle has been extensively and excellently
reviewed (144, 145) more recent experiments will be referred and
discussed. Recently (49) it has been shown that cycloendoperoxides,
time precursors of prostaglandins, PGG_2 and PGH_2 are 5-10 times more

active than $PGF_{2\alpha}$ on isolated guinea pig trachea and on increasing
its insufflation pressure. As for the structure-activity relationship, it can be suggested that a 9-hydroxyl substitution confers
spasmogenic properties to PG-compounds and the bronchoconstrictor action of PGD_2 (a structural isomer of PGE_2 having a 9-hydroxyl
group) confirms this correlation (49, 146). The role of prostaglandins in normal human respiratory physiology is not yet established. It has been suggested that endogenous PGE and PGF_α may modulate
the tone of tracheal and bronchial smooth muscle (147, 148, 149).
The diametrically opposite actions of PGE_2 and $PGF_{2\alpha}$ suggested (95)
that asthmatic attack could be induced by an overproduction of $PGF_{2\alpha}$
at the expense of the bronchodilator PGE_2 or by shift in the biosynthesis from PGE_2 to $PGF_{2\alpha}$. There are still few information available on the biosynthesis of prostaglandins in the lungs or bronchi
of asthmatic subjects. Green and coworkers (150) indicated an increase in the plasma level of prostaglandin $F_{2\alpha}$ metabolite, 15-keto-
13:14 dihydro derivative, during antigen challenge induced asthma in
man, presumably derived from the lungs. Recently Nemoto et al. (151)
have measured by radioimmunoassay the levels of serum PGE and PGF_α
in 21 asthmatic patients and in 19 normal ones. The serum concentration of PGF_α and PGE were both increased in bronchial asthmatic in
comparison with normal control subjects. However the PGF_α / PGE ratio
in asthmatic was significantly higher than in normal (respectively
2.9 and 0.9). The authors suggest that this ratio modification might
play an important role in the pathogenesis of bronchial asthma, also
in view of the unusually hypersensitivity of asthmatic patients to
$PGF_{2\alpha}$ (152). Whether or not this hyperresponsiveness could be due to
a defect in either the synthesis or metabolism of prostaglandins is
still questionable (152, 153) as well as it is the role of prostaglandins, their precursors and intermediate in the relationship of asthma.
A question not settled is also the possibility to use therapeutically
PGE type compounds or analogues in the treatment of bronchial asthma.

We try to give a sanded contribution to this problem looking for
a selective substance, able to distinguish prostaglandin synthetase
enzymes from the oxygenases which form the cyclic endoperoxides, acting
upon their breakdown to PGE_2 (139) on tracheal smooth muscle. From our
results it appears that 1-ascorbic acid antagonises the increase of
intraluminal pressure due to $PGF_{2\alpha}$ in guinea pig tracheal tube preparations. This effect was also observed in vivo experiments (139). In
more recent studies (unpublished) 1-ascorbic acid induces per se a
relaxation of guinea pig tracheal chain which is abolished by indomethacin or TYA suggesting an involvement of prostaglandin biosynthesis.
The formation and release of PGE_2 and $PGF_{2\alpha}$ from guinea pig tracheal
smooth muscle in absence and in presence of 1-ascorbic acid in the perfusing medium was measured by mass fragmentography, according to the
method of Nicosia et al. (154). The results are shown in Table 4.

TABLE 4   RELEASE OF PROSTAGLANDINS FROM GUINEA PIG
             TRACHEAL CHAIN PREPARATION

	PGE_2^a ng/g - hr	$PGF_{2\alpha}$ ng/g - hr
---	148 ± 40	678 ± 46
1-ascorbic acid 100 mg/ml	442 ± 82^b	698 ± 194

a = $\bar{x}$ $\pm$ S.E. The mean value ($\bar{x}$) has been obtained from 8 experi-
        ments.

b = p   0.01

1-Ascorbic acid induces an increase in PGE_2 formation whereas no
changes were observed on the $PGF_{2\alpha}$ production. Therefore it appears
possible that 1-ascorbic acid may act on the metabolism of cyclic endoperoxides to PGE_2. Another explanation could be related to the physiological properties of 1-ascorbic acid based on its redox system
involving the reduced glutathione (GSH). It was shown that in blood
and many tissues of scorbutic guinea pigs there is a decrease of
reduced glutathione (155). Reduced glutathione was found to stimulate
both the enzymes: the endoperoxide isomerase and the peroxidase which
form specifically PGE_2 (17, 157). On the ground of these considerations it can be possible to conclude that 1-ascorbic acid may prevent the bronchoconstriction induced by $PGF_{2\alpha}$ or other endogenous substances released during some pulmonary diseases through its effect
on PG biosynthesis. Furthermore this action of 1-ascorbic acid may
explain its therapeutic effect observed in human bronchial asthma (158,
159).

CONCLUSIONS

     Prostaglandins, cyclic endoperoxides and thromboxanes seem to have
a particular importance in the respiratory system. They are synthetized
and released by lung parenchima and by bronchial and tracheal smooth
muscles. In the lung they are rapidly metabolized as other vasoactive

substances. They affect pulmonary circulation and can induce respiratory changes. They have effect on pulmonary vascular and airway
smooth muscles and modulate through cyclic nucleotides the contraction-relaxation cycle. Therefore they may be involved in the control
of the physiological vital function of respiration. The endogenous
prostaglandins of the E series, inducing pulmonary vasodilatation
and bronchodilatation, could play such a role during increased ventilation that causes hypocapnea and thus bronchoconstriction. The F
series and the more potent endoperoxides and some metabolites, released during hypoxia, could be useful in directing blood flow to better
ventilated area of the lung. Prostaglandins, endoperoxides and
thromboxanes may also be implicated in the pathogenesis of lung
diseases such as asthma, pulmonary embolism, which is associated with
platelet aggregation and pulmonary edema. Further studies on their
synthesis, release, metabolism and pharmacological effects are necessary to ascertain their pathophysiological role.

REFERENCES

1. ELIASSON, R., (1959). Studies on Prostaglandins. Occurrence,
    Formation and Biological Actions. Acta Physiol.Scand. 45,
    suppl. 158.
2. LINN, B.O., SHUNK, C.H., FOLKERS, K., GANLEY, O. and ROBINSON, H.J.,
    (1961). SRS-S Concentrates from Normal Swine Lung without Anaphy-
    laxis. Bioch.Pharmacol. 8, 339-340.
3. BERGSTRÖM, S., DRESSLER, F., RYHAGE, R., SAMUELSSON, B. and
    SJÖVALL, J., (1962). The isolation of two further prostaglandins
    from sheep prostate glands. Ark. för Kemi 19, 563-567.
4. BERGSTRÖM, S., RYHAGE, R., SAMUELSSON, B. and SJÖVALL, J., (1962).
    The structure of prostaglandins E_1, $F_{1\alpha}$ and $F_{2\alpha}$. Acta Chem. Scand.
    16, 501-502.
5. BERGSTRÖM, S., DRESSLER, F., KRABICH, L., RYHAGE, R. and SJÖVALL, J.,
    (1962). The isolation and structure of a smooth muscle stimulating
    factor in normal sheep and pig lungs. Ark för Kemi 20, 63-66.
6. SAMUELSSON, B., (1964). Identification of Prostaglandin $F_{3\alpha}$ in bo-
    vine lung. Prostaglandin and related factors 26. Biochim.Biophys.
    Acta, 84, 707-713.
7. ÄNGGÅRD, E., (1965). The isolation and determination of prostaglan-
    dins in lung of sheep, guinea pig, monkey and man. Biochem.Pharmacol.
    14, 1507-1516.
8. KARIM, S.M.M., SANDLER, M. and WILLIAMS, E.D., (1967). Distribution
    of prostaglandins in human tissues. Brit.J. Pharmacol. Chemoth. 31,
    340 -344.
9. KARIM, S.M.M., HILLIER, K. and DEVLIN, J., (1968). Distribution of
    prostaglandins E_1, E_2, $F_{1\alpha}$ and $F_{2\alpha}$ in some animal tissues. J.Pharm.
    Pharmacol. 20, 749-753.

10. ÄNGGÅRD, E. and BERGSTRÖM, S., (1973). Biological effects of an unsaturated trihydroxiacid ($PGF_{2\alpha}$) from natural swine lung. Acta Physiol.Scand. 58, 1-12.

11. ANDERSEN, N.H. and RAMWELL, P.W., (1974). Biological aspects of prostaglandins. Arch. Int. Med. 133, 30-50.

12. CHRIST, E.J. and van DORP, D.A., (1973). Comparative aspects of prostaglandin biosynthesis in animal tissues. Adv. Biosc. 9, 35-38.

13. FLOWER, R.J. and BLACKWELL, G.J., (1976). The importance of phospholipase-A$_2$ in prostaglandin biosynthesis. Biochem.Pharmacol. 25, 285-291.

14. HAMBERG, M. and SAMUELSSON, B., (1973). Detection and isolation of an endoperoxide intermediate in prostaglandin biosynthesis. Proc.Natl. Acad. Sci. U.S.A. 70, 899-903.

I5. WLODAWER, P. and SAMUELSSON, B., (1973). On the organization and mechanism of prostaglandin synthetase. J.biol. Chem. 248, 5673-5678.

16. HAMBERG, M., SVENSSON, J., WAKABAYASHI, T. and SAMUELSSON, B., (1974). Insolation and structure of two prostaglandin endoperoxides that cause platelet aggregation. Proc. Natl. Acad. Sci. U.S.A. 71, 345-349.

17. NUGTEREN, D. H. and HAZELOF, E., (1973). Isolation and properties of intermediates in prostaglandin biosynthesis. Biochim. Biophys. Acta 326, 448-461.

18. HAMBERG, M., SVENSSON, J. and SAMUELSSON, B., (1975). Thromboxanes: a new group of biologically active compounds derived from prostaglandin endoperoxides. Proc.Natl. Acad. Sci. U.S.A. 72, 2994.

19. van DORP, D.A., (1966). The biosynthesis of prostaglandins. Mem. Soc. Endocrinol. 14, 39-47.

20. ÄNGGÅRD, E. and SAMUELSSON, B., (1965). Biosynthesis of prostaglandins from arachidonic acid in guinea pig lung. J. biol. Chem. 240, 3518-3521.

21. PIPER, P.J. and VANE, J.R., (1969). The release of prostaglandins during anaphylaxis in guinea pig isolated lungs. In: Prostaglandins, Peptides and Amines. eds. P. Mantegazza and E.W.Horton. Acad. Press, London. p.p. 15-19

22. PIPER, P.J. and VANE, J.R., (1969). Release of additional factors in anaphylaxis and its antagonism by antiinflammatory drugs. Nature (London) 223, 29-35.

23. PIPER, P.J. and WALKER, J.L., (1973). The release of spasmogenic substances from human chopped lung tissue and its inhibition. Brit. J. Pharmacol. 47, 291-304.

24. VOGT, W., MEYER, V., KUNZE, H., LUFFT, E. and BABILLI, S., (1969). Entstehung von SRS-C in der durchströmten Meerschweinchenlunge durch Phospholipase A. Archiv für Pharmac. und Exp. Pathol. 262, 124-134.

25. PALMER, M.A., PIPER, P.J. and VANE, J.R., (1970). Release of
     vasoactive substances from lungs by injection of particles.
     Br.J. Pharmacol. 40, 547P-548P.
26. LINDSEY, H.E. and WYLLIE, J.H., (1970). Release of prostaglan-
     dins from embolized lungs. Br.J. Surgery 57, 738-741.
27. PIPER, P.J. and VANE, J.R., (1971). The release of prostaglan-
     dins from lung and other tissues. Ann. N.Y. Acad. Sci 180,
     363-385.
28. BERRY, E., EDMONDS, J.F. and WYLLIE, J.H., (1971). Release of
     prostaglandin E$_2$ and unidentified factors from ventilated lungs.
     Br.J. Surgery 58, 189-192.
29. WEIR, E.K., MC.MURTRY, I.F., TUCKER, A., REVEES, F.T. and GROVER,
     R.F., (1976). Inhibition of prostaglandin synthesis or blockade
     of prostaglandin action increases the pulmonary pressor response
     to hypoxia. In: Advances in prostaglandin and thromboxane
     research. Eds. B. Samuelsson and R. Paoletti. Raven Press Vol. 2
     p.p. 914-915.
30. SAID, S.I., HOVA, N. and YOSHIDA, T., (1975). Hypoxic pulmonary
     vasoconstriction in cats, modification by indomethacin and aspi-
     rine. Fed. Proc. 34, 1229.
31. SAID, S.I., KITAMURA, S. and VREIM, C., (1972). Prostaglandins:
     release from the lung during mechanical ventilation at large tidal
     volumes. J. Clin. Invest. 51, 83a-84a.
32. BURKA, J.F. and EYRE, P., (1974). A study of prostaglandins and
     prostaglandin antagonist in relation to anaphylaxis in calves.
     Can. J. Physiol. and Pharmacol. 52, 942-951.
33. SAID, S.I., YOSHIDA, T., KITAMURA, S. and VREIM, C., (1974).
     Release of prostaglandins and other humoral mediators during
     hypoxic breathing. J. Clin. Invest. 53, 69a.
34. VANE, J.R., (1974). The use of isolated organ for detecting active
     substances in the circulating blood. Br. J. Pharmacol. Chemother.
     23, 360-363.
35. SAID, S.I., KITAMURA, S., YOSHIDA, T., PRESKITT, J. and HOLDEN, L.D.,
     (1974). Humoral control of airways. Ann. N.Y. Acad. Sci. 221, 103-
     114.
36. SVENSSON, J., HAMBER, M. and SAMUELSSON, B., (1975). Prostaglan-
     din endoperoxides IX. Characterization of rabbit aorta contract-
     ing substance (RCS) from guinea pig lung and human platelets. Acta
     Physiol. Scand. 94, 222-228.
37. ALABASTER, V.A. and BAKHLE, Y.S., (1970). The release of biological-
     ly substances from isolated lungs by 5-hydroxytryptamine and tryp-
     tamine. Br. J. Pharmacol. 40, 582P-583P.
38. BAKHLE, Y.S. and SMITH, T.W., (1972). Release of spasmogenic sub-
     stances induced by vasoactive amine from isolated lungs. Br. J.
     Pharmacol. 46, 543P-544P.
39. PIPER, P.J., VANE, J.R. and WYLLIE, J.H., (1970). Inactivation of
     prostaglandins by the lungs. Nature (London) 225, 600-604.

40. GILMORE, N.I., VANE, J.R. and WYLLIE, J.H., (1968). Prostaglandin release by the spleen. Nature (London) $\underline{218}$, 1135-1140.

41. RAMWELL, P.W. and SHAW, J.E., (1970). Biological significance of prostaglandins. In: Recent Progress in hormone research. Acad. Press. N.Y., London $\underline{26}$, 139-187.

42. HAMBERG, M. and SAMUELSSON, B., (1971). On the metabolism of prostaglandins E_1 and E_2 in man. J. biol. Chem. $\underline{246}$, 6713-6721.

43. HAMBERG, M. and SAMUELSSON, B., (1972). On the metabolism of prostaglandins E_1 and E_2 in the guinea pig. J. biol. Chem. $\underline{247}$, 3495-3502.

44. GREEN, K.and SAMUELSSON, B., (1971). Quantitative studies on the synthesis in vivo of prostaglandins in the rat. Eur. J. Biochem. $\underline{22}$, 391-395.

45. HAMBERG, M., (1972). Inhibition of prostaglandin synthesis in man. Biochem. Biophys. Res. Comm. $\underline{49}$, 720-726.

46. HAMBERG, M., (1973). Quantitative studies on prostaglandin synthesis in man. II. Determination of the major urinary metabolite of prostaglandins $F_{1\alpha}$ and $F_{2\alpha}$. Anal. Biochem. $\underline{55}$, 368-378.

47. STRANDBERG, K. and HAMBERG, M., (1974). Increased excretion of 5α, 7α dihydroxy-11-keto-tetranor-prostanoic acid on anaphylaxis in the guinea pig. Prostaglandins $\underline{6}$, 159-170.

48. GRANSTRÖM, E., (1972). On the metabolism of prostaglandin $F_{2\alpha}$ in female subjects. Eur. J. Biochem. $\underline{27}$, 462-469.

49. HAMBERG, M., SVENSSON, J., HEDQVIST, P., STRANDBERG, K. and SAMUELSSON, B., (1976). Involvement of endoperoxides and thromboxanes in anaphylactic reaction. In: Advances in prostaglandin and thromboxane research. Vol. I. p.p. 495-501. Eds. B. Samuelsson and F. Paoletti. Raven Press, N.Y.

50. HAMBERG, M., HEDQVIST, P., STRANDBERG, K., SVENSON, J. and SAMUELSSON, B., (1975). Prostaglandin endoperoxides IV. Effects on smooth muscle. Life Sci. $\underline{16}$, 451-462.

51. von EULER, U.S., (1934). Zur Kenntnis der pharmakologischen Wirkungen von Nativsekreten und Extrakten accessorischer Geschlechtsdrüsen. Arch. exp. Path. Pharmak. $\underline{175}$, 78-84.

53. HORTON, E.W., (1972). Prostaglandins.In: Mon.Endocrinol. $\underline{7}$, 1-197.

52. von EULER, U.S., (1935). A depressor substance in the vesicular gland. J. Physiol. $\underline{84}$, 21P.

54. KARIM, S.M.M. and SOMERS, K., (1972). Cardiovascular and renal action of prostaglandins. In: The prostaglandins: Progress in Research. Ed. S.M.M. Karim. MTP Med. Tech. Publ. Co. Oxford. p.p. 165-203.

55. NAKANO, J., (1973). General pharmacology of prostaglandin. In: The prostaglandins. Pharmacological and therapeutic advances. Ed. M.F. Cuthbert. W. Heinemann.Lond. p.p. 23-124.

56. FERREIRA, S.H. and VANE, J.R., (1967). Prostaglandins: their
     disappearance from and release into the circulation. Nature
     (London) 216, 868-873.

57. VANE, J.R., (1968). The release and assay of hormones. Sci.
     Basic Med. 336, 358.

58. VANE, J.R., (1969). The release and fate of vasoactive hormo-
     nes in the circulation. Br. J. Pharmacol. 35, 209-242.

59. BAKHLE, Y.S. and VANE, J.R., (1974). Pharmacokinetic function
     of the pulmonary circulation. Physiol. Rev. 54, 1007-1045.

60. Mc GIFF, J.C., TERRAGNO, N.A., STRAND, J.C., LEE, J.B. and
     LONIGRO, A.J., (1969). Selective passage of prostaglandins
     across the lung. Nature (London) 223, 742-745.

61. GRANSTRÖM, E., (1967). On the metabolism of prostaglandin
     E_1 in man. Prog. Biochem. Pharmacol. 3, 89-93.

62. BIRON, P., (1968). Vasoactive hormone metabolism by the
     pulmonary circulation. Clin. Res. 16, 112.

63. ÄNGGÅRD, E. and SAMUELSSON, B., (1966). Purification and
     properties of a 15-hydroxy prostaglandin dehydrogenase from
     swine lung. Arkiv för Kemi 25, 293-300.

64. ÄNGGÅRD, E., LARSSON, C. and SAMUELSSON, B., (1971). The
     distribution of 15-hydroxy prostaglandin dehydrogenase and
     prostaglandin- Δ^{13} - reductase in tissues of the swine.
     Acta Physiol. Scand. 81, 396-404.

65. RYAN, J.W., SMITH, V. abd NIEMEYER, S., (1972). Angiotensin
     I: metabolism by plasma membrane of lung. Science 176, 64-66.

66. LARSSON, C. and ÄNGGÅRD, E., (1970).Distribution of prosta-
     glandin metabolizing enzymes in tissues of the swine. Acta
     Pharmacol. Toxicol.28, suppl. 1, 61.

67. CHINARD, F.P., ENNS, T. and NOLAN, M.F., (1962). The perme-
     ability of characteristics of the alveolar capillary barrier.
     Trans. Ass. Am. Physiol. 75, 253-261.

68. RYAN, J. W., NIEMEYER, R.S. and GOODWIN, D.W.,(1972). Meta-
     bolic fate of bradykinin, angiotensin I, adenine nucleotides
     and prostaglandins E_1 and $F_{2\alpha}$ in the pulmonary circulation.
     Adv.Exp.Med. Biol. 21, 259-265.

69. DAWSON, C.A., COZZINI, B.O. and LONIGRO, A.J., (1975). Meta-
     bolism of (2 - ^{14}C) prostaglandin E_1 on passage through the
     pulmonary circulation. Canad. J. Physiol. Pharmacol. 53,
     610-615.

70. VONKEMAN, H., NUGTEREN, D.H. and van DORP, D.A., (1969). The
     action of prostaglandin 15-hydroxy-dehydrogenase on various
     prostaglandins. Biochim Biophys. Acta, 187, 581-583.

71. NAKANO, J., ÄNGGÅRD, E. and SAMUELSSON, B., (1969). 15-hydroxy-
     prostanoate dehydrogenase. Prostaglandins as substrates and
     inhibitors. Eur. J. Biochem. 11, 386-389.

72. BITO, L.Z., (1972). Accumulation and apparent active transport
     of prostaglandins by some rabbit tissues in vitro. J. Physiol.
     (London) 221, 371-387.

73. PIPER, P.J., (1973). Distribution and metabolism. In: The prostaglandins. Pharmacological and therapeutic advances. Ed. M.F. Cuthbert. W. Heinemann (London). p.p. 125-150.

74. ROSE, P; NEIDERHAUSER, U., PIPER, P.J., ROBINSON, E. and SMITH, A.P., (1976). Inactivation of prostaglandin $F_{2\alpha}$ in the human pulmonary circulation. Br. J. Clin. Pharmacol. 3, 342.

75. PACE-ASCIAK, C.R., (1976). Biosynthesis and metabolism of prostaglandin during animal development. In: Advances in prostaglandin and thromboxane research. Vol. 1. Eds. B. Samuelsson and R. Paoletti, Raven Press N.Y., p.p. 35-46.

76. SUN, F.F. and ARMOUR, S.B., (1974). Prostaglandin 15-hydroxy-dehydrogenase and Δ^{13} reductase levels in the lungs of maternal, fetal and neonatal rabbits. Prostaglandins 7, 327-338.

77. ÄNGGÅRD, E. and SAMUELSSON, B., (1966). The metabolism of prostaglandins in lung tissue. In: Prostaglandins (Proc. Nobel Symp. 2). Eds. S. Bergström and B. Samuelsson. Interscience Publ., N.Y. p.p. 97-105.

78. HORTON, E.W. and JONES, R.L., (1969). Prostaglandins A_1, A_2 and 19 hydroxy A_1, their actions on smooth muscle, and their inactivation on passage through the portal and pulmonary vascular bed. Br. J. Pharmacol. 37, 705-722.

79. DAWSON, W. and SWEATMAN, W.J.F., (1975). Possible role of prostaglandins in asthma. Int.Arch.Allergy Appl. Immun. 49, 213-216.

80. DAWSON, W., LEWIS, R.L., Mc. MAHON, R.E. and SWEATMAN, W.J.F., (1974). Potent bronchoconstrictor activity of 15-keto prostaglandin $F_{2\alpha}$. Nature (London) 250, 331-332.

81. VANE, J.R., (1971). Inhibition of prostaglandin synthesis as a mechanism of aspirin like drugs. Nature, New Biol. 231, 232-235.

82. FLOWER, R.J. and VANE, J.R., (1974). Inhibition of prostaglandin biosynthesis. Biochem. Pharmacol. 23, 1439-1450.

83. DOWNINGS, D.T., AHERN, D.G. and BACHTA, M., (1970). Enzyme inhibition by acetylenic compounds. Biochem. Biophys. Res. Comm. 40, 218-223.

84. NIJKAMP, F.P., FLOWER, R.J., MONCADA, S. and VANE, J.R., (1976). Partial purification of rabbit aorta contracting substance-releasing factor and inhibition of its activity by anti-inflammatory steroids. Nature. 263, 479-482.

85. LEWIS, G.P. and PIPER, P.J., (1975). Inhibition of release of prostaglandins as an explanation of some of the action of antiinflammatory corticosteroids. Nature. 254, 308-311.

86. LEWIS, G.P. and PIPER, P.J., (1976). Inhibition of prostaglandin release by antiinflammatory steroids. In: Advances in prostaglandins and thromboxane research. Vol. 1. Eds. B. Samuelsson and R. Paoletti. Raven Press. p.p. 121-124.

87. GRYGLEWSKI, T.J., PANCZENKO, B., KORBUT, R., GRODZINSKA, L. and OCETKIEWCZ, A., (1975). Corticosteroids inhibit prostaglandin release from perfused mesenteric blood vessels of rabbit and from perfused lung of sensitized guinea pig. Prostaglandins 10, 343-355.

88. CRUTCHLEY, D.J. and PIPER, P.J., (1974). Prostaglandin inactivation in guinea pig lung and its inhibition. Br. J. Pharmacol. $\underline{52}$, 197-203.

89. EAKINS, K.E., KARIM, S.M.M. and MILLER, J.D., (1970). Antagonism of some smooth muscle action of prostaglandins by polyphloretin phosphate. Br. J. Pharmacol. $\underline{39}$, 556-563.

90. MATHE, A.A., STRANDBERG, K and ASTRÖM, A., (1971). Blockade by poliphloretin phosphate of the $PGF_{2\alpha}$ action on isolated human bronchi. Nature New Biol. $\underline{230}$, 215-216.

91. MATHE, A.A., STRANDBERG, K and FREDHOLM, B., (1972). Antagonism of prostaglandin $F_{2\alpha}$ induced broncho-constriction and blood pressure changes by polyphloretin phosphate in guinea pig and cat. J. Pharm. Pharmacol. $\underline{24}$, 378-382.

92. VILLANUEVA, R., HINDS, L., KATZ, R. and EAKINS, E.A., (1972) The effect of polyphloretin phosphate on some muscle action of prostaglandins in the cat. J. Pharmacol. Exp. Ther. $\underline{180}$, 78-85.

93. BITO, L.Z. and BAROODY, R.A., (1975). Inhibition of pulmonary prostaglandin metabolism by inhibitors of prostaglandin bio-transport (Probenecid and Bromcresol Green). Prostaglandins. $\underline{10}$, 633-639.

94. BITO, L.Z., (1975). Saturable, energy-dependent, transmembrane transport of prostaglandins against concentration gradient. Nature. $\underline{256}$, 134-136.

95. HORTON, E.W., (1969). Hypothesis on physiological roles of prostaglandins. Physiol. Rev. $\underline{49}$, 122-161.

96. LEWIS, A.J. and EYRE, P., (1972). Some cardiovascular and respiratory effects of prostaglandins E_1, E_2 and $F_{2\alpha}$ in the calf. Prostaglandins $\underline{2}$, 55-64.

97. OKPAKO, D.T., (1972). The action of histamine and prostaglandins $F_{2\alpha}$ and E_2 on pulmonary vascular resistance of the lung of the guinea pig. J. Pharm. Pharmacol. $\underline{24}$, 40-46.

98. KADOWITZ, P.J., GEORGE, W.J., JOINER, P.D. and HYMAN, A.L., (1973). Effect of prostaglandins E_1 and $F_{2\alpha}$ on adrenergic responses in the pulmonary circulation. Adv. in Bioscience $\underline{9}$, Ed. S. Bergström. Pergamon Press. p.p. 501-506.

99. BURKA, J.E. and EYRE, P., (1974). Studies of prostaglandins and prostaglandin-antagonists on bovine pulmonary vein $\underline{in}$ $\underline{vitro}$. Prostaglandins $\underline{6}$, 333 341.

100. STARLING, M.B. and ELLIOTT, R.B., (1974). The effects of prostaglandins, prostaglandin-inhibitors and oxygen on the closure of the ductus arteriosus, pulmonary arteries and umbilical vessels "in vitro". Prostaglandins $\underline{8}$, 187-203.

101. KADOWITZ, P.J., JOINER, P.D. and HYMAN, A.L., (1975). Physiological and pharmacological roles of prostaglandins. Ann. Rev. Pharmacol. $\underline{15}$, 285-306.

102. STRONG, C.G. and BOHR, D.F., (1967). Effects of prostaglandins E_1, E_2, A_1 and $F_{2\alpha}$ on isolated vascular smooth muscle. Am. J. Physiol. $\underline{219}$, 725-733.

103. HAUGE, A., LUNDE, P.K.M. and WALER, B.A., (1967). Effects of prostaglandin E_1 and adrenaline on the pulmonary vascular resistance (PVR) in isolated rabbit lungs. Life Sci. 6, 637-680.

104. NAKANO, J. and Mc CURDY, J.R., (1967) Effects of prostaglandins E_1 (PGE_1) and A_1 (PGA_1) on the systemic venous return and pulmonary circulation. Clin. Res. 15, 409.

105. MAXWELL, G.M., (1967). The effect of prostaglandin E_1 upon the general and coronary hemodynamics and metabolism of the intact dog. Br. J. Pharmacol. 31, 162.

106. SAID, S.I., (1968). Some respiratory effects of prostaglandins E_2 and $F_{2\alpha}$. In: Prostaglandin Symposium of the Worcester Foundation. Eds. P.W. Ramwell and J. E. Shaw. Intersciences N.Y.. p. 207.

I07. DU CHARME, D.W., WEEKS, J.R. and MONTGOMERY, R.G., (1968). Studies of the hypertensive effect of prostaglandin $F_{2\alpha}$. J. Pharm. Exp. Ther. 160, 1.

108. GILES, T.D., QUIROZ, A.C. and BURCH, G.E., (1969). The effects of prostaglandin E_1 on the systemic and pulmonary circulation of intact dogs. The influence of urethane and pentobarbital in anesthesia. Experientia 25, 1056.

109. NAKANO, J. and COLE, B., (1969). Effects of prostaglandins E_1 and $F_{2\alpha}$ on systemic, pulmonary and splanchnic circulation in dogs. Am. J. Physiol. 217, 222.

110. KUIDA, H., (1971). Effects of prostaglandin. Bovine pulmonary circulation. Fed. Proc. 30, 380.

111. ALPERT, J.S., HAINES, F.W., KNUTSON, P.A., DALEN, J.E. and DEXTER, L., (1973). Prostaglandins and pulmonary circulation. Prostaglandins 3, 759-765.

112. KADOWITZ, P.J., JOINER, P.D. and HYMAN, A.L., (1975). Effect of prostaglandin E_2 on pulmonary vascular resistance in intact dog, swine and lamb. Eur. J. Pharmacol. 31, 72-80.

113. HYMAN, A.L., (1969). The active responses of pulmonary veins in intact dogs to prostaglandins $F_{2\alpha}$ and E_1. J. Pharmacol. Exp. Ther. 165, 267-273.

114. HYMAN, A.L., WOOLVERTON, W.C., GUTH, P.S. and ICHINOSE, H., (1971) Pulmonary vasopressor response to decrease in blood pH in intact dogs. J. Clin. Invest. 50, 1028-1043.

115. BERGSTRÖM, S, CARLSON, L.A. and WEEKS, J., (1968). The prostaglandins: a family of biologically active lipids. Pharmacol. Rev. 20, 1-48.

116. NAKANO, J., (1973). General pharmacology of prostaglandins In: The prostaglandins pharmacological and therapeutics advances. Ed. M.F. Cuthbert, W. Heinemann, Ltd. London, p.p. 23-124

117. LESLIE, C.A. and LEVINE, L., (1973). Evidence for the presence of a prostaglandin E_2 - 9 - ketoreductase in rat organs. Biochem. Biophys. Res. Comm. 52, 717.

118. SALZMAN, E.W., (1976). Prostaglandins and platelet function.
      In: "Advances in prostaglandin and thromboxane research.
      Vol. 2. Eds. B. Samuelsson and R. Paoletti. Raven Press.
      N.Y., p.p. 767-780.

119. Mc QUEEN, D.S., (1972). The effects of some prostaglandins
      on respiration in rats and cats. Br. J. Pharmacol. $\underline{45}$, 147P
      148P.

120. BROOKES, L.G. and MARSHALL, R.C., (1974). The effects of some
      prostaglandins on respiration in the rabbit. J. Pharm.
      Pharmacol. $\underline{26}$, suppl. 80P-81P.

121. SAID, S.I., MUREN, O. and KIRBY, B.J., (1968). Some respira-
      tory effects of prostaglandins E_2 and $F_{2\alpha}$. Clin. Res. $\underline{16}$, 90.

122. BERGSTRÖM, S., (1967). Prostaglandins, member of a new hormon-
      al system. Science $\underline{157}$, 382-391.

123. CARLSON, L.A., EKELUND, L.G. and ORÖ, L., (1969). Circulato-
      ry and respiratory effects of different doses of PGE_1 in man.
      Acta Physiol. Scand. $\underline{75}$, 161-169.

124. Mc QUEEN, D.S., (1974). The effects of some prostaglandins
      on respiration in anaesthetized cats. Br. J. Pharmacol. $\underline{50}$
      559-568.

125. WASSERMAN,M.A., (1975). Bronchopulmonary responses to prosta-
      glandin $F_{2\alpha}$, histamine and acetylcholine in the dog. Eur.J.
      Pharmacol. $\underline{32}$, 146-155.

126. WASSERMAN, M.A., (1975). Bronchopulmonary effects of prosta-
      glandin $F_{2\alpha}$ and three of its metabolites in the dog.
      Prostaglandins $\underline{9}$, 959-973.

127. WASSERMAN, M.A., (1976). Bronchopulmonary pharmacology of
      some prostaglandin endoperoxide analogs in the dog. Eur. J.
      Pharmacol. $\underline{36}$, 103-114.

128. PARKER, C.W. and SNIDER, D.E., (1973). Prostaglandins and
      asthma. Ann. Intern. Med. $\underline{78}$, 963.

129. CUTHBERT, M.F., (1975). Prostaglandins and asthma. Br. J.
      Clin. Pharmacol. $\underline{2}$, 293-295.

130. ROBISON, G.A. and SUTHERLAND, E.W., (1970). Sympathin E,
      sympathin I and the intracellular level of cyclic AMP.
      Circulation Res. Suppl. I, 26 and 27: 1-147 to 1-161.

131. BÄR, H.P., (1974). Cyclic nucleotides and smooth muscle. In:
      Advances in cyclic nucleotide research. Vol. 4. Eds. P.
      Greengard and G. A. Robison, Raven Press N.Y.. p.p. 195-237.

132. ANDERSSON,R., KÖVESI, G. and ERICSSON, E., (1975). Influence
      of contracting and relaxing drugs on cyclic AMP level in
      tracheal smooth muscle. Acta Physiol. Scand. (quoted by
      no. 135).

133. KALINER, M.A., ORANGE, R.P., KOOPMAN, W.J., AUSTEN, K.F. and
      LARAIA, P.J., (1971). Cyclic adenosin 3', 5', monophosphate
      in human lung. Biochim. Biophys. Acta, $\underline{252}$, 160-164.

134. KALINER, M.A., ORANGE, R.P. and AUSTEN, K.F., (1972). Immuno-
     logical release of histamine and slow-reacting substance of
     anaphylaxis from human lung: IV. Enhancement by cholinergic
     and α - adrenergic stimulation. J. Expt. Med. $\underline{136}$, 556-567.

135. ANDERSSON, R., NILSSON, K., WIKBERG, J., JOHANSON, S., MOHME-
     LUNHOLM, E. and LUNDHOLM, L., (1975). Cyclic nucleotides
     and the contraction of smooth muscle. In: Advances in cyclic
     nucleotides research. Vol. 5. Eds. G.I. Drummond, P. Greengard
     and G.A. Robison. Raven Press N.Y. p.p. 491-518.

136. ANDERSSON, R., (1972). Cyclic AMP and calcium ions in mechanic-
     al and metabolic responses of smooth muscle; influence of some
     hormones and drugs. Acta Physiol. Scand. suppl. 382, 1-59.

137. KADOWITZ, P.J., JOINER, P.D., HYMAN, A.L. and GEORGE, W.J.,
     (1975). Influence of prostaglandins E_1 and $F_{2\alpha}$ on pulmonary
     vascular resistence, isolated lobar vessels and cyclic nucleo-
     tide levels. J. Pharmacol. Exp. Ther. $\underline{192}$, 677-687.

138. GRODZINSKA, L., PANCZENKO, B and GRYGLEWSKI, R.J., (1975).
     Generation of prostaglandin E-like material by the guinea pig
     trachea contracted by histamine. J. Pharm. Pharmacol. $\underline{27}$, 88-91.

139. PUGLISI, L., BERTI, F., BOSISIO, E., LONGIAVE, D. and NICOSIA,
     S., (1976). Ascorbic acid and $PGF_{2\alpha}$ antagonism on tracheal
     smooth muscle. In: Advances in prostaglandin and thromboxane
     research. Vol. 1, Eds. B. Samuelsson and R. Paoletti. Raven
     Press N.Y. p.p. 503-506.

140. YEN, S.S., MATHE, A.A. and DUGAN, J.J., (1976). Release of
     prostaglandins from healthy and sensitized guinea pig lung
     and trachea by histamine. Prostaglandins $\underline{11}$, 227-239.

141. MAIN, I.H.M., (1964). The inhibitory action of prostaglandins
     on respiratory smooth muscle. Br. J. Pharmacol. $\underline{22}$, 511-519.

142. HORTON, E.W. and MAIN, I.H.M., (1965). A comparison of the
     actions of prostaglandins $F_{2\alpha}$ and E_1 on smooth muscle. Br. J.
     Pharmacol. $\underline{24}$, 470.

143. SWEATMAN, W.J.F. and COLLIER, H.O.J., (1968). Effects of
     prostaglandins on human bronchial muscle. Nature $\underline{217}$, 69.

144. CUTHBERT, M.F., (1973). Prostaglandins and respiratory smooth
     muscle. In: The prostaglandins, pharmacological and therapeutic
     advances. Ed. M.F. Cuthbert. W. Heinemann Ltd., London,
     p.p. 253-285.

145. SMITH, A.P., (1976). Prostaglandins and the respiratory system.
     In: Advances in prostaglandins research. Prostaglandins: phy-
     siological, pharmacological and pathological aspects. Ed.
     S.M.M. Karim, M.T.P. Press. Ltd., pp. 83-102.

146. ROSENTHALE, M. E., DERVINIS, A. and STRIKE, D., (1976). Actions
     of prostaglandins on the respiratory tract of animals. In:
     Advances in prostaglandin and thromboxane research. Vol. 1.
     Eds. B. Samuelsson and R. Paoletti. Raven Press N.Y. p.p. 477-
     493.

147. OREHEK, J., DOUGLAS, J.S., LEWIS, A.J., BOUHUYS, A., (1973).
      Prostaglandin regulation of airway smooth muscle tone.
      Nature New Biol. 245, 84-85.

148. FARMER, J.B., FARRAR, D.G. and WILSON, J., (1974). Antago-
      nism of tone and prostaglandin mediated responses in a
      tracheal preparation by indomethacin and SC-19220. Br. J.
      Pharmacol. 52, 559-565.

149. LAMBLEY, J. and SMITH, A.P., (1975). The effect of arachi-
      donic acid, indomethacin and SC-19220 on guinea pig tracheal
      tone. Eur. J. Pharmacol. 30, 148.

150. GREEN, K., HEDQVIST, P. and SVANBORG, N., (1974). Increased
      plasma levels of 15-keto-13:14 dihydro-prostaglandin $F_{2\alpha}$
      after allergen-provoked asthma in man. The Lancet 14, 1419-
      1421.

151. NEMOTO, T., AOKI, H., IKE, A., YAMADA, K., KONDO, T.,
      KOBAYASHI, S. and INAGAWA, T., (1976). Serum prostaglandin
      levels in asthmatic patients. J. Allergy Clin. Immunol. 57,
      89-94.

152. MATHE, A.A., HEDQVIST, P., HOLMGREN, A. and SVANBORG, N.,
      (1973). Bronchial hyperreactivity to prostaglandin $F_{2\alpha}$ and
      histamine in patients with asthma. Br. Med. J. 1, 193-196.

153. HORTON, E.W., (1975). Prostaglandins. A short review. Scot.
      Med. J. 20, 155-160.

154. NICOSIA S. and GALLI, G., (1974). A rapid gaschromatographic-
      mass spectrometric method for prostaglandin analysis at pi-
      comole levels. Anal. Biochem. 61, 192.

155. BANERJER, S., DEB, C. and BELAVADY, B., (1952) Effect of
      scurvy on glutathione and dehydroascorbic acid in guinea
      pig tissues. J. Biol. Chem. 195, 271-276.

156. SAMUELSSON, B and HAMBERG, M, (1974). Role of endoperoxides
      in the biosynthesis and action of prostaglandins. In: Prosta-
      glandin synthetase inhibitors. Eds. H.J. Robinson and J.R.
      Vane, Raven Press N.Y. p.p. 107-119.

157. NUGTEREN, D.H., BEERTHUIS, R.K. and van DORP, D.A., (1966).
      The enzymic conversion of all-cis 8,11, 14-eicosatrienoic
      acid into prostaglandin E_1. Rec. Trav. Chim. Pays Bas
      85, 405-419.

158. MIYARES, C.M., REYES DIAZ, J.M., GUEVARA, F.S., COLLAZO, A.A.
      and AVILA, F., (1973). Altas dosis de vitamin C en el asma
      bronquial. Rev. Cub. Med. 12, 327.

159. ZUSKIN, E., LEWIS, A.J. and BOUHUYS, A., (1973). Inhibition
      of histamine induced airway constriction by ascorbic acid.
      J. Allergy Clin. Immunol. 51, 218-226.

GENERATION OF PROSTAGLANDIN AND THROMBOXANE-LIKE

SUBSTANCES BY LARGE AIRWAYS AND LUNG PARENCHYMA

R.J. Gryglewski, A. Dembińska-Kieć and
L. Grodzińska
Department of Pharmacology, Copernicus
Academy of Medicine in Cracow
31-531 Cracow, 16 Grzegórzecka, Poland

BIOSYNTHESIS OF PROSTAGLANDINS
AND THROMBOXANES IN LUNGS

Anggard and Samuelsson (1965) were the first researchers to study the biosynthesis of prostaglandins in lungs. They found that the low speed supernatant (900 x g) of a homogenate of guinea pig lungs incubated with arachidonic acid converted 10% of the substrate to more polar products than arachidonic acid. These products were identified as $PGF_{2\alpha}$, PGE_2 and their metabolites. Guinea pig lung homogenates were used by Vane (1971) in his work that lead to the discovery of the anti-prostaglandin synthetase activity of aspirin like drugs. In rat lung homogenates dihomo- γ -linolenic acid is mainly converted to PGD_1, similarily as prostaglandin endoperoxides PGH_1 and PGH_2 are chiefly metabolized to PGD_1 and PGD_2, respectively (Nugteren and Hazelhof, 1973)[1]. Further studies on metabolism of radioactive arachidonic acid by whole homogenates of guinea pig lungs brought to unexpected results (Hamberg and Samuelsson, 1974). The formation of PGF_2 and PGE_2, both in 2 - 3% yield was confirmed. However, the major products formed were thromboxane B_2 and 12L-hydroxy-5,8,10-heptadecatrienoic acid (HHT), each in about 20% of the recovered radioactivity. Thromboxane B_2 cochromatographed with PGE_2 but could be distinguished from this prostaglandin by its lack of conversion into PGB_2 by alkali. Thromboxane B_2, prostaglandins and HHT originate from the same intermediates of arachidonic acid metabolism, namely from prostaglandin endoperoxides (PGG_2 and PGH_2).

Thus it has been demonstrated that guinea pig lungs
similarily to human platelets (Hamberg et al., 1975)
convert arachidonic to thromboxanes in a considerably
greater extent than to prostaglandins.
    Thromboxane B_2 has little or no biological activity
on isolated organs, however its unstable precursor
thromboxane A_2 has a powerful contractile action on a
strip of rabbit aorta sharing this property with prostaglandin endoperoxides (Svensson et al., 1975). As the
matter of fact thromboxane A_2 is about 50 times more
potent than PGG_2 in contracting rabbit aorta (Bunting
et al., 1976, Needleman et al., 1976). Piper and Vane
(1969) were first workers who demonstrated that perfused
guinea pig lungs during anaphylactic shock released a
substance that contracted rabbit aorta. Piper and Vane
(1969) named this substance RCS (a rabbit aorta contracting substance). RCS is released from perfused
guinea pig lungs by a variety of chemical agents which
include bradykinin, histamine slow reacting substances
A and C, releasing factor for RCS (RCS-RF) and arachidonic acid. (Vargaftig and Dao Hai, 1971; Piper and Vane,
1971; Palmer et al., 1973). RCS is also generated by
vibrated rabbit splenic slices (Gryglewski and Vane,
1972a), by dog spleen microsomes incubated with arachidonic acid (Gryglewski and Vane, 1972b), by aggregating
platelets (Vargaftig and Zirinis, 1973) and by ram seminal vesicle microsomes incubated with arachidonic acid
(Hamberg and Samuelsson, 1973). Any RCS can comprise
either prostaglandin endoperoxides or thromboxane A_2 or
it can be a mixture of both unstable products of oxydation
of arachidonic acid. Until recently the only way to
differentiate biologically between prostaglandin endoperoxides and thromboxane A_2 was measuring a half life
time of RCS. In aqueous solution at $37^{\circ}C$ half lives for
prostaglandin endoperoxides and thromboxane A_2 are 5 min
and 30 sec respectively (Hamberg and Samuelsson, 1973,
Hamberg et al., 1975, Svensson et al., 1975). However,
recently Vane's group has discovered (Bunting et al.,
1976) that strips of rabbit coeliac or mesenteric
arteries are contracted by thromboxane A_2 and relaxed
by prostaglandin endoperoxides. Thus a strip of rabbit
mesenteric artery (or coeliac artery) directly differ-
entiates between these two potential components of RCS.
    Perfused guinea pig lungs (Piper and Vane , 1969,
1971; Palmer et al., 1973; Gryglewski et al., 1975) and
chopped guinea pig lungs (Palmer et al., 1970, 1973;
Fjalland , 1974) when stimulated immunologically chemically or mechanically release simultaneously RCS and
prostaglandins as well as 15-Keto,13-14-dihydro prostaglandin metabolites (Mathé and Levine, 1973). These

findings encourage to put several questions. Is RCS
composed of prostaglandin endoperoxides or thromboxane
A_2? What type of prostaglandins does prevail in perfusate or superfusate from lung tissues? Is any common
mechanism which triggers the release of RCS and prostaglandins from lungs by various stimuli? Are RCS and
prostaglandins generated by the same structural elements
in lungs?

Some evidence has already accumulated that RCS from
lungs mainly comprises thromboxane A_2. Using physicochemical techniques Hamberg and Samuélsson (1974) have
found that perfused guinea pig lungs release thromboxane
B_2 (654 - 2304 ng), HHT (192 - 387 ng), PGE_2 (15 - 93 ng)
and PGF_2 (93 - 171 ng) following an injection of 30 μg
of arachidonic acid. Using a bioassay technique (a
superfused strip of rabbit aorta) Svensson et al.,
(1975) have detected that only 1 - 3 ng/ml of prostaglandin endoperoxides is present in the perfusate of
guinea pig lungs after a challenge with 30 μg of arachidonic acid. The major part of RCS from lungs consists
of a material with the half life of 30 seconds at $37^{\circ}C$,
which corresponds to the half life of thromboxane A_2
(Hamberg et al., 1975, Needleman et al., 1976). There
is also interesting ex post evidence. In 1973 Palmer et
al. have tried the effect of RCS from lungs on various
vascular strips which included a strip of rabbit coeliac
artery. This vascular strip has been contracted by RCS
from lungs indicating the presence of thromboxane A_2 in
the lung perfusate (Bunting et al., 1976). Other questions
still remain to be answered.

RELEASE OF PROSTAGLANDINS FROM CONTRACTING TRACHEA

The release of prostaglandins and thromboxanes
from lungs by any agent, perhaps except for exogenous
arachidonic acid, seems to be preceded by a distortion
of cellular membranes (Vane and Piper, 1971) which
enables activation of phospholipase A_2 and liberation
of endogenous arachidonic acid. That "distortion of
cellular membranes" might be easily quantified when the
contractile elements of lungs are exposed to known
concentrations of spasmogens. It is important to point
out that in lungs there exist at least two types of
contractile elements: a) smooth muscles of large airways
and b) contractile proteins in alveolar interstitial
cells (Kapanci and Gabbiani, 1976). These two types of
lung contactile cells differ in their physiological
function, and in their responsiveness to immunological
and pharmacological agents. In aim to spot the site of

generation of prostaglandins and thromboxanes in lungs
one should separate these two types of contractile elements and expose them to the same spasmogen. It is much
easier to separate smooth muscles than contractile
elements of parenchyma. Spirally cut guinea pig trachea
or a chain of tracheal rings are frequently used in
pharmacological experiments.

Bouhuys' group (Orehek et al., 1973, 1975; Bouhuys,
1975; Bouhuys et al., 1976) have elegantly designed a
series of experiments with isolated spirals of guinea
pig trachea contracted by histamine or acetylcholine.

The authors have shown that contractions of tracheae
are accompanied by a release of prostaglandins, and this
release can be blocked by prostaglandin synthetase
inhibitors (indomethacin, aspirin and 5,8,11,14-eicosa-
tetraynoic acid). We have essentially confirmed the
results of Bouhuys' group (Grodzińska et al., 1975).
The only discrepancy is that in our bioassay system we
have detected mainly prostaglandins of the E type,
whereas Orehek et al.,(1973,1975) were able to spot also
$PGF_{2\alpha}$ by thin layer chromatography. Yen et al. (1976)
have found that incubated slices of guinea pig trachea
release both PGE_2 and $PGF_{2\alpha}$ although PGE_2 is released
to a greater extent. The same authors have reported that
histamine infusion into the perfused lungs causes several-
fold increase in the outflows of both prostaglandins.
Therefore there is no doubt that guinea pig trachea
has a capacity to generate PGE_2 and $PGF_{2\alpha}$. However,
we believe that contracting guinea pig trachea releases
mainly PGE_2. Except for our direct evidence (Grodzińska
et al., 1975) there is also an indirect but crucial
evidence. Orehek et al.,(1973, 1975) and Grodzińska et
al. (1975) have found that indomethacin at a low concentration (0.6 - 1 μg/ml) increases by 100% the contractile
response of trachea to a high dose of histamine. At this
low concentration indomethacin is expected to inhibit
specifically prostaglandin biosynthesis without any
side-effects. The result of this experiment may be
explained in terms of blocking by indomethacin of generation of a bronchodilator PGE_2 but it cannot be
explained in terms of blocking of generation of a
bronchoconstrictor $PGF_{2\alpha}$. Orehek et al. (1975) observed
that indomethacin at a high concentration of 60 μg/ml
supressed the contractile response of trachea to histamine. This may be explained by non-specific depressive
action of high concentrations of indomethacin. This
effect of indomethacin has been described for other
smooth muscles (Sorrentino et al., 1972). Certain prostaglandin synthetase inhibitors have dual action on prostaglandin system in the guinea pig trachea. Meclofenamic

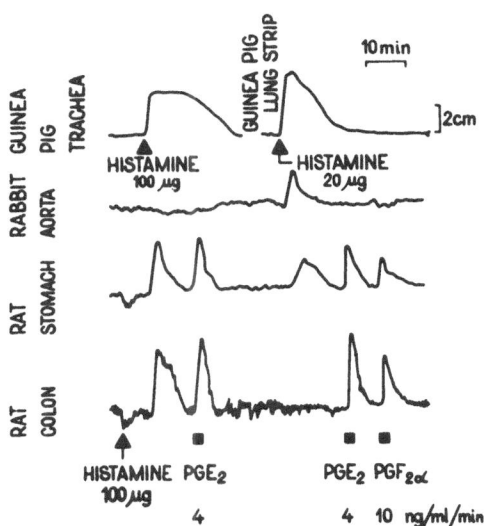

Fig. 1. Release of prostaglandins and rabbit aorta contracting substance by guinea pig trachea and guinea pig
trachea and guinea pig lung strip contracted by histamine. Both preparations were separately mounted over the
same bank of assay organs, which were rabbit aorta, rat
stomach and rat colon. The assay tissues were superfused
with Krebs bicarbonate (1.5 ml/min, at 37°C) which
contained diphenhydramine (10 µg/ml) and other receptor
antagonists and indomethacin. Histamine (100 µg) did
not contract the assay tissues, however, it contracted
trachea and released from trachea a substance which in
turn contracted the rat stomach and rat colon but not
rabbit aorta. These contractions were matched by an
infusion of PGE$_2$ at a concentration of 4 ng/min. Lung
strip when contracted by histamine (20 µg) released a
substance which contracted the rabbit aorta and rat
stomach but not rat colon. These contractions could not
be matched by infusions of PGE$_2$ or PGF$_{2\alpha}$. (According
to Gryglewski et al., 1976).

acid at low concentrations (1 µg/ml) inhibits prostaglandin synthetase in trachea and potentiates its contractile responses to PGF$_{2\alpha}$ whereas maclofenamic acid
at high concentrations (10 µg/ml) blocks receptor
sites for prostaglandins. (Panczenko et al., 1975).
    Orehek et al. (1975) put forward an interesting
hypothesis that prostaglandin biosynthesis in trachea

probably occurs in epithelial or subepithelial tissues since gentle scratching of its mucosal surface was associated with the release of a large amount of prostaglandins, On the other hand Coburn et al. (1974) have found that purified airway smooth muscles from dogs still generate and release prostaglandins.

It well might be that the release of PGE_2 by the contracting airways smooth muscles is a self-defensive mechanism protecting airways against over constriction. We have postulated (Szczeklik et al., 1975) that this prostaglandin mechanism dominates over the beta-adrenergic mechanism in airways of aspirin-sensitive patients, and therefore a great number of prostaglandin synthetase inhibitors precipitate severe attacks of bronchospasm in those patients.

In the paper of Orehek et al. (1975) there is short but meaningful statement: "The effluent of contracting tracheae did not contain slow-reacting substance nor rabbit aorta contracting substance". We have fully confirmed this observation. (Gryglewski et al., 1976). In 16 superfused tracheal preparations contracted with histamine (100 μg) the release of prostaglandins was 4.0 ± 0.8 ng/ml of PGE_2 equivalents (mean $\pm$ S.E.) but no RCS appeared. In additional similarily designed six experiments the effluent from superfused trachea dripped over a strip of rabbit mesenteric artery that is relaxed by prostaglandin endoperoxides and contracted by thromboxane A_2 (Bunting et al., 1976). This strip is also relaxed by PGE_2. The effluent from histamine-contracted trachea resulted in a relaxation of mesenteric artery. This relaxation strictly corresponded to the amount of the released PGE_2 as measured by contractions of rat stomach strip and rat colon that were superfused in cascade with rabbit mesenteric artery.

We conclude that superfused guinea pig trachea when contracted by histamine releases prostaglandins, mainly of the E type. There is no detectable release of either prostaglandin endoperoxides or thromboxane A_2 as measured by the appearance of RCS in the effluent.

RELEASE OF THROMBOXANE A_2 BY CONTRACTING LUNG STRIPS

The extensive studies of Kapanci and Gabbiani (1976) have shown that contractile proteins in alveolar interstitial cells are organized into dense fibrillar bundles, representing "intra-cytoplasmic muscles". Interstitial cells extend between, and are anchored on, the basement membranes of two adjacent alveoli. These

cells are contracted by histamine, hypoxia and other
agents. Owing to their privilaged spatial situation
the contractile interstitial cells may fold the alveolo-
capillary membrane into the capillary space or close
the alveolus and thus change the alveolo-capillary configuration (Kapanci and Gabbiani, 1976). It should be
stressed that the contractile interstitial cells differ
in many respects from large airways smooth muscles. For
instance, regular smooth muscles are in vitro relaxed
by adrenaline, whereas interstitial cells are contracted
by adrenaline (Kapanci and Gabbiani, 1976). We have
demonstrated that parenchymal lung strips which contain
interstitial cells generate thromboxane A_2-like substance,
whereas tracheal smooth muscles generate prostaglandins
(Gryglewski et al., 1976).

In the above experimental set we prepared thin
(2 x 25 mm) subpleural strips of guinea pig lungs that
under the microscopic examination were free from bronchial
smooth muscle. These strips were superfused with Krebs
bicarbonate (37°C 1.5 ml/min) and the effluent was en-
riched with a mixture of combined antagonists (Gilmore
et al., 1968) plus indomethacin (1 µg/ml) plus diphenhydramine hydrochloride (10 - 20 µg/ml). The effluent
superfused in cascade a bank of assay organs that were
capable to detect thromboxane A_2, prostaglandin endoperoxides and prostaglandins, but were made insensitive
to histamine by the presence of diphenhydramine. The
bank of assay organs consisted of rabbit aortic strip,
rabbit vena cava strip, mesenteric artery strip (2
experiments), rat stomach strip and rat colon. All
vascular strips are contracted by thromboxane A_2 (Bunting et al., 1976). Prostaglandin endoperoxides also
contract vascular strips except for mesenteric artery
that is relaxed by PGG_2 and PGH_2 as well as by PGE_2
(Bunting et al., 1976). Rat stomach strip is contracted
by PGE_2, $PGF_{2\alpha}$, prostaglandin endoperoxides and thromboxane A_2 in a decreasing order of potency, whereas rat
colon is usually more sensitive to $PGF_{2\alpha}$ than to PGE_2
and is not contracted by prostaglandin endoperoxides
and thromboxane A_2 (up to 100 ng).

When lung strips were contracted with histamine
(10 - 20 µg) there was observed the release of a substance that contracted all vascular strips and a rat
stomach strip but did not contract rat colon. (Fig. 2).
This substance had a half life shorter than 60 seconds
(Fig. 2). These findings indicate that contractile
elements of the lung strip generate thromboxane A_2.
Prostaglandins and prostaglandin endoperoxides (if any)
are generated at amounts that are not biologically
detectable (see discussion).

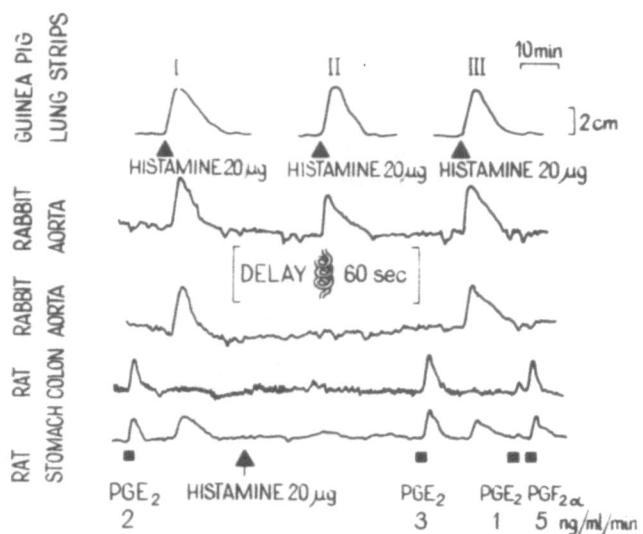

Fig. 2. Histamine-contracted strips of guinea pig lungs
release a rabbit aorta contracting substance (RCS) which
disappears from the Krebs bicarbonate (37°C) after 60
seconds of incubation without the subsequent appearance
of PG-like activity, strongly indicating that the RCS
from lung strips is identical with thromboxane A_2.
Three strips from the same lungs (I, II, III) were
mounted separately over one bank of superfused tissues
which consisted of two rabbit aortas, the rat stomach
and rat colon. During the histamine challenge of the
second lung strip (II) a delay coil (60 sec) was in-
serted between two superfused aortic strips. Sixty
seconds of incubation inside the delay coil was sufficient do destroy any biological activity of the RCS
released from the second lung strip (II) and no PG-like
activity appeared. Therefore we assume that contractions
of the rat stomach (control lung preparations I and II
in this figure, see also Fig. 1) appearing during the
histamine challenge of lung strips are associated with
the intrinsic biological activity of an unstable RCS
(thromboxane A_2) released by lung parenchyma and are
not due to the concomitant or subsequent appearance of
PGs. (According to Gryglewski et al., 1976).

        Yen et al. (1976) have incubated lung fragments
of guinea pig lungs and found that there is the spontaneous release of prostaglandins and their metabolites
in following amounts (ng/g wet weight, mean $\pm$ S.E, n =6):

PGEs 4.5 + 0.5, PGF$_{2\alpha}$ 11.7 + 1.5 and metabolites
17.0 + 4.0. This baseline release prostaglandins is not
very high taking into account the effect of the trauma
of mincing. In very similar conditions the fragments
of human polypous mucosa generate PGEs in amounts of
74 + 16 ng/g wet weight (our unpublished observation).
In our experiments we have used thin strips of parenchymal tissue which were superfused with fresh Krebs bicarbonate for several hours and no baseline release of
any biologically active metabolites of arachidonic acid
could be detected, possibly because of the limits of
sensitivity of bioassay (see discussion). However, when
this small piece of parenchyma was challenged with histamine it released only one biologically detectable
substance, i.e. thromboxane A$_2$. Because of the scanty
amount of the "generator tissue" that has been used in
our experiments we cannot exclude the possibility that
lung parenchyma may release also prostaglandin endoperoxides and prostaglandins, however, we have proved
that thromboxane A$_2$ is the major product that is formed
by lung parenchyma. Another point is that Yen et al.
(1976) have used in their experiments chopped lungs
after removing of trachea, large bronchi and blood
vessels. Their biological material contained smooth
muscles of small bronchi. Our subpleural parenchymal
strips were free of any smooth muscles.

We conclude that the RCS released by histamine
from lung parenchyma (probabely from interstitial cells)
comprises mainly thromboxane A$_2$. We also postulate ana-
tomical compartmentization for biochemical transformations of arachidonic acid in lungs. Main airways seem
to generate predominantly prostaglandins whereas the
lung parenchyma mainly thromboxane A$_2$.

DISCUSSION

Perfused guinea pig lungs , and lung homogenate
generate from arachidonic acid chiefly thromboxanes and
small amounts of prostaglandins. (Hamberg and Samuelsson,
1974; Svensson et al., 1975). We have shown that histamine-contracted tracheal spirals release mainly prostaglandins of the E type and no detectable amounts of
either thromboxane A$_2$ or prostaglandin endoperoxides.
Whereas histamine-contracted lung strips release chiefly
thromboxane A$_2$ and no detectable amounts of either
prostaglandin endoperoxides or prostaglandins. In these
experiments the threshold sensitivity of bioassay organs
to PGE$_2$ and PGF$_{2\alpha}$ was at a range of 0.5 - 1.0 ng (for
either prostaglandin) (Gryglewski et al., 1976).

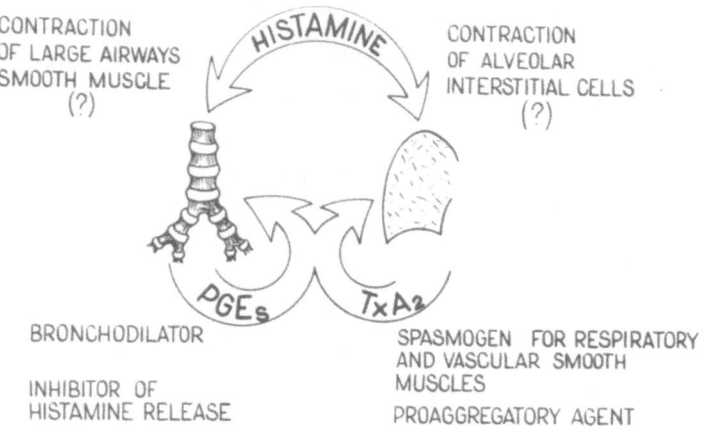

Fig. 3. Diagrammatic representation of the effect of histamine on the release of prostaglandins and thromboxane A_2 from guinea pig tracheal spirals and strips of lung parenchyma.

PGH_2 could be detected at the lowest concentration of $6 - 12$ ng as standarized by the threshold responses of aortic strips to $1 - 2$ ng of 9,11-methanoepoxy analog of PGH_2 (U 46619). The threshold sensitivity of aortic strip to PGH_2 was calculated according to the ratio of potency PGH_2 : U 46619 that has been reported by Malmsten (1976).

Airways smooth muscle can produce both PGE_2 and $PGF_{2\alpha}$ (Orehek et al., 1973, 1975 ; Yen et al., 1976). We have a direct and an indirect evidence that contracting tracheal spirals release mainly prostaglandins of the E series (Grodzińska et al., 1975; Gryglewski et al., 1976). This phenomenon can be teleologically explained as the self-defense of smooth muscles against the overdone bronchoconstriction. Bronchodilator PGEs locally attenuate the contractile action of histamine and may also inhibit histamine release from mast cells by activation of adenylcyclase in these cells (Fig. 3). The specificity of histamine as the releaser of prostaglandins from airways smooth muscle still remains to be elucidated. Not only histamine (Bakhle and Smith, 1972 ; Yen et al., 1976) but also many other agents (Palmer et al., 1973) release prostaglandins when in-

jected into perfused guinea pig lungs. Orehek et al.,
(1973, 1975) observed the release of prostaglandins
from guinea pig trachea contracted either by histamine
or acetylcholine, whereas Dunlop and Smith (1975) did
not detect a prostaglandin release from strips of human
bronchi contracted by acetylcholine.

Thromboxane A_2 is the main product that is released
from histamine-contracted lung parenchyma. Our parenchymal preparation contained no smooth muscles and
probably few migrating cells, since a thin strip of
subpleural lung parenchyma was superfused with Krebs
bicarbonate for a long period of time. The pathophysiological significance of formation of thromboxane A_2 by
lung parenchyma remains unknown. Although extremely unstable thromboxane A_2 has a powerful contractile action
on smooth muscles of vascular and respiratory systems
and it is the most potent pro-aggregatory agent so far
known (Hamberg et al., 1975). In all respects thromboxane A_2 differs from PGEs that are formed in large
airways (Fig. 3). One can hardly offer an explanation
for a positive feed-back mechanism by which histamine
liberates from lung parenchyma even a more dangerous
agent than histamine ifself. Anyway it seems that the
"protective action" of PGEs and catecholamines in lungs
can be antagonized not only by acetylcholine, histamine,
slow reacting substance A and PGFs (Orehek, 1973, 1975;
Kaliner and Austen, 1975) but also thromboxane A_2.

ACKNOWLEDGEMENTS

I gratefully acknowledge the generous grant of
equipment from the Trustees of The Wellcome Trust,
London, Great Britain.

REFERENCES

Anggard, E., and Samuelsson B., 1965. Biosynthesis of
    prostaglandins from arachidonic acid in guinea pig
    lung. J. Biol. Chem. 240: 3518-3521.
Bakhle, Y.S., and Smith, T.W., 1972. Release of spasmo-
    genic substances induced by vasoactive amines from
    isolated lungs. Br. J. Pharmac. 46: 543-544.
Bouhuys, A., 1975. Prostaglandins and asthma. Lancet
    1: 399.
Bouhuys, A., Douglas, J.S., and Brink, C., 1976. Actions
    of chemical mediators and drugs on effector cells.
    Pneumonologie 153: 148-149.

Bunting, S., Moncada, S., and Vane, J.R., 1976. The
    effects of prostaglandin endoperoxides and thromb-
    oxane A_2 on strips of rabbit coeliac artery and
    certain other smooth muscle preparations. Proc. Br.
    Pharmacol. Soc. 1 - 2 April, p. 48.
Coburn, R.F., Hizig, B., and Yamaguchi, T., 1974. Prosta-
    glandin (PG) synthesis in canine trachealis muscle
    (CTM) and guinea pig taenia coli (TC). Fed. Proc.
    33: 451.
Dunlop, L.S., and Smith, A.P., 1975. Reduction of anti-
    gen-induced contractions of sensitized human bron-
    chus in vitro by indomethacin. Br. J. Pharmac. 54:
    495-497.
Fjalland, B., 1974, Inhibition by non-steroidal anti-
    inflammatory agents of the release of rabbit aorta
    contracting substance and prostaglandins from chop-
    ped guinea pig lungs. J. Pharm. Pharmacol. 26: 448-
    451.
Gilmore, N., Vane, J.R., and Wyllie, J.H., 1968, Prosta-
    glandins released by the spleen. Nature 218: 1135-
    1140.
Grodzińska, L., Panczenko, B., and Gryglewski, R.J.,
    1975, Generation of prostaglandin E-like material
    by the guinea-pig trachea contracted by histamine.
    J. Pharm. Pharmac. 27: 88-91.
Gryglewski, R.J., Dembińska-Kieć, A., Grodzińska, L.,
    and Panczenko, B., 1976, Differential generation of
    substances with prostaglandin-like and thromboxane-
    like activities by guinea pig trachea and by lung
    strips, in:Lung cells in disease (A. Bouhuys, ed.),
    in press. Elsevier Pub. Amsterdam.
Gryglewski, R.J., Panczenko, B., Korbut, R., Grodzińska,
    L., and Ocetkiewicz, A., 1975, Corticosteroids in-
    hibit prostaglandin release from perfused mesenteric
    blood vessels of rabbit and from perfused lungs of
    sensitized guinea pig. Prostaglandins 10: 343-355.
Gryglewski, R., and Vane, J.R., 1972a, The release of
    prostaglandins and rabbit aorta contracting sub-
    stance (RCS) from rabbit spleen and its antagonism
    by anti-inflammatory drugs. Br. J. Pharmac. 45:
    37-47.
Gryglewski, R., and Vane, J.R., 1972b, The generation
    from arachidonic acid of rabbit aorta contracting
    substance (RCS) by microsomal enzyme preparation
    which also generates prostaglandins. Br. J. Pharmac.
    46: 449-457.
Hamberg, M., and Samuelsson, B., 1973, Detection and
    isolation of an endoperoxide intermediate in prosta-
    glandin biosynthesis. Proc. Nat. Acad. Sci. USA 70:
    899-903.

Hamberg, M., and Samuelsson, B., 1974, Prostaglandin endoperoxides VII. Novel transformations of arachidonic acid in guinea pig lung. Biochem. Biophys. Res. Commun. 61; 942-949.

Hamberg, M., Svensson, J., and Samuelsson, B., 1975, Thromboxanes: A new group of biologically active compounds derived from prostaglandin endoperoxides. Proc. Nat. Acad. Sci. USA, 72: 2994-2998.

Kaliner, M., and Austen, K.F., 1975, Immunologic release of chemical mediators from human tissues in: Annual Review of Pharmacology (H.W. Elliot, R. George and R. Okum, eds.), pp. 177-189. Annual Review Inc. California.

Kapanci, Y., and Gabbiani, G., 1976, Location and function of contractile interstitial cells of the lung. Pneumonologie 153: 141.

Malmsten, C., 1976, Some biological effects of prostaglandin endoperoxide analogs. Life Sci. 18: 169-176.

Mathe, A.A., and Levine, L., 1973, Release of prostaglandins and metabolites from guinea pig lungs: inhibition by catecholamines. Prostaglandins 4: 877-890.

Needleman, P., Moncada, S., Bunting, S., Vane, J.R., Hamberg, M., and Samuelsson, B., 1976, Identification of an enzyme in platelet microsomes which generates thromboxane A_2 from prostaglandin endoperoxides. Nature 261: 558-560.

Nugteren, D.H., Hazelhop, E., 1973, Isolation and properties of intermediates in prostaglandin biosynthesis. Biochim. Biophys. Acta 326: 448-461.

Orehek, J., Douglas, J.S., and Bouhuys, A., 1975, Contractile responses of the guinea-pig trachea in vitro: modification by prostaglandin synthesis-inhibiting drugs. J. Pharmacol. 194: 554-564.

Orehek, J., Douglas, J.S., Lewis, A.J., and Bouhuys, A., 1973, Prostaglandin regulation of airway smooth muscle tone. Nature New Biology 245: 84-85.

Palmer, M.A., Piper, P.J., and Vane, J.R., 1970, The release of rabbit aorta contracting substance from chopped lungs and it antagonism by anti-inflammatory drugs. Br. J. Pharmac. 40: 581P - 582P.

Palmer, M.A., Piper, P.J., and Vane J.R., 1973, Release of rabbit aorta contracting substance (RCS) and prostaglandins induced by chemical or mechanical stimulation of guinea pig lungs. Br.J.Pharmac. 49: 226-242.

Panczenko, B., Grodzińska, L., and Gryglewski, R.J., 1975, The dual action of meclofenamate on the contractile response to PGF_2 in the guinea pig trachea. Pol. J. Pharmacol. 27: 273-276.

Piper, P.J., and Vane, J.R., 1969, Release of additional
    factors in anaphylaxis and its antagonism by anti-
    inflammatory drugs. Nature 233: 29-35.
Piper, P.J., and Vane, J.R., 1971, The release of
    prostaglandins from lung and other tissues. N.Y.
    Acad. Sci. 180: 363-385.
Sorrentino, L., Capasso, F., Di Rosa, M., 1972, Indo-
    methacin and prostaglandins. Europ. J. Pharmac. 17:
    306-308.
Svensson, J., Hamberg, M., Samuelsson, B., 1975,
    Prostaglandin endoperoxides IX. Characterization
    of rabbit aorta contracting substance (RCS) from
    guinea pig lung and human platelets. Acta Physiol.
    Scand. 94: 222-228.
Szczeklik, A., Gryglewski, R.J., and Mysik-Czerniawska,
    G., 1975, Relationship of inhibition of prostaglan-
    din biosynthesis by analgesics to asthma attacks
    in aspirin-sensitive patients. Br. Med. J. 1:67-69.
Vane, J.R., 1971, Inhibition of prostaglandin synthesis
    as a mechanism of action for aspirin-like drugs.
    Nature New Biol. 231: 232-235.
Vargaftig, B.B., and Dao Hai, N., 1971, Release of
    vasoactive substances from guinea-pig lungs by slow
    reacting substance C and arachidonic acid. Pharma-
    cology 6: 99-108.
Vargaftig, B.B., and Zirinis, P., 1973, Arachidonic
    acid induced platelet aggregation is accompanied
    by release of potential inflammatory mediators
    distinct from PGE_2 and PGF_2 Nature New Biol.
    244: 114-116.
Yen, S.S., Mathe, A.A., and Dugan, J.J., 1976, Release
    of prostaglandins from healthy and sensitized guinea-
    pig lung and trachea by histamine. Prostaglandins
    11: 227-239.

CLINICAL EFFECTS OF THE PROSTAGLANDINS ON THE RESPIRATORY SYSTEM

MAURICE F CUTHBERT

DEPARTMENT OF PHARMACOLOGY & THERAPEUTICS
LONDON HOSPITAL MEDICAL COLLEGE LONDON E 1
&
CHEST UNIT
DEPARTMENT OF MEDICINE
KINGS COLLEGE HOSPITAL   LONDON S E 5 ENGLAND

The prostaglandins were first isolated from the lungs by Bergström and his coworkers more than 14 years ago (1). It is now established that in animal studies all prostaglandins have powerful effects on both respiratory smooth muscle and on pulmonary vascular smooth muscle and that the lungs have an important function in their metabolic degradation, since some 90-95% of infused E and F prostaglandins are removed predominantly by the enzyme 15-prostaglandin dehydrogenase (15-PGDH) on a single passage through the pulmonary circulation. It is of interest that the A prostaglandins, which have a similar but less potent effect than the E prostaglandins, are less readily degraded by 15-PGDH in the lungs of experimental animals. Similar results have been obtained with prostaglandin F_{2a} (PGF_{2a}) in man and it is suggested that degradation occurs in the pulmonary endothelial cells (2,3).

These findings have led to considerable speculation on the possible physiological and pathological role of the prostaglandins in relation to the function of the respiratory system and have raised the possibility that certain prostaglandins, or substances which interfere with their synthesis or action, might prove of therapeutic value in the treatment of certain respiratory diseases, notably in bronchial asthma.

In this article it is proposed to review the actions
of the naturally-occurring prostaglandins on respiratory
function, both in healthy subjects and in asthmatics,
and to consider their possible physiological involvement in the control of normal and abnormal respiratory
function. Finally, some speculations will be made on
their possible therapeutic application.

EFFECTS OF PROSTAGLANDIN AEROSOLS ON RESPIRATORY FUNCTION

Animal Studies

The actions of the prostaglandins on respiratory smooth
muscle in experimental animals have already been reviewed by Professor Puglisi who has referred to the funde-
mental observation of Sweatman & Collier in 1968 (4)
who demonstrated that isolated human bronchial muscle
has intrinsic tone and that PGF_{2a} causes contraction
and both prostaglandins E_1 and E_2 (PGE_1 & PGE_2) relaxation of this preparation. These effects are independent
of vagal innervation or histamine release and are not
influenced by adrenoceptor blocking agents. Present
evidence suggests that the prostaglandins act directly
on smooth muscle, relaxation being mediated through the
cyclic AMP system. A more detailed in vitro study see (4a).

By far the most important animal experiment which stimulated the author's interest in the effects of prostaglandin
aerosols in man was that of Large, Leswell & Maxwell in
1969 (5). In a comparative study, these workers showed
that PGE_1 and isoprenaline had similar bronchodilator
potency in anaesthetised guinea-pigs when given intravenously, but when given by aerosol PGE_1 had 50-100
times the activity of isoprenaline. The duration of the
bronchodilatation was similar for both agents but in
contrast to isoprenaline, inhalation of PGE_1 produced no
tachycardia in high dosage and the bronchodilatation
was not inhibited by beta-adrenoceptor blockers. Another
interesting observation was that an F prostaglandin,
namely PGF_{2b}, was reported to be an effective bronchodilator when given by aerosol to experimental animals
(6) althogh this compound is considerably less active
than PGE_1.

Human Studies

Preliminary studies in healthy subjects (7) showed that
the inhalation of PGE_1 (free acid) had no effect on the
blood pressure, heart rate, electrocardiogram or forced
expiratory volume in 1 second (FEV_1). This result on the

FEV_1 was not unexpected since the respiratory smooth muscle in the smaller airways of healthy subjects is normally almost fully relaxed and it is not possible to demonstrate the small amount of possible further relaxation by spirometric methods.

The higher doses of PGE_1 caused coughing and retrosternal soreness and for this reason this study was repeated with the neutral triethanolamine salt which was better tolerated. Subsequently it was shown that in asthmatic subjects, the inhalation of PGE_1 (triethanolamine salt) 55ug had a bronchodilator effect of comparable degree and duration to that of isoprenaline 550ug. Maximal bronchodilatation occurred within 5 minutes of isoprenaline inhalation while the effect of PGE_1 was less rapid, maximal values being obtained some 30 minutes after inhalation. In this and in subsequent studies inhalations were from metered aerosols either of prostaglandins or isoprenaline both with freon propellents. No significant cardiovascular effects were noted with PGE_1 and only 2 of the 6 asthmatics complained of irritation of the upper respiratory tract. In one subject, persistent coughing led to the gradual onset of bronchospasm; this appeared to be the result of reflex cholinergic stimulation rather than a direct effect of PGE_1. Other workers have confirmed the bronchodilator effect of PGE_1 in asthmatics (8.9).

More recently, a double-blind study has been completed to compare the effects of metered aerosols of PGE_1, PGE_2 & isoprenaline with placebo using total body plethysmography to investigate changes in bronchial muscle tone (10,11). In both groups, PGE_1 and PGE_2 were effective bronchodilators causing a fall in airway resistance and a rise in specific airway conductance. In healthy subjects, the bronchodilator effects were more variable than in the asthmatics but were statistically significant. Changes in lung volume brought about by either of the prostaglandins and isoprenaline were insignificant. In both healthy subjects and asthmatics the bronchodilator effects of PGE_1 and PGE_2 were comparable in degree and duration to that of isoprenaline using the doses in the previous study.

Prostaglandin F_{2a} (PGF_{2a}) has been shown to be a potent bronchoconstrictor of short duration (12) in healthy subjects. In our own study (10) the inhalation of PGF_{2a} 40-80ug caused a maximal fall in specific airway conductance 4-6 minutes and there was a gradual return to normal over 20-30 minutes. Both PGE_2 (55ug) and isoprenaline (550ug) by metered aerosol effectively reversed the PGF_{2a}-induced

bronchoconstriction. Inhalation of PGF_{2a} had a similar
irritant effect to that of the E prostaglandins. Pretreatment with several possible prostaglandin antagonists
including flufenamic acid, disodium cromoglycate and
atropine failed to modify the PGF_{2a}-induced bronchoconstriction. For a recent of PGF_{2a} on lung mechanics (12a)

In a recent study (9), Mathé and his coworkers have drawn
attention to the possible role of PGF_{2a} in bronchial asthma.
Using total body plethysmography, asthmatics were found to
be about 10 times more sensitive than healthy subjects to
histamine inhalation but approximately 8000 times more
sensitive than healthy subjects to PGF_{2a} inhalation.
The authors suggest that hypersensitivity to endogenous
PGF_{2a} may significantly contribute to increased airways
resistance in bronchial asthma and that local release of
PGF_{2a} might explain the precipitation of the asthmatic
attack by a variety of stimuli. In our own study (11) we
have confirmed the hypersensitivity of asthmatics to
inhaled PGF_{2a} but have shown marked individual variation,
some asthmatics being no more sensitive than healthy
subjects while others showed the marked sensitivity reported by Mathé et al. This variation was not correlated with
any particular feature of the nature or duration of the
asthma or its treatment. Differences in the technical
administration of the aerosolised PGF_{2a} may account for the
differences in the results of these two studies.

The widespread use of PGE_2 and PGF_{2a} in therapeutic
abortion and induction of labour has presented an opport-
unity to study the effects of intravenous prostaglandins
in women with no previous history of respiratory disease.
The infusion of PGF_{2a} caused a modest increase in airways
resistance; this was not sufficient to cause symptoms
but could present a hazard in asthmatics(13) and bronchospasm has been precipitated in one asthmatic patient (14).
PGE_2 infusion paradoxically also caused a modest increase
in airway resistance in these female patients.

EFFECTS ON RESPIRATION AND PULMONARY VASCULATURE

Intravenous infusion of PGE_1 in healthy subjects causes a
dose-dependent increase in total and alveolar ventilation
(15). Respiratory rate is increased and there is a fall in
arterial oxygen saturation due to disturbance in ventilation
/perfusion relationships. These effects may be due to stimulation of the respiratory centre or secondary to cardiovascular effects.

Animal studies have shown that PGE_1 and PGE_2 intravenously increase pulmonary arterial pressure but reduce pulmonary vascular resistance; these changes appear to be secondary to changes in cardiac output since the pulmonary perfusion pressure falls when cardiac output or pulmonary arterial blood flow are kept constant (16,17) Similar results have been obtained with PGA_1 showing that both A & E prostaglandins have a dilator effect on human pulmonary vessels.

In contrast, PGF_{2a} produces a marked increase in pulmonary arterial pressure and pulmonary venous pressure in experimental animals and it has been suggested that the arterial effect may be secondary to venoconstriction (18,19). Prostaglandin isomer 8-iso-PGE_1 has some interesting properties; it has similar effects as PGF_{2a} on the pulmonary circulation while retaining the properties of the E prostaglandins on the systemic circulation (17)

PHYSIOLOGICAL INVOLVEMENT

Prostaglandins E_2 and F_{2a} are normally present in human lungs and bronchi (20,21) and it is clear that they have mutually antagonistic actions on human bronchial muscle both in vitro (4) and in vivo (10,11). There is also good evidence from animal experiments that prostaglandins can be released from the lungs by a variety of stimuli (22) including challenge of sensitised lungs, embolisation, mechanical stimulation, histamine or 5-hydroxytryptamine infusion and by ventilation. The common factor would appear to be disturbance of the cell membrane (22). The quantities of prostaglandins which can be released from tissues compared with those which can be extracted suggest that stimulation of release involves immediate synthesis from precursors.

In view of these findings it is possible that naturally-occurring PGE_2 and PGF_{2a} act as local hormones and contribute to the control of normal and possibly abnormal bronchial smooth muscle tone. This would require a specific and differential release of prostaglandins to account for the normal variations in this tone. At present there is little evidence to support this concept.

Since both PGE_2 and PGF_{2a} have powerful effects on both respiratory and vascular smooth muscle it is also possible that they may produce proportional changes in ventilation and pulmonary blood flow and thus contribute to the close relationship known to exist between ventilation and perfusion of specific regions of the lung (23).

PATHOLOGICAL INVOLVEMENT

The discovery by Vane in 1971 (24) that the non-steroidal anti-inflammatory drugs, for example, indomethacin, aspirin and salicylate, are effective inhibitors of the synthesis of both E and F prostaglandins has led to several attempts to elucidate the role of the prosta- glandins in the maintenance of bronchial smooth muscle tone. Experiments in animal and human in vitro prepar- ations (25,26) have shown that indomethacin is remarkably effective in reducing bronchial muscle tone but it is not certain whether this is due to inhibition of PG synthetase of antagonism of PGF_{2a} released from damaged tissues. The effects of indomethacin in asthmatics has recently been investigated (27). In a dosage of 200mg day, indo- methacin failed to produce any improvement in exercise- induced bronchospasm, antigen challenge or measurement of peak flow rate. Since this indomethacin dose is known to almost completely reduce the urinary excretion of prostaglandin metabolites (28),serum prostaglandin levels are elevated in asthmatics (29) and prostaglandin metab- olites appear in the blood following antigen challenge (30), these results are interpreted as throwing doubt on the hypothesis that prostaglandins are important mediators in bronchial asthma.

This hypothesis is attractive since only some asthmatics are exquisitely sensitive to PGF_{2a} (11) but it does dep- end on certain assumptions: Firstly, that doses of indo- methacin which inhibit the urinary excretion of prosta- glandin metabolites also significantly reduce the bio- synthesis in the lung parenchyma. Moreover, since indo- methacin inhibits the biosynthesis of both E and F prostaglandins it is assumed that the effect of the constrictor PGF2a is not simply offset or otherwise influenced by a comparable reduction in the dilator E prostaglandins

THERAPEUTIC POSSIBILITIES

Although the prostaglandins have most promising broncho- dilator properties, it seems unlikely that any naturally- occurring prostaglandin will prove to be of value as a treatment for bronchial asthma. There are two reasons for this statement. Firstly, all the prostaglandin aerosols so far tested in man are irritant to the upper respiratory tract; this may be an intrinsic property of the prostaglandins or could be limited to the naturally- occurring compounds. The other difficulty is that the E

prostaglandins are not stable in solution so their form-
ulation as freon-propelled aerosols would be inpractical.
The report that PGF_{2b} is an effective bronchodilator in
experimental animals (6) was therefore of great interest
since the F compounds are stable. Unfortunately, the activity of this compound in man was so low that it proved
impossible to determine whether it was a bronchodilator
in healthy subjects using total body plethysmography
(Cuthbert & Smith, unpublished observations). However,
the search for a stable, non-irritant PG analogue with
a bronchodilator activity of acceptable therapeutic duration is being pursued with great activiy and it is
expected that a number of analogues suitable for evaluation in man will shortly become available after the
appropriate aerosol toxicity tests have been completed
in experimental animals.

From the view-point of the clinical pharmacologist, the
outstanding advantage of the body plethysmograph is that
it is capable of measuring changes in the larger of the
small airways (the important resistance bronchioles)
with such sensitivity that it can determine whether a
potential new bronchodilator is likely to be effective
in asthmatics simply by carrying out studies in healthy
consenting subjects.

In respect of remarks made in earlier sections of this
paper, it will be evident that an equally promising
approach which is being pursued with similar enthusiasm,
is the synthesis of compounds which will specifically
inhibit the intracellular synthesis of PGF_{2a} or a specific
antagonist to its action on the smooth muscle receptor.
One of the most interesting developments in this field
is poly-phloretin phosphate, which can block the effect
of PGF2a on isolated human bronchial muscle while
leaving the relaxant effect of PGE_2 intact (31).
However, both this and other prostaglandin antagonists
at present available are either relatively unselective
or are unsuitable for administration to man.

CONCLUSION

Despite the difficulties in the synthesis of compounds
of potential therapeutic value, there is no doubt that
the relationship of the prostaglandins to the function
of the respiratory system will remain an area of intense
interest and vigorous research over the next few years.
It is hoped that the result of such activities will lead

not only to a better understanding of the role of the
prostaglandins in the maintenance of normal and abnormal
bronchial muscle tone but also to the development of
compounds of value in the treatment of certain respiratory diseases, notably in bronchial asthma of both
allergic and non-allergic aetiology.

ABSTRACT

Prostaglandin F_{2a} (PGF_{2a}) causes a contraction while both
prostaglandins E_1 and E_2 (PGE_1 and PGE_2) usually cause a
relaxation of both animal and human respiratory smooth
muscle. In most animal species, intravenous PGF_{2a} causes
bronchoconstriction and PGE_1 and PGE_2 are effective in
inhibiting the bronchoconstriction induced by various
agonists and by anaphylaxis. A study in which the relative bronchodilator properties of PGE_1 and isoprenaline
were compared in anaesthetised ventilated guinea-pigs
showed that while these substances were approximately
equipotent when given intravenously, PGE_1 was some 50-
100 times more potent than isoprenaline by the aerosol
route. This finding may be related to the rapid enzymatic
inactivation of prostaglandins by the lung.

Studies on the pulmonary vasculature showed that the
effects of the prostaglandins parallel those on respiratory smooth muscle, PGF_{2a} causing an increase and both
PGE_1 and PGE_2 a decrease in pulmonary vascular resistance.

In healthy subjects the inhalation of PGE_1 did not affect
the forced expiratory volume in one second (FEV_1) but in
5 of 6 asthmatics with reversible airways obstruction
the inhalation of PGE_1 55ug had a bronchodilator effect
comparable in both degree and duration to that of isoprenaline 550ug. These findings have been confirmed in
both healthy and asthmatic subjects using total body
plethysmography in which both PGE_1 and PGE_2 caused a fall
in airway resistance and an increase in specific airway
conductance. The effects on lung volume were insignificant.

In contrast, PGF_{2a} inhalation caused bronchoconstriction
in normal healthy subjects which could be readily reversed by both PGE_2 or isoprenaline. Pretreatment with
flufenamic acid, atropine and disodium cromoglycate
had no modifying action. The finding that some asthmatics
are extremely sensitive to inhaled PGF_{2a} raises the
possibility of its involvement in the aetiology of

bronchial asthma although the significance of PG synthetase inhibition in this respect is at present unclear.

Intravenous infusions of prostaglandins have also been shown to have effects on respiratory smooth muscle tone when used in otherwise healthy women undergoing termination of pregnancy. Both PGF_{2a} and PGE_2 cause a modest degree of bronchoconstriction. These changes were not sufficient to cause symptoms in these patients but could represent a hazard in asthmatics.

Since both PGE_2 and PGF_{2a} occur naturally in the lung, have mutually antagonistic pharmacological effects, are release by anaphylaxis and other stimuli and are also rapidly inactivated ezymatically it has been suggested that these potent substances may act as local hormones in the control of bronchial smooth muscle tone. Experiments with inhibitors and antagonists of endogenous prostaglandin are in accord with this hypothesis although there is as yet no convincing data due to the unselectivity of the available compounds and difficulties in their administration to man.

The naturally-occurring prostaglandins seem unlikely to prove to be bronchodilators of therapeutic value since all those tested to date are irritant to the upper respiratory tract and the E compounds are chemically unstable. However, in the search for a non-irritant, stable PG analogue it is certain that the body plethysmograph, which can detect small changes in bronchial muscle tone in healthy subjects with great sensitivity, will play an important part since current experience indicates that this technique is of predictive value in the identification of bronchodilators of therapeutic value. An equally promising approach lies in the development of synthetic compounds which specifically inhibit the synthesis of endogenous PGF_{2a} or block its action at the smooth muscle receptor.

REFERENCES

1. BERGSTROM S et al (1962) The isolation and structure of a smooth-muscle stimulating factor in normal pig and sheep lungs. Arkiv fur Kemi **20** 63

2. PIPER P J, VANE J R & WYLLIE J H (1970) Inactivation of prostaglandins by the lung. Nature (Lond) **225** 600

3. JOSE P, NIEDERHAUSER U, PIPER P J, ROBINSON C & SMITH P A (1976) Inactivation of prostaglandin F_{2a} in the human pulmonary circulation. British Journal of Clinical Pharmacology **3** 342

4. SWEATMAN W J F & COLLIER HOJ (1968) Effects of prostaglandins on human bronchial muscle. Nature(Lond) **217** 69

4a GARDINER P J (1975) The effects of some natural prostaglandins on isolated human circular bronchial muscle. Prostaglandins **10** 607

5. LARGE B J, LESWELL P F & MAXWELL D R (1969) Bronchodilator activity of prostaglandin E_1 in experimental animals. Nature (Lond) **224** 78

6. ROSENTHALE M E, DERVINIS A & GLUCKMAN J (1973) Comparative studies on the bronchodilating properties of prostaglandin F_{2b}. Advances in Biosciences **9** 229

7. CUTHBERT M F (1969) Effect on airways resistance of prostaglandin E_1 given by aerosol to healthy and asthmatic volunteers. British Medical Journal **4** 723

8. HERXHEIMER H & ROETSCHER I (1971) Effects of prostaglandin E_1 on lung function in bronchial asthma. European Journal of Clinical Pharmacology **3** 123

9. MATHÉ A A, HEDQVIST P, HOLMGREN A & SVANBORG N (1973) Bronchial hyperreactivity to prostaglandin F_{2a} and histamine in patients with asthma. British Medical Journal **1** 193

10. SMITH A P & CUTHBERT M F (1972) The antagonistic action of prostaglandins F_{2a} and E_2 aerosols on bronchial muscle tone in man. British Medical Journal **2** 212

11. SMITH A P, CUTHBERT M F & DUNLOP L S (1975) Effects of inhaled prostaglandins E_1, E_2 and F_{2a} on the airway resistance of healthy and asthmatic man. Clinical Science **48** 421

12. HEDQVIST P, HOLMGREN A & MATHÉ A A (1971) Effect of prostaglandin F_{2a} on airways resistance in man. Acta Physiologica Scand **82** 29a

13. SMITH A P (1973) The effects of intravenous infusion of graded doses of prostaglandins F_{2a} and E_2 on lung resistance in patients undergoing termination of pregnancy. Clinical Science **44** 17

14. FISHBURNE J I, BRENNER W E, BRAAKSMA J T, STAUROVSKY L G, MUELLER R A, HOFFER J L & HENDRICKS C H (1972) Cardiovascular and respiratory responses to intravenous PGF_{2a} in the pregnant woman. Obstetrics &

Gynecology <u>39</u> 892

15. CARLSON L A, EKELUND L & ORO L (1969) Circulatory and
    respiratory effects of different doses of prosta-
    glandin E_1 in man. Acta Physiologica Scand <u>75</u> 161

16. NAKANO J & COLE B (1969) Effects of prostaglandins E_1
    and F_{2a} on systemic, pulmonary and splanchnic circ-
    ulations in dogs. American Journal of Physiology <u>217</u>
    222

17. NAKANO J & KESSINGER J M (1970) Effects of 8-iso PGE_1
    on the systemic and pulmonary circulations in dogs.
    Proceedings of the Society for Experimental Biology
    & Medicine <u>133</u> 1314

18. DUCHARME D W & WEEKS J R (1967) Cardiovascular pharma-
    cology of PGF_{2a}, a unique pressor agent. Proceedings
    of the Nobel Symposium 2 'Prostaglandins' Bergstrom
    S & Samuelsson B Eds. Interscience New York p 173

19. SAID S I (1968) Some respiratory effects of prostaglan-
    dins E_2 and F_{2a}. Prostaglandin Symposium of the
    Worcester Foundation. Ramwell P W & Shaw J E. Eds
    Interscience New York p 207

20. ÄNGGARD E (1965) The isolation and determination of
    prostaglandins in lungs of sheep, guinea pig, monkey
    and man. Biochemical Pharmacology <u>14</u> 1507

21. KARIM S M M, SANDLER M & WILLIAMS E D (1967) Distribution
    of prostaglandins in human tissues. British Journal of
    Pharmacology <u>31</u> 340

22. PIPER P J & VANE J R (1971) The release of prostaglandins
    from lung and other tissues. Annals of the New York
    Academy of Sciences. <u>180</u> 363

23. CLARKE S W, GRAF P D & NADEL J A (1970) The in vivo
    visualisation of small-airway constriction after
    pulmonary microembolism in cats and dogs. Journal
    of Applied Physiology <u>29</u> 646

24. VANE J R (1971) Inhibition of prostaglandin synthesis
    as a mechanism of action for aspirin-like drugs.
    Nature New Biology <u>231</u> 646

25. FARMER J B, FARRAR D G & WILSON J (1974) The effect of
    indomethacin on the tracheal smooth muscle of the
    guinea pig. British Journal of Pharmacology <u>52</u> 559

26. DUNLOP L S & SMITH A P (1975) Reduction of antigen-
    induced contraction of sensitised human bronchus in
    vitro by indomethacin. British Journal of Pharmacology
    <u>54</u> 495

27. SMITH A P (1975) Effect of indomethacin in asthma: Evidence against a role for prostaglandins in its pathogenesis. British Journal of Clinical Pharmacology **2** 307

28. SAMUELSSON B (1973) Quantitative aspects of prostaglandin synthesis in man. Advances in Biosciences, International Conference on Prostaglandins. Bergstrom S Ed. **9** 7

29. NEMOTO T, KONDO T & TOSHIO I (1976) Serum prostaglandin levels in asthmatic patients. Journal of Allergy and Clinical Immunology **57** 89

30. GREEN K, HEDQVIST P & SVANBORG N (1974) Increased plasma levels of 15-keto-13,14-dihydro-prostaglandin F_{2a} after allergen-provoked asthma in man. Lancet **2** 1419

31. MATHÉ A A, STRANDBERG K & ASTROM A (1971) Blockade by polyphloretin phosphate of the prostaglandin F_{2a} action on isolated human bronchi. Nature New Biology **230** 215

12a. PATEL K R (1976) Effect of prostaglandin F_2 alpha on lung mechanics in extrinsic asthma. Postgraduate Medical Journal **52** 275

ASPIRIN-SENSITIVE ASTHMA: ITS RELATIONSHIP

TO INHIBITION OF PROSTAGLANDIN BIOSYNTHESIS

R.J. Gryglewski, A. Szczeklik,
and E. Niżankowska
Departments of Pharmacology and Allergology
and Clinical Immunology, Copernicus
Academy of Medicine in Cracow, Poland

INTRODUCTION

Patients with bronchial asthma do not constitute a homogenous population. One group which can be clearly distinguished on clinical grounds is formed by patients who do not tolerate aspirin. Although such patients have been described soon after introduction of aspirin to pharmacotherapy, it was only recently that aspirin-sensitive asthma has been recognized as a distinct clinical syndrome. In fact the sequence of symptoms and the natural history of this syndrome are so characteristic that, based on large clinical observations, a typical or "classic" case has been constructed (Samter and Beers, 1968).

There are no unique characteristics of these patients during childhood or adolescence. Beginning with the third or fourth decade of life, however, the typical patient may start to experience intense vasomotor rhinitis. Over the period of months chronic nasal congestion appears and physical examination reveals nasal polyps. Bronchial asthma and intolerance to aspirin develop during subsequent stages of the illness. The intolerance presents itself under a unique picture: within minutes to hours following ingestion of aspirin acute asthmatic attacks occur which are sometimes accompanied by rhinorrhea.

Varying figures for incidence of aspirin-sensitive asthma have been reported depending on methods used. When oral challenge studies coupled with measurement of airway resistance were used, the percentage of adult

asthmatic patients who have aspirin-induced bronchospasm
has been estimated as high as 10 - 20% (Van Leeuwen,
1928; McDonald et al., 1972). Other surveys that have
not employed oral challenge techniques or the use of
spirometry have reported lower prevalence of aspirin
intolerance among asthmatic patients, e.g. 4.3% (Chafee
and Settipane, 1974). In our experience, 5 to 10 per
cent of the adult asthmatics in Poland suffer from aspirin-induced asthma. Whatever the precise prevalence figure may be, aspirin intolerance among asthmatic subjects is a clinical association common enough to be of
considerable practical importance (Maur et al., 1974).

        Allergic mechanism of the disorder has been exclu-
ded by numerous and extensive immunological studies
(see Schlumberger et al., 1975). Furthermore, in aspirin-
sensitive patients asthmatic attacks may be precipitated
by certain other analgesics with various chemical structures, which makes immunological cross-activity most
unlikely. Non-immunological concepts which still remain
to be proved are that there is an injury of kinin receptors (Samter and Beers, 1967), activation of complement
system by aspirin (Yurchak et al., 1970), lack of a hypothetical enzyme inhibitor permitting aspirin to activate tissue enzymes (Yurchak et al., 1970), and α/β
adrenergic imbalance (Fischerman and Cohen, 1973). We
put forward a concept that in the sensitive patients
induction of asthmatic attacks by aspirin-like drugs is
due to inhibition of prostaglandin biosynthesis in their
tissues (Szczeklik et al., 1974, 1975a). The present
article evaluates this hypothesis in view of the data
recently accumulated.

BRONCHIAL REACTIVITY TO PGE_2, $PGF_{2\alpha}$ AND HISTAMINE

        Although several substances with potent biological
activity are produced in the respiratory tract, their
importance in regulating the bronchomotor tone in physiology and disease in not well known. Partial information on this subject can be obtained from studies of
bronchial reactivity. In principle, these studies depend
on having a patient inhale an aerosolized substance,
while the effects of this procedure on lung function
are being recorded. Histamine and acetylcholine have
been most often used for demonstration of bronchial
hyperreactivity, a leading feature of asthma, and recently $PGF_{2\alpha}$ has been added to this group (Mathé et
al., 1973). Little attention, however, was given to the
relation between type of asthma and bronchial response
to various stimulants, and no such data were available

for aspirin-sensitive subjects.

One of us has studied the effects of inhaled $PGF_{2\alpha}$, PGE_2 and histamine on bronchial reactivity in 32 aspirin-sensitive asthmatics and compared with 41 asthmatics without this sensitivity (Niżankowska, 1976). The later group was further divided into 22 patients with atopic (extrinsic) and 19 with infectious (intrinsic) type of asthma.

In majority of patients inhalation of PGE_2 in a single dose of 60 μg resulted in improvement of pulmonary function. Better response was often observed in aspirin-sensitive subjects than in the other patients studied; it did not reach, however, the level of statistical significance.

The sensitivity to bronchoconstricting agents as well as their tolerance in individual patients was assessed by the minimum dose of either agent causing 40% reduction in peak expiratory flow from absolute baseline values (Szczeklik et al., 1976c). Atopic patients were characterized by vivid reactivity to low doses of both $PGF_{2\alpha}$ and histamine. In patients with infectious asthma significantly higher doses of both $PGF_{2\alpha}$ and histamine were necessary to induce bronchoconstriction as compared to atopics. Aspirin-sensitive patients responded quickly with bronchial spasm to similar doses of histamine as atopics, but tolerated significantly higher doses of $PGF_{2\alpha}$. The ratio of $PGF_{2\alpha}$/histamine differentiated aspirin-sensitive from the non-sensitive asthmatics ($p < 0.01$) (Fig. 1). Similarly, significant difference was obtained when group of aspirin-sensitive patients was compared separately with either infectious or atopic asthma groups. These results point to differences in bronchial reactivity to $PGF_{2\alpha}$ and histamine depending on type of asthma and suggest that the regulatory role played by these substances might not be the same in various types of asthma.

THE EFFECT OF ANALGESICS ON PULMONARY FUNCTION
IN ASPIRIN-SENSITIVE ASTHMATICS

Over the past 3 years we have performed oral challenge tests with 13 different non-steroidal anti-inflammatory drugs in 90 patients with aspirin-sensitive asthma. These tests, details of which had been described elsewhere (Szczeklik et al., 1975a), consisted of clinical observation and registration of peak expiratory flow (index of bronchial patency) over 4 hours period following an ingestion of the drug studied. Only one drug in one dose was tested on any one day. The tests began with

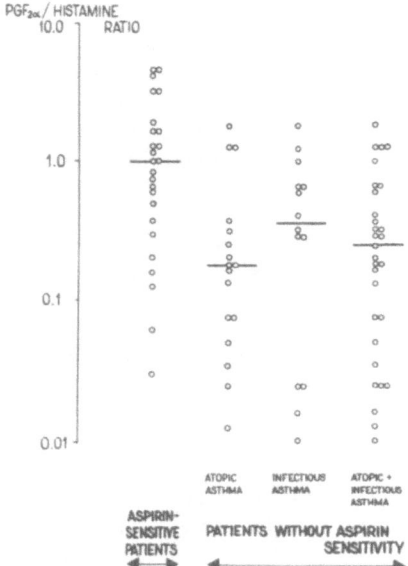

Fig. 1. Bronchial reactivity to $PGF_{2\alpha}$ and histamine in 32 aspirin-sensitive and 41 aspirin-non-sensitive asthmatics. The later group was composed of 22 patients with atopic and 19 with infectious type of asthma. For calculating $PGF_{2\alpha}$ /histamine ratio the minimum doses in ug of either compound causing 40% reduction in peak expiratory flow from absolute baseline values were used. The $PGF_{2\alpha}$ /histamine ratio was significantly higher in aspirin-sensitive patients than in any other group of asthmatic patients.

the smallest dose of the drug, and were repeated with the increasing doses of the same drug on the following days until the reaction become positive or a dose equivalent to that in one tablet was reached. The placebo day varied. Any reactions causing subjective symptoms were relieved immediately with β-adrenergic atimulating drugs as an aerosol inhalation or intravenous aminophylline injection.

    The results obtained revealed that, additionally to indomethacin (Vaneslow, 1967) and mefenamic acid (Samter, 1973) several other non-steroidal anti-inflammatory drugs are able to induce bronchospasm in aspirin-sensitive asthmatic patients. The complete list includes: aspirin, indomethacin, mefenamate, flufenamate, fenoprofen, ibuprofen, diclofenac, naproxen an phenylbutazone. The intensity of bronchospasm induced by these agents varied from patient to patient (vide infra) and

was usually accompanied by profuse rhinorrhea, lacrima-
tion and conjunctival injection.

   The bronchoconstricting property was not, however,
common to all the analgetics studied. Four of them in
a dose equivalent to one tablet failed to cause adverse
reactions in any one patient. These were: salicylamide,
benzydamine, chloroquine, and dextropropoxyphene. In
few aspirin-sensitive patients these drugs were later
on administered over the period of a week in a dose of
3 tablets a day, without any untoward reaction. Thus,
if necessary, these four drugs can be used safely in
patients with aspirin-sensitive asthma.

   The ability of these twelve drugs to inhibit PG
biosynthesis was studied in vitro using bovine seminal
vesicle microsomes (Gryglewski, 1974; Gryglewski et al.,
1976). Aspirin, indomethacin, fenamates, fenoprofen,
ibuprofen, naproxen and phenylbutazone - all were found
PG synthetase inhibitors. Their concentrations which
inhibited the enzymic activity by 50% (IC_{50}) ranged
from 0.1 µM (indomethacin) to 189 µM (aspirin). Only
diclofenac was not studied by us in vitro, however, it
had been found by others (Ku et al., 1975) to be a
strong PG synthetase inhibitor. On the contrary, the
remaining 4 drugs, which were inactive in aspirin-sensitive patients, namely salicylamide, chloroquine, benzydamine and dextropropoxyphene did not inhibit prostaglandin synthetase activity in our experiments ($IC_{50} >$
1.000 µM). Thus, 9 of 13 studied anti-inflammatory
drugs inhibited PG biosynthesis in vitro and only these
nine drugs induced obstruction to airflow and other adverse reactions in aspirin-sensitive patients.

   For accurate estimation of anti-enzymic potency
of the drugs studied, the Q ratio was calculated (Gryglewski, 1974). This ratio is a better approximation of
the expected pharmacological activity in vivo than the
anti-prostaglandin synthetase potency of a drug calculated as its IC_{50} in vitro. A ratio represents the potency of the drug to inhibit PG synthetase corrected
by its affinity to bind to plasma albumin (Gryglewski
et al., 1976). Comparison of the Q ratios with the results of clinical challenge tests revealed pretty good
correlation between the ability of the drugs to induce
bronchoconstriction with their estimated power to inhibit PG synthesis in vivo. Thus, the highest Q ratio
for indomethacin (1400) correlated with the most powerful bronchoconstrictor action of indomethacin in the
clinical setting. On the other hand, phenylbutazone, as
expected from its Q ratio (6.8), showed the lowest degree
of activity in the aspirin-sensitive patients. Q ratios
indicated also the expected order of magnitude of the

threshold doses of aspirin, fenoprofen and ibuprofen
which precipitated asthmatic attacks in aspirin-sensitive patients (Szczeklik et al., 1975a, 1976a).

PATTERNS OF SENSITIVITY TO PG INHIBITORS
AMONG PATIENTS WITH ASPIRIN-INDUCED ASTHMA

The intensity of the adverse reactions induced by
the same PG synthetase inhibitor varied greatly in different patients. Thus, in some patients ingestion of
30 mg aspirin precipitated an open asthmatic attack,
while in the others doses 10 times higher caused only
moderate fall in spirometric values. In order to study
this phenomenon in more detail, minimal doses of aspirin, ibuprofen and fenoprofen causing greater than 15%
reduction in peak expiratory flow were determined in
18 aspirin-sensitive patients (Szczeklik et al., 1976a).
The procedure of the challenge tests was the same as
described above, except that the increase in dosage of
a drug tested proceded at lower rate. In other words,
the drugs were administered over a period of several
days in slightly increasing doses until the exact threshold causing positive reaction was achieved. Individual
pattern of sensitivity to the threshold doses of the
3 drugs tested has been found for each aspirin-sensitive
patient. Thus, patients in whom low doses of aspirin
precipitated adverse reactions also responded to low
doses of fenoprofen and ibuprofen, while those who developed obstruction to airflow only after relatively high
doses of aspirin, also needed high doses of the other
two drugs to respond with bronchospasm. Statistical
analysis confirmed these observations revealing highly
significant positive correlation between the minimum
doses of the three drugs causing positive reactions.
These results, together with similar observations concerning indomethacin and fenamates (Szczeklik et al.,
1975a) indicate that the degree of enzymic inhibition
which is sufficient to precipitate bronchoconstriction
is an individual hallmark of each patient. Knowing the
threshold dose for any of PG synthetase inhibitors in
a patient, one can predict the threshold doses for the
rest of aspirin-like drugs in this particular patient.

GENERATION OF PROSTAGLANDINS BY NASAL POLYPS

Studies were designed to investigate PG generating
system in respiratory tissue of patients with aspirin-
sensitive asthma. To this aim, nasal polyps were used.

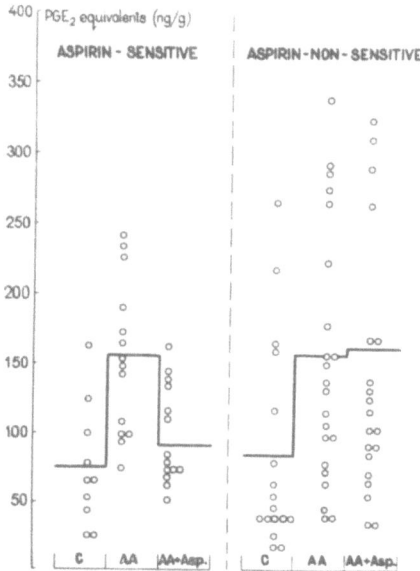

Fig. 2. Release of prostaglandins from polyp fragments in 14 aspirin-sensitive and 22 aspirin-non-sensitive patients with asthma. In each group, the first column (C) represents spontaneous release, (ng PGE_2 equivalents/ g wet tissue) the second (AA) - release following incubation with arachidonic acid (33 µM), and third (AA + Asp) - release following incubation with arachidonic acid (33 µM) and aspirin (167 µM). Horizontal bars represent mean values. Note that the used concentration of aspirin inhibited the AA-stimulated PG-release only in aspirin sensitive patients, but was ineffective in control group.

Several factors spoke in favor of this choice. Nasal polyps, which represent the most available tissue of the respiratory tract, occur frequently in aspirin-sensitive patients, a phenomenon which does not seem to be incidental (Samter, 1973). Moreover, they were proved to be a good biological material to study the release of mediators such as histamine, slow reacting substance of anaphylaxis and eosinophil chemotactic factor (Kaliner et al., 1973).

     Polyps obtained from 14 aspirin-sensitive asthmatics and 22 patients without this sensitivity were studied. The procedure, described in details elsewhere (Szczeklik et al., 1976d), could be summarized as follows: nasal polyps removed without anaesthesia were fragmented with scissors, divided into 3 parts and sus-

pended in Tyrode's solution. The first sample contained
plain Tyrode, the second sodium arachidonate at a concentration of 33 µM, while the third contained aspirin
at a concentration of 167 µM and sodium arachidonate at
a concentration of 33 µM. Following incubation PGs were
bioassayed (Vane, 1964; Gilmore et al., 1968) in the
medium, while the polyp slices were removed, and homo-
genized separately in the solutions composed identically
as those in which incubation was previously carried out.
After 3 min of incubation, PGs concentration was also
bioassayed in boiled homogenates. Thin layer chromatography revealed that PGs produced from arachidonic acid
by nasal polyps comprised mainly of PGE_2 (Szczeklik et
al., 1976d).
     The spontaneous release of PGEs from polyp frag-
ments and PG biosynthesis by polyp homogenates were
essentially similar in aspirin-sensitive and non-sensitive patients (Fig. 2). Arachidonic acid stimulated PG
biosynthesis to the same extent in both groups. However,
the response to aspirin was different. Aspirin was used
in a single concentration of 167 µM, which inhibits PG
synthesis in microsomal preparations less than by 50%
(Gryglewski, this book). In polyp fragments of aspirin-
sensitive patients the above concentration of aspirin
completely inhibited the arachidonate-stimulated release
of prostaglandins, whereas in the control group no inhibition occurred (Fig. 2). Aspirin also more effectively
inhibited PG biosynthesis in polyp homogenates from
aspirin-sensitive patients than in the control group,
although the difference between the groups did not reach
the level of significance. Thus, the enzymic system generating PGEs in polyp fragments of aspirin-sensitive
patients has an increased susceptibility to the inhibitory action of aspirin. It might well be that this phenomenon is not restricted to nasal mucosa, but occurs
also in further part of respiratory tract, namely in
bronchi.

                         DISCUSSION

     We believe the results here presented strongly
suggest that precipitation of the asthmatic attacks by
aspirin-like drugs in sensitive patients is due to the
inhibition of PG biosynthesis in tissues of the respiratory tract of these patients. Several questions, then
come to one's mind, which still await the answer.
     Why does the inhibition of PG biosynthesis brings
about bronchoconstriction? One should assume that the
consequences of the removal of bronchodilator PGEs

dominate over the consequences of the removal of bronchoconstrictor PGFs and TXA_2. Human bronchi were reported to generate predominantly PGEs (Cuthbert, 1973),
also histamine-contracted guinea pig trachea released
mainly PGEs (Grodzińska et al., 1975). The removal of
endogenous PGEs by PG synthetase inhibitors leaves the
effects of endogenous bronchoconstrictor unopposed.
However, if this was the only reason for bronchial obtu-
ration, one could expect that PG-synthetase inhibitors
in doses high enough to block the bronchial enzyme would
lead to bronchoconstriction in healthy and in asthmatic
subjects without aspirin hypersensitivity, which is not
the case. Our general concept (Szczeklik et al., 1975a)
is that PGEs along with β-adrenergic system play the
main defensive role in bronchial asthma, however, aspirin-sensitive patients differ from other asthmatic patients as well as from healthy subjects by relying more
on the PGEs mechanism than on the β-adrenergic one.
In vitro, PGEs stabilize membranes of mastocytes, inhibit their degranulation, and block the release of histamine (Lichtechstein, 1973; Walker, 1973). Similar modulatory action of the PGEs in vivo would consist of attenuation of histamine action on bronchi and of the inhibition of further histamine release from its stores. We
speculate that this modulatory action is of the first
importance for aspirin-sensitive patients with asthma,
whereas aspirin-non-sensitive asthmatics possibly rely
more on their β-adrenergic regulatory mechanisms. Therefore, in aspirin-sensitive patients PG synthetase inhibitors by removing PGEs leave their respiratory tract
defenceless against the action of spasmogens. Our concept
is supported by recent finding (Arroyave et al., 1976)
of significant rise in plasma histamine following aspirin oral challenge in aspirin-sensitive asthmatics; on
the contrary, under similar challenge conditions plasma
histamine level remained unchanged in aspirin-non-sensitive asthmatics and in control subjects.

What could be the reason for greater susceptibility of PG generating system to inhibitory action of aspirin-like drugs in sensitive patients? One explanation
for this phenomenon is an increased cell membrane permeability for aspirin resulted in by a specific, but unknown, infectious agent. Such a concept is in accordance
with clinical picture of this disorder, in which aspirin
reaction is the culminating and often most dramatic feature of the preexisting disease (Samter, 1973). Aspirin
intolerance in many instances develops in people who
have never had any difficulty with aspirin before. Symptoms like rhinorrhea and polyposis usually precede the
onset of aspirin reactions, and clinical course leading

to aspirin intolerance even if aspirin was carefully
avoided (Maur et al., 1974). Also our polyp studies
support this concept.

Another possibility is that in aspirin-sensitive
asthmatics the microsomal prostaglandin synthetase is
more susceptible to inhibition by aspirin. Our polyp
studies do not provide convincing evidence to prove
this point. Such possibility implies abnormality of the
cyclo-oxygenase, which would react more easily with aspirin. The enzymopathy could be the local one (caused
by infectious agent in the respiratory tract) or generalized, due to genetic disorder. It is, therefore, interesting to note that a possible defect in another system caused by aspirin has been suggested recently. In
a small group of aspirin-sensitive patients studied,
low doses of aspirin ranging from 10 to 100 mg were reported to cause a definite prolongation of bleeding
time, while no such effects were noted in aspirin-non-
sensitive subjects (Fisherman and Cohen, 1974). We were,
however, unable to detect any changes in template bleeding time following ingestion of 40 mg aspirin in 21
patients with aspirin-sensitivity as compared to 17
normal subjects (Szczeklik et al., 1975b). Furthermore,
in another study (Schwartz and Bennet, 1973) no difference in platelet aggregation induced by collagen were
found in 7 aspirin-sensitive patients in comparison
with normal subjects. The issue, however, is far from
being settled. Recently Snider and Parker,(1976)have
found an increased susceptibility to aspirin of arachidonic acid oxydizing enzymes in platelets of 3 aspirin
sensitive patients.

This short review of possible mechanisms through
which prostaglandins participate in precipitation of
asthmatics attacks in hypersensitive patients, has
gathered a few, often contradictory hypotheses. Testing
of these hypotheses might provide not only an extended
explanation for pathogenesis of aspirin hypersensitivity,
but also help in better understanding of basic pathoge-
nic mechanisms operating in bronchial asthma.

ACKNOWLEDGEMENTS

This study was supported by NIH Special Foreign
Currency Agreement USA No. 05-082 N, and by The
Wellcome Trust.

REFERENCES

Arroyave, C.M., Stevenson, D.D., Bhat, K.N., and Tan, E.M., 1976, Oral challenge in asthmatic patients: a study of plasma mediators (abstr.). J. Allergy Clin. Immunol. 57: 206.

Chafee, F.H., and Settipane, G.A., 1974, Aspirin intolerance. I. Frequency in an allergic population. J. Allergy Clin. Immun. 53: 193-204.

Cuthbert, M.F., 1973, Prostaglandins and respiratory smooth muscle, in: Prostaglandins: Pharmacological and Therapeutic Advances, (M.F. Cuthbert and J.B. Lippincott, eds.), pp. 252-265, Philadelphia.

Fisherman, E.W., and Cohen, G.N., 1974, Alpha-beta-adrenergic imbalance in intrinsic-intolerance rhinitis or asthma. Ann. Allergy 33: 86-101.

Gilmore, N., Vane, J.R., and Wyllie, J.H., 1968, Prostaglandins released by the spleen. Nature 218: 1135-1140.

Grodzińska, L., Panczenko, B., and Gryglewski, R.J., 1975, Generation of prostaglandin E-like material by the guinea-pig trachea contracted by histamine. J. Pharm. Pharmac. 27: 88-91.

Kaliner, M., Wasserman, S.I., and Austen, K.F., 1973, Immunologic release of chemical mediators from human nasal polyps. N. Engl. J. Med., 289: 277-281.

Ku, E.C., Wasvary, J.M., and Cash, W.D., 1975, Diclofenac sodium (GP 45840, Voltaren), a potent inhibitor of prostaglandin synthetase. Biochem. Pharmac. 24: 641-643.

Lichtenstein, L.M., 1973, The control of IgE-mediated histamine release, in: Asthma, (K.F. Austen, and L.M. Lichtenstein, eds.), p. 91. Academic Press, New York-London.

McDonald, J.R., Mathison, D.A., and Stevenson, D.D., 1972, Aspirin intolerance in asthma. I. Detection by oral challenge. J. Allergy Clin. Immunol. 50: 198-207.

Mathé, A.A., Hedqvist, P., Holmgren, A., and Svanborg, N., 1973, Bronchial hyperreactivity to prostaglandin F_2 and histamine in patients with asthma. Br. Med. J. 1: 193-196.

Maur, K., Adkinson, N.F., Van Metre, T.E., Marsch, D.G., and Norman, P.S., 1974, Aspirin intolerance in a family . J. Allergy Clin. Immun. 54: 380-395.

Niżankowska, E., 1976, The effect of prostaglandins on bronchial reactivity in asthma. Doctors thesis, Copernicus Academy of Medicine, Kraków.

Samter, M., and Beers, R.J., 1967, Concerning the natu-
    re of intolerance to aspirin. J. Allergy 40: 281-
    293.
Samter, M., and Beers, R.J.,Jr., 1968, Intolerance to
    aspirin: clinical studies and consideration to its
    pathogenesis. Ann. Int. Med. 68: 975-983.
Samter, M., 1973, Intolerance to aspirin. Hosp. Pract.
    (Dec.), 8: 85-90.
Schlumberger, H.D., Löbbecke, E.A., and Kallós, P.,
    1974, Acetylsalicylic acid intolerance. Acta Med.
    Scand. 196: 451-458.
Schwartz, H.J., and Bennet, B., 1973, The differential
    effect of acetylsalicylic acid on in vitro aggrega-
    tion of platelets from normal asthmatic and aspi-
    rin-sensitive subjects. Int. Arch. Allergy, 45:
    899-903.
Snider, D.E., and Parker, C.W., 1976, Altered arachido-
    nic acid metabolism in platelets from aspirin-sensi-
    tive subjects. Clin. Res. 450A.
Szczeklik, A., Gryglewski, R.J., and Czerniawska-Mysik,
    G., 1974, Is aspirin-sensitive asthma due to the
    inhibition of prostaglandin biosynthesis? 1974,
    Abstracts of IXth Congress of European Academy of
    Allergy and Clin. Immunology, London, 34.
Szczeklik, A., Gryglewski, R.J., and Czerniawska-Mysik,
    G., 1975a, Relationship of inhibition of prostaglan-
    din biosynthesis by analgesics to asthma attacks in
    aspirin-sensitive patients. Br. Med. J. 1: 67-69.
Szczeklik,A., Musiał, J., and Serwońska, M., 1975b, The
    effect of aspirin on bleeding time and platelet
    aggregation in patients hypersensitive to non-ste-
    roidal anti-inflammatory drugs. Abstracts of XI
    Congress of Pol. Soc. Hematol., Gdańsk, p.64.
Szczeklik, A., Gryglewski, R.J., Czerniawska-Mysik, G.,
    and Zmuda, A., 1976a, Aspirin-induced asthma: hyper-
    sensitivity to fenoprofen and ibuprofen in relation
    to their inhibitory action on prostaglandin genera-
    tion by different microsomal enzymic preparations.
    J. Allergy Clin. Immunol. 58: 10-18.
Szczeklik, A., and Czerniawska-Mysik, G., 1976b, Prosta-
    glandins and aspirin-induced asthma, Lancet 1: 488.
Szczeklik, A., Niżankowska, E., and Niżankowski, R.,
    1976c, Bronchial reactivity to prostaglandins F_2,
    E_2 and histamine in different types of asthma.
    Respiration-subjected for publication.
Szczeklik, A., Gryglewski, R.J., Olszewski, E., Dembińs-
    ka-Kieć, A., and Czerniawska-Mysik, G., 1976d,
    Aspirin sensitive asthma: the effect of aspirin on
    the release of prostaglandins from nasal polyps.
    New. Eng. J. Med. - subjected for publication.

Vane, J.R., 1964, The use of isolated organs for detec-
    ting active substances in the circulating blood.
    Br. J. Pharmac. Chemother. 23: 360-373.
Van Leeuwen, W.S., 1928, Pathognomonische Bedeutung der
    Uebermpfindlichkeit gegen Aspirin bei Asthmatikeren.
    Münch. Med. Woch. 75: 1588-1592.
Vaneslow, N.A., and Smith, J.R., 1967, Bronchial asthma
    induced by indomethacin. Ann. Int. Med. 66: 568-572.
Walker, J.L., 1973, The regulatory function of prosta-
    glandins in the release of histamine and SRS-A from
    passively sensitized human lung tissue, in: Advances
    in Biosciences, Vol. 9, (S. Bergström and S. Bern-
    hard, eds.), p. 235, Pergamon Press, Vieweg-Oxford,
Yurchak, A.M., Wicher, K., and Arbesman, C.E., 1970,
    Immunologic studies on aspirin, J. Allergy 46:
    245-252.

PHYSIOLOGICAL ROLES OF PROSTAGLANDINS IN REPRODUCTION

N. L. Poyser

Department of Pharmacology, University of
Edinburgh, 1 George Square, Edinburgh, EH8 9JZ

MALE

It is now over 40 years since prostaglandins were
detected in secretions of human and sheep vesicular
glands (von Euler, 1936).   Bergström, Samuelsson and
co-workers have since found seminal plasma to contain
many different prostaglandins, with PGE_1 and PGE_2 being
the most abundant in both human and sheep semen.   The
physiological significance of seminal prostaglandins
in reproduction is still not clear.   However, two
reports have shown that seminal plasma from infertile
men, for whom no obvious reason can be found for their
infertility, contains significantly less PGE than
normal (Bygdeman, Fredricsson, Svanborg and Samuelsson,
1970; Collier, Flower and Stanton, 1975).   The
picture has become more complicated recently by the
findings that 19-OH PGE_1 and 19-OH PGE_2 are the major
components of human semen, being present in larger
quantities than PGE_1 and PGE_2 (Taylor and Kelly, 1974;
Jonsson, Middleditch and Desiderio, 1975).   Studies in
hypogonodal men have shown the levels of 19-OH PGE in
semen to be low, but to increase greatly following
treatment with testosterone (Shakkebaek, Kelly and
Corker, 1976).   Prostaglandin levels in human semen
are, therefore, under hormonal control.   19-OH PGE
is also a major component in the semen of other
primates, but is lacking in the semen of sheep and
other domestic and laboratory species (Kelly, Taylor,
Hearn, Short, Martin and Marston, 1976).   With the
exception of the sheep, sub-primate species contain

very little of any prostaglandin in their semen,
(von Euler, 1936; see Poyser, 1974), but they reproduce
successfully despite this lack.

NON-PREGNANT FEMALE

(1)   Menstruation

Pickles and his co-workers established the
presence of prostaglandins in human endometrium and
menstrual fluid.   Endometrial levels and the ratio of
PGF to PGE were higher in the luteal phase than in the
proliferative phase (Pickles, Hall, Best and Smith,
1965).   More recent studies have extended these findings.   Downie, Poyser and Wunderlich (1974) and Singh,
Baccarini and Zuspan (1975) reported mean endometrial
levels of $PGF_{2\alpha}$ and PGE_2 between 10 and 25 ng/100 mg
tissue in the proliferative phase. $PGF_{2\alpha}$ levels rose
during the luteal phase up to 60 to 75 ng/100 mg tissue,
but PGE_2 levels remained lower only increasing at
menstruation.   The ratio of PGF to PGE was near unity
during the proliferative phase but increased to between
2 and 3.5 during the luteal phase.   Levitt, Tabon and
Josimovich (1975) reported levels about 8 times higher,
though they used dry tissue.   Again the ratio of PGF
to PGE was near unity during the proliferative phase,
but increased to 4 during the luteal phase.   Willman,
Collins and Clayton (1976) reported mean levels of
$PGF_{2\alpha}$ of 8.9 and 18.7 ng/100 mg tissue during the
proliferative and luteal phases respectively, with a
large increase at menstruation up to 63.5 ng/100 mg
tissue.   Surprisingly, PGE levels were slightly higher
at the 3 stages studied so that the ratio of PGF to PGE
was always below unity.   In general endometrial prostaglandin levels are higher in the luteal than in the
proliferative phase and, with the exception of one
study, the ratio of PGF to PGE is higher also during
the luteal phase, thus confirming the earlier observations of Pickles et al. (1965).

More recent studies in our laboratory using
radioimmunoassays on individual endometrial samples have
revealed that a wide range of values can be found (Downie,
Simon and Poyser, unpublished results).   This has made
it impossible for us to compare results of samples
obtained from normal patients and patients suffering
from disorders of menstruation to see if any differences
exist.   Willman et al. (1976) report, however, that
endometrial prostaglandin levels are higher than normal
in patients suffering from fibryomata with mennorhagia,

dysmenorrhoea and irregular uterine bleeding.   Flufenamic acid does alleviate the symptoms of severe
primary dysmennorrhoea (Schwartz, Zor, Lindner and Naor,
1974), and reduces menstrual blood loss in patients
with menorrhagia due to dysfunctional bleeding (Anderson,
Haynes, Guillebaud and Turnbull, 1976).   Indomethacin
was also found to alleviate the symptoms of dysmenorrhoea
and to reduce the abnormally high levels of PGF which
occurred in uterine washings (Halbert, Demers, Fontana
and Jones, 1975).   Several disorders of menstruation
are apparently associated with an over production of
prostaglandins and symptomatic relief can be obtained
by the use of prostaglandin synthesis inhibitors, though
the precise role of prostaglandins in menstruation has
still to be elucidated.

(2)   Luteinizing Hormone (LH) Release

Orczyk and Behrman (1972) and Armstrong and
Grinwich (1972) first suggested that prostaglandins
may be involved in LH release from their studies using
prostaglandin synthetase inhibitors.   Further studies
have revealed that:

i) PGE_1 or PGE_2 treatment of rats causes LH release
by acting on the hypothalamus rather than directly on
the anterior pituitary (Harms, Ojeda and McCann, 1973;
Tsafriri, Koch and Lindner, 1973; Spies and Norman,
1973).

ii) $PGF_{2\alpha}$ causes LH release in sheep (Carlson,
Barcikowski and McCrackan, 1973).

iii) Indomethacin decreases plasma LH levels in
ovariectomised rats, and inhibits the release of LH
induced by ovarian steroids in ovariectomised rats and
oestradiol in anoestrous sheep (Sato, Jyujo, Hiroma and
Iesaka, 1975; Ojeda, 1976; Roberts, Carlson and
McCracken, 1976).

iv) This inhibition of LH release by indomethacin is
overcome by LH-releasing hormone (LH-RH) (Ojeda, Harms
and McCann, 1975; Roberts et al., 1976).

v) PGE_2 releases LH-RH in rats (Ojeda, Wheaton and
McCann, 1975; Eskay, Warberg, Mical and Porter, 1975).

vi) Oestradiol-induced LH release from the sheep is
associated with increased $PGF_{2\alpha}$ release from the brain
(Roberts et al., 1976).

All these observations indicate a physiological role for PGE_2 in the rat and $PGF_{2\alpha}$ in the sheep, the prostaglandin acting at the hypothalamic level to cause LH-RH release. This in turn stimulates the release of LH from the anterior pituitary. Follicle stimulating hormone (FSH) release is apparently not controlled in a similar manner.

(3)   Ovulation

The evidence for a physiological role of prostaglandins in ovulation is summarised:

i)   Indomethacin treatment of rats and rabbits blocks ovulation (Orczyk and Behrman, 1972;   Armstrong and Grinwich, 1972;   O'Grady, Caldwell, Auletta and Speroff, 1972).

ii)   This blockade in rats could be overcome by the administration of PGE_2 (Tsafriri, Lindner, Zor and Lamprecht, 1972).

iii)   Amounts of PGF and PGE in rabbit graafian follicles increase significantly after an ovulatory dose of HCG or LH, and after mating, up to the time of ovulation (Yang, Marsh and LeMaire, 1973;   Armstrong, Grinwich, Moon and Zamechik, 1974).

iv)   These increases in follicular prostaglandin levels were prevented if the rabbits were pretreated with indomethacin (Yang et al., 1973).

v)   Ovulation in rabbits was blocked after the intra-follicular injection of indomethacin of $PGF_{2\alpha}$ antiserum (Armstrong et al., 1974).

Although these observations strongly support a physiological role of PGE_2 and/or $PGF_{2\alpha}$ in ovulation, the precise role they play has still to be elucidated. Although indomethacin blocks LH-induced ovulation, it does not prevent luteinization of the follicle and progesterone secretion, indicating that prostaglandins are not involved in these other processes.

(4)   Uterine Luteolytic Hormone

In many sub-primate species, corpus luteum function is terminated (luteolysis) by a hormone secreted from the uterus. This hormone acts in a local manner and is responsible for regulating the length of the oestrous

cycle or pseudopregnancy.   There is much evidence that
this hormone is $PGF_{2\alpha}$ and the whole subject has recently
been extensively reviewed, the main evidence being,
(see Horton and Poyser, 1976, for references):

i) $PGF_{2\alpha}$ is luteolytic in many species of mammal, a
notable exception being the human.

ii) Levels of $PGF_{2\alpha}$ in the uterine venous blood
increase towards the end of the oestrous cycle, just
prior to luteal regression, in the sheep, guinea-pig,
cow and pig, and also the mare (Douglas and Ginther,
1976).

iii) $PGF_{2\alpha}$ reaches the ovary in a local manner due to
a counter-current exchange mechanism which operates
between the uterine vein and ovarian artery, in the
sheep and cow.   How it reaches the ovary in a local
manner in other species has still to be elucidated.

iv) Levels of $PGF_{2\alpha}$ in the endometrium or whole
uterus are higher at the end of the oestrous cycle or
pseudopregnancy than on earlier days in the sheep,
guinea-pig, cow and hamster. $PGF_{2\alpha}$ synthesising
capacity of the guinea-pig uterus is also higher at
the end of the oestrous cycle.

v) Various treatments which lead to premature luteal
regression also result in an earlier release of $PGF_{2\alpha}$
from the uterus.

vi) Indomethacin treatment prevents luteal regression
and extends oestrous cycle length in the guinea-pig
and pseudopregnancy in the rat and rabbit.

vii) Active immunisation against $PGF_{2\alpha}$ extends luteal
function and oestrous cycle length in sheep and guinea-
pigs.   Passive immunisation against $PGF_{2\alpha}$ extends
luteal function and pseudopregnancy in rabbits.

viii) The physiological stimulus for $PGF_{2\alpha}$ synthesis
and release from the uterus is probably a combination
of oestrogen and progesterone released from the ovaries.
Oestradiol also increases the $PGF_{2\alpha}$ synthesising
capacity of the uterus by increasing the amount of
prostaglandin synthetase enzyme present.   Recently an
oxygenated derivative of $PGF_{2\alpha}$ (6-keto-$PGF_{1\alpha}$) has been
identified in incubates of the guinea-pig uterus,
(Poyser, unpublished results).   It was synthesised in
quantities less than $PGF_{2\alpha}$, but its possible involvement
in luteolysis merits investigation.

PREGNANT FEMALE

(1)   Implantation

Most of the studies to date on implantation have
involved work on mice, and can be summarised:

i)   Indomethacin treatment in early pregnancy prevents
implantation.

ii)   This effect is reversed about 60% by the administration of PGE_2 and/or $PGF_{2\alpha}$ (Lau, Saksena and Chang, 1973).

iii)   Implantation can be induced in pregnant mice
ovariectomised on day 3 by replacement therapy with
progesterone and oestradiol.   This can be blocked by
the administration of indomethacin, but this inhibition
can be overcome by treating the mice additionally with
$PGF_{2\alpha}$ and histamine (Saksena, Lau and Chang, 1976).
Saksena et al., (1976) conclude 'it is likely that in
the progesterone dominated uterus an injection of
oestradiol triggers the release of $PGF_{2\alpha}$ and histamine
which are involved in the chain of events leading to
implantation of the blastocyst.'

In rats also, it has been reported that indomethacin
treatment on days 2 or 3 of pregnancy prevents implantation (Chatterjee, 1973).   In addition, the prostaglandin
synthesising capacity of the pseudopregnant rat uterus
is highest on day 5, the day of implantation in the
pregnant rat (Fenwick, Jones, Naylor, Poyser and Wilson,
1976).   Although more PGF than PGE was assayed as being
present, the major compound identified was $6\text{-keto-}PGF_{1\alpha}$.
This raises the question which prostaglandin is actually
involved in the implantation process, although much more
evidence is needed in the mouse, rat and other species
to establish a physiological role for prostaglandins
in implantation.

(2)   Maintenance of Luteal Function

It is important for the luteolytic influence of
the uterus to be negated if an animal becomes pregnant,
since ovarian progesterone is necessary for the
maintenance of pregnancy throughout the whole of
gestation in some species and during the first one-third
in others.   Mechanisms by which this may be achieved
have also been reviewed recently, and the main
when administered near term.

ii)   Indomethacin or aspirin treatment prolongs
gestation in rat, hamster, rabbit and monkey (Aiken,
1972;   Chester, Dukes, Slater and Walpole, 1972;   Lau,
Saksena and Chang, 1975;   Challis, Davies and Ryan,
1975;   Novy, Cook and Manaugh, 1975).

iii)   The passive immunisation of pregnant rats against
$PGF_{2\alpha}$ also prolongs gestation (Dunn, Humphries, Judkins,
Kendall and Knight, 1973).

iv)   The prostaglandin synthesising capacity of the
pregnant rat uterus increases at the end of gestation.

v)   Peripheral plasma levels of 15-keto-13,14-dihydro-
$PGF_{2\alpha}$ increase in women 10 to 30 times during labour.
(Green, Bygdeman, Toppozoda and Wiqvist, 1975).

vi) $PGF_{2\alpha}$ levels in uterine venous plasma increase at
term in the sheep, goat, cow and rat.   The stimulus
for this increase is probably oestrogen, of placental
origin as in the sheep and goat, or from the maternal
ovaries as in the rat.   However, oxytocin may be of
importance in maintaining $PGF_{2\alpha}$ output once labour
has started.

   $PGF_{2\alpha}$ has probably more than one role in the
initiation of parturition.   The cow, pig, rabbit, rat,
hamster and goat are dependent upon corpora lutea for
progesterone throughout gestation, though progesterone
levels must fall just prior to parturition if labour
is to ensue.   It is probable, therefore, that $PGF_{2\alpha}$
released from the uterus acts on the corpora lutea to
terminate their function so to reduce progesterone
output.   Indeed, indomethacin treatment of pregnant
rats near term does prevent the rapid fall in progesterone levels.   In addition, $PGF_{2\alpha}$ may also act directly
on the uterus to stimulate contractions and/or lower
the threshold of the uterus for the stimulant action of
oxytocin in these and other species (see Fig. 1).

   (4)   Other Possible Physiological Roles

   Reported observations in the literature may lead
to the establishment of other physiological roles for
prostaglandins in reproduction:

(i)   Studies in rabbits have shown that prostaglandins
may be involved in egg transport through the oviduct,

observations can be summarised (see Horton and Poyser, 1976 for references):

i)   The early pregnant guinea-pig uterus synthesises and releases much less $PGF_{2\alpha}$ than the non-pregnant uterus around day 15.

ii)   In the sheep, $PGF_{2\alpha}$ release from the uterus after day 13 of pregnancy is very much reduced when compared to non-pregnant sheep at the equivalent time.   These findings were achieved using chronic sampling techniques, but other workers using acute sampling dispute these findings.

iii)   In the non-pregnant cow, pulses of 15-keto-13,14, dihydro-$PGF_{2\alpha}$ occur in the peripheral plasma during the last days of the normal oestrous cycle.   These pulses coincide with a decrease in progesterone levels and no doubt reflect the pulsatile release of $PGF_{2\alpha}$ from the uterus.   In the pregnant animal this increase in $PGF_{2\alpha}$ metabolite level in peripheral plasma is not seen and progesterone levels remain high.   This indicates that $PGF_{2\alpha}$ release from the early pregnant bovine uterus is greatly reduced, thereby allowing luteal maintenance (Kindahl, Edqvist, Ban and Granström, 1976).

In these 3 species, it appears that the embryo or conceptus secretes an anti-luteolytic factor which acts locally on the uterus to inhibit prostaglandin synthesis and release.   Part of its action may be also to inhibit oestradiol output from the ovary. However, other mechanisms may also be involved, including the secretion of a luteotrophic factor by the embryo or conceptus.   There is evidence for this in the sheep and guinea-pig.   It has been known for several years that the placenta of the rat, rabbit and hamster produces a luteotrophin which is essential for luteal maintenance.   Whether any other mechanism is involved in luteal maintenance in these species needs further investigation.

(3)   Parturition

$PGF_{2\alpha}$ has been suspected of having an important role in parturition for several years.   Observations supporting this view are, (see Horton and Poyser, 1976, for references unless otherwise stated):

i) $PGF_{2\alpha}$ will induce parturition in several species

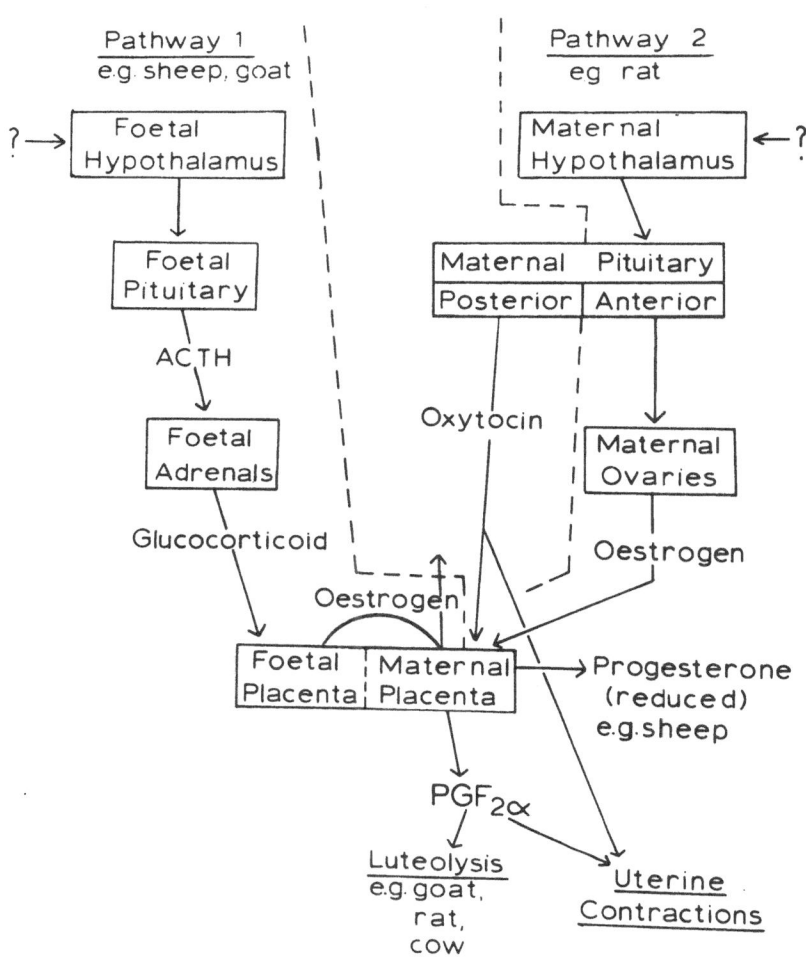

Figure 1.   Mechanisms involved in the release of $PGF_{2\alpha}$ at the end of gestation in several species.   Each pathway is species specific, though oxytocin may be common to both.   Also shown are the possible roles of $PGF_{2\alpha}$ in the initiation of parturition.   Note that the reduction in progesterone output in the pregnant sheep is independent of $PGF_{2\alpha}$.

(Saksena and Harper, 1975; Hodgson, 1976).

(ii) Treatment of pregnant rabbits with indomethacin results in the degeneration and resorption of implanted foetuses, suggesting that prostaglandins may be vital to foetal survival and growth (O'Grady et al., 1972).

(iii) Active immunisation of female rabbits against PGE or PGF did not interfere with ovulation, fertilisation or implantation, but did cause foetal death and abortion.   Apparently the primary effect was directed towards the placenta suggesting that prostaglandins have a role in placental growth, (Elzayat and Stylos, 1974).   A model of a role for prostaglandins in the regulation of placental blood flow has recently been proposed (Rankin, 1976).

(iv) PGE may have a role in the maintenance of patency of the ductus arteriosus during foetal life (Coceani, Olley and Bodach, 1975).

(v) PGG_2, PGH_2 and/or thromboxane A_2 produced locally may be involved in the closure of the umbilical artery at birth (Tuvemo, Strandberg, Hamberg and Samuelsson, 1976).

REFERENCES

AIKEN, J.W. (1972).   Aspirin and Indomethacin prolong parturition in rats: Evidence that prostaglandins contribute to expulsion of foetus.   Nature, (Lond.) 240, 21-25.

ANDERSON, A., HAYNES, P.J., GUILLEBAUD, J. and TURNBULL, A.C. (1976).   Reduction of menstrual blood-loss by prostaglandin synthetase inhibitors.   Lancet 1, 774-776.

ARMSTRONG, D.T. and GRINWICH, D.L. (1972).   Blockade of spontaneous and LH-induced ovulation in rats by indomethacin, an inhibitor of prostaglandin synthesis. Prostaglandins 1, 21-28.

ARMSTRONG, D.T., GRINWICH, D.L., MOON, Y.S. and ZAMECHIK, J. (1974).   Inhibition of ovulation in rabbits by intrafollicular injection of indomethacin and prostaglandin F antiserum.   Life Sci. 14, 129-140.

BYGDEMAN, M., FREDRICSSON, B., SVANBORG, K. and
SAMUELSSON, B. (1970).   The relation between
fertility and prostaglandin content of seminal fluid
in man.   Fert. Steril. 21, 622-9.

CARLSON, J.C., BARCIKOWSKI, B. and McCRACKEN, J.A.
(1973).   Prostaglandin $F_{2\alpha}$ and the release of LH in
sheep.   J. Reprod. Fert., 34, 357-362.

CHALLIS, J.R.G., DAVIES, I.J. and RYAN, K.J. (1975).
The effects of dexamethasone and indomethacin on the
outcome of pregnancy in the rabbit.   J. Endocr. 64,
363-370.

CHATTERJEE, A. (1973).   Some studies on the effects of
prostaglandin $F_{2\alpha}$ and indomethacin in the physiology
of pseudopregnancy and pregnancy in rats.   Proc.
Ind. Nat. Sci. Acad. 39, 408-419.

CHESTER, R., DUKES, M., SLATER, S.R. and WALPOLE, A.L.
(1972).   Delay of parturition in the rat by anti-
inflammatory agents which inhibit the biosynthesis
of prostaglandins.   Nature (Lond.) 240, 37-38.

COCEANI, F., OLLEY, P.M. and BODACH, W. (1975).   Lamb
ductus arteriosus: effect of prostaglandin synthesis
inhibitors on the muscle tone and the response to
prostaglandin E_2.   Prostaglandins 9, 299-308.

COLLIER, J.G., FLOWER, R.J. and STANTON, S.L. (1975).
Seminal prostaglandins in infertile men.   Fert.
Steril. 26, 868-871.

DOUGLAS, R.H. and GINTHER, O.J. (1976).   Concentration
of prostaglandins F in uterine venous plasma of
anaesthetized mares during the estrous cycle and
early pregnancy.   Prostaglandins 11, 251-260.

DOWNIE, J., POYSER, N.L. and WUNDERLICH, M. (1974).
Levels of prostaglandins in human endometrium during
the normal menstrual cycle.   J. Physiol. (Lond.)
236, 465-472.

DUNN, M.V., HUMPHRIES, N.G., Judkins, G.R., KENDALL,
J.Z. and KNIGHT, G.W. (1973).   The effect of
prostaglandin $F_{2\alpha}$ antibody on gestation length in
the rat.   Prostaglandins 3, 509-514.

ELZAYAT, S. and STYLOS, W.A. (1974). The effect of circulating prostaglandin antibodies on reproduction in rabbits with special reference to the placenta. Endocrinology 95, 1642-1648.

ESKAY, R.L., WARBERG, J., MICAL, R.S. and PORTER, J.C. (1975). Prostaglandin E_2-induced release of LH-RH into hypophysial portal blood. Endocrinology 97, 816-824.

FENWICK, L., JONES, R.L., NAYLOR, B., POYSER, N.L. and WILSON, N.H. (1976). Production of prostaglandins by the pseudopregnant rat uterus, in vitro, and the effect of tamoxifen, with the identification of 6-keto-prostaglandin $F_{1\alpha}$ as a major product formed. Br. J. Pharmac. In press.

GRÉEN, K., BYGDEMAN, M., TOPPOZODA, M. and WIQVIST, N. (1975). The role of prostaglandin $F_{2\alpha}$ in human parturition. Amer. J. Obstet. Gynec. 120, 25-31.

HALBERT, D.R., DEMERS, L.M., FONTANA, J. and JONES, D.E.D. (1975). Prostaglandin levels in endometrial jet wash specimens in patients with dysmenorrhoea before and after indomethacin therapy. Prostaglandins, 10, 1047-1056.

HARMS, P.G., OJEDA, S.R. and McCANN, S.M. (1973). Prostaglandin involvement in hypothalamic control of gonadotrophin and prolactin release. Science 181, 760-761.

HODGSON, B.J. (1976). Effects of indomethacin and ICI 46474 administered during ovum transport on fertility in rabbits. Biol. Reprod. 14, 451-457.

HORTON, E.W. and POYSER, N.L. (1976). Uterine luteolytic hormone: A physiological role for prostaglandin $F_{2\alpha}$. Physiol. Rev. 56. In press.

JONSSON, H.T., JR., MIDDLEDITCH, B.S. and DESIDERIO, D.M. (1975). Prostaglandins in human seminal fluid: two novel compounds. Science 187, 1093-1094.

KELLY, R.W., TAYLOR, P.L., HEARN, J.P., SHORT, R.V., MARTIN, D.E. and MARSTON, J.H. (1976). 19-hydroxy-prostaglandin E_1 as a major component of the semen of primates. Nature (Lond.) 260, 554-545.

KINDAHL, H., EDQVIST, L.-E., BAN, A. and GRANSTRÖM, E. (1970). Blood levels of progesterone and 15-keto-13,14-dihydro-prostaglandin $F_{2\alpha}$ during the normal oestrous cycle and early pregnancy in heifers. Acta Endocr. (Copenh.) 82, 134-149.

LAU, I.F., SAKSENA, S.K. and CHANG, M.C. (1973). Pregnancy blockade by indomethacin, an inhibitor of prostaglandin synthesis; its reversal by prosta-glandin and progesterone in mice. Prostaglandins 4, 795-804.

LAU, I.F., SAKSENA, S.K. and CHANG, M.C. (1975). Effects of indomethacin and prostaglandin $F_{2\alpha}$ on parturition in the hamster. Prostaglandins 10, 1011-1018.

LEVITT, M.J., TABON, H. and JOSIMOVICH, J.B. (1975). Prostaglandin content of human endometrium. Fert. Steril. 26, 296-300.

NOVY, M.J., COOK, M.J. and MANAUGH, L. (1974). Indomethacin block of normal onset of parturition in primates. Amer. J. Obstet. Gynec. 118, 412-416.

OJEDA, S.R. (1976). PG role in control for gonadotropin secretion. Prostaglandins and Therapeutics, 2, 3.

OJEDA, S.R., HAMS, P.G. and McCANN, S.M. (1975). Effect of inhibitors of prostaglandin synthesis on gonadotropin release in the rat. Endocrinology 97, 843-854.

OJEDA, S.R., WHEATON, J.E. and McCANN, S.M. (1975). Prostaglandin E_2-induced release of luteinizing hormone - releasing factor. Neuroendocrinology 17, 283-287.

O'GRADY, J.P., CALDWELL, B.V., AULETTA, F.J. and SPEROFF, L. (1972). The effects of an inhibitor of prostaglandin synthesis (indomethacin) on ovulation, pregnancy and pseudopregnancy in the rabbit. Prostaglandins 1, 97-106.

ORCZYK, G.P. and BEHRMAN, H.R. (1972). Ovulation blockade by aspirin or indomethacin. In vivo evidence for a rold of prostaglandin in gonadotrophin secretion. Prostaglandins 1, 3-20.

PICKLES, V.R., HALL, W.J., BEST, F.A. and SMITH, G.N. (1965). Prostaglandins in endometrium and menstrual fluid from normal and dysmenorrhoeic subjects. J. Obst. Gynaec. Br. Commonw. 72, 185-192.

POYSER, N.L. (1974). Some aspects of prostaglandins in reproduction. Biochem. Soc. Trans. 2, 1196-1200.

RANKIN, J.H.G. (1976). A role for prostaglandins in the regulation of placental blood flows. Prostaglandins 11, 343-352.

ROBERTS, J.S., CARLSON, J.C. and McCRACKEN, J.A. (1976). Prostaglandin $F_{2\alpha}$ production by the brain and its role in LH secretion. Advances in Prostaglandin and Thromboxane Research 2, pp. 609-619, ed. B. Samuelsson and R. Paoletti, New York: Raven Press.

SAKSENA, S.K. and HARPER, J.K. (1975). Relationship between concentration of prostaglandin F (PGF) in the oviduct and egg transport in rabbits. Biol. Reprod. 13, 68-76.

SAKSENA, S.K., LAU, I.F. and CHANG, M.C. (1976). Relationship between oestrogen, prostaglandin $F_{2\alpha}$ and histamine in delayed implantation in the mouse. Acta endocrin. (Copenh.) 81, 801-807.

SATO, T., JYUJO, T., HIRONO, M. and IESAKA, T. (1975). Effects of an inhibitor of prostaglandin synthesis on the hypothalamic pituitary system in rats. J. Endocr. 64, 395-396.

SCHWARTZ, A., ZOR, U., LINDNER, H.R. and NAOR, S. (1974) Primary dysmenorrhoea. Alleviation by an inhibitor of prostaglandin synthesis and action. Obstet. Gynec. 44, 709-712.

SHAKKEBAEK, N.E., KELLY, R.W. and CORKER, C.S. (1976). Prostaglandin concentrations in the semen of hypogonadal men during treatment with testosterone. J. Reprod. Fert. 47, 119-121.

SINGH, E.J., BACCARINI, I.M. and ZUSPAN, F.P. (1975) Levels of prostaglandins $F_{2\alpha}$ and E_2 in human endometrium during menstrual cycls. Amer. J. Obstet. Gynec. 121, 1003-1007.

SPIES, H.G. and NORMAN, R.L. (1973).   Luteinizing
  hormone release and ovulation induced by the intraventricular infusion of prostaglandin E_1 into
  pentobarbital blocked rats.   Prostaglandins 4,
  131-141.

TAYLOR, P.L. and KELLY, R.W. (1974).   19-hydroxylated
  E prostaglandins as the major prostaglandins of
  human semen.   Nature (Lond.) 250, 665-667.

TSAFRIRI, A., LINDNER, H.R., ZOR, A. and LAMPRECHT,
  S.A. (1972)   Physiological role of prostaglandins
  in the induction of ovulation.   Prostaglandins 2,
  1-10.

TSAFRIRI, A., KOCH, Y. and LINDNER, H.R. (1973).
  Ovulation rate and serum LH levels in rats treated
  with indomethacin or prostaglandin E_2. Prostaglandins
  3, 461-468.

TUVEMO, T., STRANDBERG, K., HAMBERG, M. and SAMUELSSON,
  B. (1976).   Formation and action of prostaglandin
  endoperoxides in the isolated human umbilical
  artery.   Acta. Physiol. Scand. 96, 145-149.

VON EULER, U.S. (1936).   On the specific vasodilating and plain muscle stimulating substances
  from accessory glands in man and certain animals
  (prostaglandin and vesiglandin).   J. Physiol.
  (Lond.) 88, 213-234.

WILLMAN, E.A., COLLINS, W.P. and CLAYTON, S.G. (1976)
  Studies in the involvement of prostaglandins in
  uterine symptomatology and pathology.   Br. J. Obstet.
  Gynaec. 83, 337-341.

YANG, N.S.T., MARSH, J.M. and LEMAIRE, W.J. (1973).
  Prostaglandin changes induced by ovulatory stimuli
  in rabbit graafian follicles.   The effect of
  indomethacin.   Prostaglandins 4, 395-404.

CLINICAL USE OF PROSTAGLANDINS FOR TERMINATION OF PREGNANCY AND

INDUCTION OF LABOUR

Marc Bygdeman

Department of Obstetrics and Gynecology

Karolinska Hospital, Stockholm, Sweden

INTRODUCTION

Prostaglandin $F_{2\alpha}$ or E_2 are at present used routinely in many countries for termination of second trimester pregnancy. Recent studies indicate that some prostaglandin analogues may be practicable by more convenient routes not only in the second trimester but also during the first trimester of pregnancy. Considerable experience is also available on the clinical usefulness of intravenous administration of $PGF_{2\alpha}$ or PGE_2 and of oral administration of PGE_2 for induction of labour.

FIRST TRIMESTER OF PREGNANCY

Menstrual Induction

Treatment for this purpose has been restricted to the first two to four weeks following the first missed menstrual period. The main reason being the believe that during this period the treatment will result in a high percentage of complete abortion and that the abortion process will not be associated with a heavy bleeding, important prerequisites for a non-surgical and out-patient procedure. Effective doses of prostaglandin will result in an abortion also during the remaining part of the first trimester but an increasing frequency of incomplete expulsion of the conceptus during this period of gestation necessitates surgical intervention anyway in the majority of the patients (1).

Several studies have shown that intravenous or intravaginal administration of PGE_2 or $PGF_{2\alpha}$ for termination of early pregnancy will result in an abortion but that effective doses are associated with a high frequency of side effects which prevented their use. The only clinically useful application of a primary prostaglandin compound for this purpose has been the intrauterine administration of either PGE_2 or $PGF_{2\alpha}$. Further studies in this area have arrived at using prostaglandin analogues to reduce side effects with the intrauterine route and to explore other routes which would allow selfadministration (e.g. oral and vaginal routes). Csapo (2) has reported that intrauterine instillation of 5 mg $PGF_{2\alpha}$ to 100 pregnant patients in whom the mean menstrual delay was 11 days resulted in termination of pregnancy in 97. A similar degree of efficacy was also found by Ragab and Edelman (3) using the same type of treatment and by Karim (4) who injected a mixture of the methyl esters of 15-methyl PGE_2 and 15-methyl $PGF_{2\alpha}$ into the uterine cavity. Equally effective was repeated vaginal administration (3 or 4 times with 3 hours interval) of suppositories containing either 15-methyl $PGF_{2\alpha}$ methyl ester or 16,16-dimethyl PGE_2 (5,6,7) (Table I). The clinical events following the treatment by either route was also similar. Bleeding generally started three to six hours after the initiation of therapy and lasted for one to two weeks. Although some patients described the bleeding as heavy the mean hematocrite did not change significantly (3,6,7).

The differences were more pronounced with regard to side effects. All patients who received intrauterine administration of $PGF_{2\alpha}$ obtained pretreatment with meperedine hydrochloride, atropin and diazepan to minimize side effects. Such pretreatment was not necessary following intrauterine administration of the mixture of the two analogues. Although most of these patients experienced uterine cramps they were painful enough to require analgesics in only 27 out of 142 cases. In most patients oral or rectal analgesics were sufficient to control uterine discomfort following vaginal administration of either 15-methyl $PGF_{2\alpha}$ methyl ester or 16,16-dimethyl PGE_2. Intramuscular injection of meperedine hydrochloride was, however, necessary in approximately 10 per cent of the patients for effective alleviation of pain. Gastrointestinal side effects were encountered by both routes of administration. Despite pretreatment 30 per cent of the patients experienced one or more episodes of vomiting following intrauterine administration of $PGF_{2\alpha}$ (3). The frequency of vomiting and diarrhea seemed significantly reduced if the analogues were used by the same route (4). Approximately 50 per cent of the patients who received vaginal suppositories containing 15-methyl $PGF_{2\alpha}$ methyl ester suffered no gastrointestinal side effects if simultaneous treatment with diphenoxylate chloride was given. The frequency was further reduced if 16,16-dimethyl PGE_2 was used. Clinical signs of pelvic infection or incomplete abortion was only occasionally seen with both routes of administration.

TABLE I

Interruption of very early pregnancy by intrauterine or vaginal administration of prostaglandin

Author & ref.	Compound	Type of administration	Total dose (mg)	No. of pat.	Stage of gestation (weeks)	Bleeding Onset (hrs)	Bleeding Duration (days)	Per cent abortion	Pre-treatment
Csapo (2)	$PGF_{2\alpha}$	single intra-uterine instillation	5	100	5-6	4	8	97	yes
Ragab et al (3)	$PGF_{2\alpha}$	single intra-uterine instillation	5	100	5-8	5.7	14.5	100	yes
Karim et al (4)	methyl ester of 15-methyl PGE_2 and 15-methyl $PGF_{2\alpha}$	single intra-uterine instillation	0.01 + 0.1	142	5-6	2-8	6.2	96	no
Bygdeman et al (6)	15-methyl $PGF_{2\alpha}$ methyl ester	repeated vaginal suppositories	4	63	5-8	4	11.4	100	no
Lundström et al (7)	16-16-di-methyl PGE_2	repeated vaginal suppositories	3	87	5-8	5.2	10.5	95	no

Intrauterine administration of prostaglandins technically resembles that of vacuum aspiration and both methods need premedication with meperedine hydrochloride (8). It therefore appears that intrauterine administration does not offer any major clinical advantages over the operative procedure for termination of early pregnancy. The situation is somewhat different for the vaginal route of administration. The simplicity of the administration and the possibility of selfadministration may under certain circumstances outweigh the disadvantages of the method in comparison to vacuum aspiration.

Preoperative Cervical Dilatation

Different prostaglandins and different routes of administration have been used to achieve partial or full cervical dilatation prior to vacuum aspiration. The time interval between the start of therapy and the operation has varied. One important reason is the normal clinical management of the patient. Late first trimester patients are in Sweden, for instance, admitted to the hospital in the afternoon the day before the operation, a condition which allows a long pretreatment period. In other countries it is preferable if both the pretreatment and the operation could be performed on the same day.

The efficacy of extra-amniotic, intramuscular or vaginal administration of 15-methyl $PGF_{2\alpha}$ free acid or methyl ester has been studied by Toppozada et al (9) and Borell et al (1). In both these studies the vacuum aspiration was performed 18 hours after the start of treatment.

A single extra-amniotic instillation (mean dose 400 μg) or 3 intramuscular injection (300-800 μg per injection) in 67 patients in the 11th to 13th week of gestation induced a satisfactory outcome in 81 per cent of the patients (cervix dilated to 10 mm or more). In the remaining cases the cervix was found at operation to be open for 7-9 mm. Extra-amniotic administration was associated with a low incidence of vomiting (mean of 1 episode per patient) but there was a transient and moderate degree of uterine pain reaction. The intramuscular route was technically more simple and caused less uterine pain but a higher frequency of vomiting and diarrhea constituted a clinical disadvantage (9). Borell et al. (1) administered 15-methyl $PGF_{2\alpha}$ methyl ester vaginally to 58 patients. The mean gestational age of the women was 11 weeks and the majority of the patients was nulliparous. Each patient was given diphenoxylate chloride with each of the first two prostaglandin treatments. All but two patients had cervical dilatation to 10 mm or more at the time of operation 18 hours after the start of treatment. The amount of bleeding measured at operation was in all cases less than 100 ml. The major side effect was pain which could be controlled by the administration of oral or intramuscular analgesics (Table II).

It is natural that a longer period of treatment will be more efficient since the dilatation of the cervix is dependent on the increased uterine contractility. It is, however, possible to achieve at least partial dilatation of the cervix even if the pretreatment period is reduced to three to four hours. Brenner et al. (10) administered one 50 mg $PGF_{2\alpha}$ vaginal suppository to 40 first trimester nulliparae three hours prior to vacuum aspiration. When subjects of similar gestational age were compared $PGF_{2\alpha}$-treated subjects were dilated sufficiently to perform the operation in 55 per cent. The corresponding figure for the control group was 5 per cent. Vomiting and diarrhea did not appear severe enough to limit the clinical practicability of the method. The efficacy of intracervical injections of 200 µg of 16,16-dimethyl PGE_2 p-benzaldehyde semicarbazone has also been investigated. The compound was given to 120 patients in the 6th to 12th week of gestation three to four hours prior to vacuum aspiration. In 75 per cent of the patients adequate cervical dilatation (8-12 mm) was achieved to enable evacuation of the uterus without mechanical dilatation. Eighty-five per cent of the patients were free from any side effects. In the remaining 15 per cent the side effects were diarrhea, vomiting, cold and shivering (4)(Table II).

It is a wellknown fact that primary instrumental evacuation of the uterus in the 10th to 12th week of gestation is associated with an increased complication rate. Late sequele in terms of an increased prematurity rate following forceful mechanical dilatation cannot be disregarded. Preoperative dilatation of prostaglandin will reduce or eliminate the need for mechanical dilatation and be a clinically valuable method especially in primigravidae patients. The choice of drug, dose, route of administration and treatment interval will depend on the clinical routine preferred in different countries.

SECOND TRIMESTER OF PREGNANCY

The first attempt to terminate pregnancy in the second trimester were accomplished by continous intravenous administration of PGE_2 or $PGF_{2\alpha}$. This procedure was fairly successful, yet it was unfortunate that efficacious doses were associated with a high incidence of side effects such as vomiting and diarrhea. To avoid these side effects, prostaglandin was administered into the uterus between the uterine wall and the fetal membranes (extra-amniotic route) aiming at a direct local action and thus a reduction of side effects. A similar local effect could be obtained with the intra-amniotic route of administration.

Further clinical advantages could be obtained if prostaglandin analogues, i.e. 15-methyl $PGF_{2\alpha}$ or 16,16-dimethyl PGE_2 were used, due to the higher potency and longer duration of action of the analogue compared to the parent compound.

TABLE II

Dilatation of the cervix by prostaglandins prior to vacuum aspiration

Author & ref.	Compound	Route of administration	Dose (mg)	No. of pat.	Time from start of treatment to operation (hours)	Degree of cervical dilatation (mm)		Amount of operative bleeding (ml)		
						7-9	>10	≤100	100-500	≥500
Toppozada et al (9)	15-methyl PGF$_{2\alpha}$	Extra-amniotic	0.4	45	18	7	38	32	11	2
		Intra-muscular	3 x 0.3-0.8	22	18	6	16	18	4	0
Borell et al (1)	15-methyl PGF$_{2\alpha}$ methyl ester	Vaginal	4 x 1.0 4 x 1.5	58	18	2	56	58	0	0
Karim (4)	16,16-di-methyl PGE$_2$ ester	Intra-cervical	0.2	120	3-4	87	22	Not indicated		
Brenner et al (10)	PGF$_{2\alpha}$	Vaginal	50	40	3	25	15	Not indicated		

The availability of prostaglandin analogues has also made possible the development of less invasive methods for second trimester abortion while maintaining safety with a high success rate and minimum side effects. Both 16,16-dimethyl PGE_2 and 15-methyl $PGF_{2\alpha}$ methyl ester have been found effective to induce abortion following vaginal administration.

Extra-amniotic Administration

Prostaglandins possess the unusual property of acting locally, not only after endogenous release, but also unlike the majority of other drugs after exogenous administration. A radiological study revealed that a small amount of $PGF_{2\alpha}$ solution injected into the lower uterine segment induced local contractions, which forced the pool of solution upwards to spread between the uterine wall and the fetal membranes (11). This mechanism is believed to facilitate the development of forceful, reasonably coordinated contractions that result in expulsion of the conceptus and may explain why fundal (=high intrauterine) administration does not offer any particular advantage over instillation of the drug into the isthmic region via a Foley catheter.

In the initial clinical trials, two-hourly doses of 750 μg $PGF_{2\alpha}$ or 200 μg PGE_2 were instilled via a transcervically placed extra-amniotic Foley catheter. The balloon of the cathether was filled with 20-30 ml saline to reduce leakage of prostaglandin via the cervix and expulsion of the catheter before completion of the abortion. Embrey et al reported an 85% success rate with $PGF_{2\alpha}$ and 93% with PGE_2 within a period of 36 hours (12). The mean induction abortion interval was 19.4 hours and 24.9 hours respectively with the two compounds. Bygdeman et al obtained 46 abortions out of 50 patients (92%) in the second trimester of pregnancy with $PGF_{2\alpha}$ and a mean induction abortion interval of 24.2 hours (13)(Table III). The frequency of side effects following extra-amniotic administration is low, also compared to that following intra-amniotic injection. Occasional episodes of vomiting, diarrhea and slight temperature elevation were noted. Similar results have been obtained by several investigators using slightly different doses and time intervals between the injections.

The major limitations of the method are the need for repeated instillations and the inconvenience of an indwelling catheter which also harbours the risk for potential intrauterine infection. These drawbacks could be eliminated by the use of a single injection technique allowing for removal of the catheter after the instillation of a suitable dose of a prostaglandin analogue which has a more long lasting stimulatory effect on uterine contractility.

TABLE III

Results of extra-amniotic administration of prostaglandins for termination of midtrimester pregnancy in selected studies

Author and Reference	No. of Patients	Compound	Dose per Injection (μg)	Injection Interval (hours)	Mean Induction Abortion Interval (hours)	Success Rate (%)
Bygdeman et al (13)	50	$PGF_{2\alpha}$	(250)–750	2	24.2	92 [x]
Embrey et al (12)	93	$PGF_{2\alpha}$	(250)–750	2	24.9	85 [x]
Embrey et al (12)	70	PGE_2	200	2	19.3	93 [x]
Wiqvist et al (15)	55	15-methyl $PGF_{2\alpha}$	730	Single dose	13.6	84 [xx]
Karim (4)	316	2a,2b-dihomo 15-methyl $PGF_{2\alpha}$ methyl ester	1000	Single dose	14.3	82 [xx]

[x]   Abortion within 36 hours
[xx] Abortion within 24 hours

A prostaglandin analogue seems more suitable and safer than a high single dose of $PGF_{2\alpha}$ or PGE_2 for this purpose. However, a recent report indicates that the disadvantage of a single dose technique using PGE_2 may be overcome by incorporating the compound into a high viscosity solution (14).

The results of two analogue studies are presented in Table III to exemplify the clinical characteristics of this method. One single injection of 500-780 µg 15-methyl $PGF_{2\alpha}$ (mean dose 730 µg) induced abortion in 46 out of 55 second trimester patients. The mean induction abortion interval was 13.6 hours (15). Karim found that 1 mg of the analogue 2a, 2b dihomo 15-methyl $PGF_{2\alpha}$ methyl ester given as one single extra-amniotic injection resulted in an abortion in 82% of the patients with a mean induction abortion interval of 14.3 hours (4). It was found in both studies that in the majority of those patients who failed to abort within 24 hours the cervix had dilated sufficiently to enable the evacuation of the uterus by vacuum aspiration or to finalize the abortion process by one or two intramuscular injections of 15-methyl $PGF_{2\alpha}$. The frequency of side effects were low and comparable to that following repeated administration of the primary compounds.

It could be concluded that although the extra-amniotic route of administration could be used at any gestational age, it appears most valuable during the late first and early second trimester (11-15 weeks gestation) when the uterus has become to large for saction curettage yet is still to small for intra-amniotic injection.

Intra-amniotic Administration

The intra-amniotic route of administration is the method mostly used for interruption of pregnancy after 14th week of gestation when the amniotic cavity could easily be punctured. The initial metabolism of intra-amniotically administered $PGF_{2\alpha}$ is similar to that following intravenous administration although considerably slower. The prostaglandins are also slowly transferred across the fetal membranes into the maternal plasma. The half life time of $PGF_{2\alpha}$ in the amniotic fluid ranged from 13.5-20 hours (16;17). One of the analogues, 15-methyl $PGF_{2\alpha}$ disappeared more slowly and its half life time was 27-31 hours. The slow disappearence rate following intra-amniotic administration is most likely the explanation for the prolonged duration of uterine stimulation following intra-amniotic injection of prostaglandin in comparison with e.g. extra-amniotic instillation.

Repeated intra-amniotic injections of natural prostaglandins requires the maintenance of a catheter for administration of the second and subsequent doses, but is highly efficacious without causing a disturbing range of side effects. In a multicentre study performed by the WHO Prostaglandin Task Force, intra-amniotic administr-

TABLE IV

Results of intra-amniotic administration of prostaglandins for termination of midtrimester pregnancy in selected studies

Author and Reference	No. of Patients	Compound	Dose per Injection (mg)	No. of Doses	Injection Interval (hours)	Mean Induction Abortion Interval (hours)	Success Rate (%)
WHO Prostaglandin Task Force (18)	717	$PGF_{2\alpha}$	25	2	6	19.7	85.4
Brenner et al (20)	40	$PGF_{2\alpha}$	50	1	–	19.1	95.0
Karim et al (19)	40	PGE_2	5	1–4	10	14.0	95.0
Wiqvist et al (21)	50	15-methyl $PGF_{2\alpha}$	2.5	1	–	18.8	98.0
Karim (4)	290	15-methyl $PGF_{2\alpha}$	1.0	1	–	20.2	90.0

ation of 25 mg $PGF_{2\alpha}$ repeated after six hours resulted in an abortion in 85.4% of the cases within 48 hours (18). The mean induction
abortion interval was 19.7 hours (Table IV). Several individual investigators have reported the success rate of 90-95% with this dose
schedule.

PGE_2 has not been as widely used as $PGF_{2\alpha}$ by the intra-amniotic
route. Karim et al could interrupt pregnancy in 38 out of 40 women
given 10-hourly intra-amniotic injections of 5 mg PGE_2 for a maximum of four doses. The mean time to abortion was 14 hours.

Single dose schedules allow for immediate withdrawal of the
catheter used for the injection. However, in order to attain an
acceptable rate of efficacy, 90% or more abortions within 48 hours,
high doses of natural prostaglandin (40-50 mg $PGF_{2\alpha}$) need to be
given. Brenner et al (20) injected 50 mg $PGF_{2\alpha}$ as a single intra-
amniotic dose in 40 patients. Without supplementary therapy the
success rate was 95% within 48 hours and the mean induction abortion interval 19.1 hours.

Due to the longer duration of effect, the 15-methyl analogues
of $PGF_{2\alpha}$ seem to be more suitable for a single injection procedure
also by the intra-amniotic route. Wiqvist et al (21) administered
intra-amniotically a single dose of 1.0, 2.5 or 5 mg 15-methyl $PGF_{2\alpha}$
and achieved abortion in 46, 98 and 95% of the patients. The authors
felt that 2.5 mg of the analogue was the optimal dose and that this
dose was significantly more effective and caused fewer side effects
than a single dose of 40 mg $PGF_{2\alpha}$. Karim (4) has also reported
high   efficacy and low frequency of side effects with a single
intra-amniotic injection of 15-methyl $PGF_{2\alpha}$ free acid or methyl
ester. Karim claimed, however, that in contrast to the results presented by Wiqvist et al even 1 mg of the analogue will result in
a high success rate (90%) and that the lower dose is associated with
a reduced frequency of side effects.

Comparison between Hypertonic Saline and Prostaglandin

Hypertonic saline injected intra-amniotically has for many years
been the common method for termination of second trimester abortion.
The WHO Prostaglandin Task Force finished in 1975 a multicentre
randomized investigation comparing 200 ml of 20% hypertonic saline
with two intra-amniotic injections of 25 mg $PGF_{2\alpha}$ given at six hour
intervals (18). The study comprised a total of 1513 patients. In
the prostaglandin group 614 of 717 or 85.4% aborted within 48 hours
following the initial prostaglandin injection. With hypertonic saline the corresponding figures were 641 of 716 patients or 80.5%
($p < 0.05$). The mean induction abortion interval was significantly
shorter in the prostaglandin group compared with the saline group,
19.7 versus 30.4 hours ($p < 0.001$). The difference in the cumulative
abortion rate was most pronounced after 24 hours when approximately

65% of the prostaglandin treated patients had aborted compared with
some 20% of the patients who had received hypertonic saline. The
frequency of complete abortion was the same in the two groups
(66.4% for prostaglandin and 68.2% for hypertonic saline). The
average number of episodes of vomiting per patient was 1.5 for
prostaglandin and 0.4 for hypertonic saline. The corresponding
figures for diarrhea were 1.4 and 0.1 respectively. Both differences
were highly significant.

    Based on this study and information available in the literature
(for ref. see 22;23) the advantages and disadvantages of intra-
amniotic administration of $PGF_{2\alpha}$ in comparison with hypertonic
saline:
Advantages with $PGF_{2\alpha}$
Higher success rate within 48 hours
Shorter induction abortion interval
Lower risk for coagulation disorders
Lower risk for myometrial necrosis
Rapid metabolism if injected intravenously by mistake
Reduced frequency of maternal deaths.

Disadvantages with $PGF_{2\alpha}$
Higher frequency of gastrointestinal side effects
Increased risk for cervical lacerations.

Vaginal Administration

    Efforts have been directed towards the development of new
methods that are less invasive and therefore easier to administer
for midtrimester abortion while maintaining safety with a high
success rate and a minimum of side effects. Therefore the oral and
vaginal routes of administration have been investigated. By the oral
route 16,16-dimethyl PGE_2 and its methyl ester have so far not given
acceptable clinical results in that the success rate was only 60%
(24;25).

    The uterine stimulatory effect of prostaglandins following
vaginal administration is largely depending on their absorption into
the systemic circulation and is therefore usually associated with a
high frequency of side effects mainly in terms of vomiting and diarrhea similar to that following intravenous infusion.

    Most investigators agree that vaginal administration of repeat-
ed doses of $PGF_{2\alpha}$ or PGE_2 results in satisfactory rates of abortion
and that serious complications are rare. However, some investigators
have concluded that the incidences of side effects are too high for
the method to be preferred method of abortion (22;26;27;28).
Administration of vaginal PGE_2 suppositories (20 mg every three to
five hours) may be useful for inducing abortion in patients with fetal death in utero, late missed abortion or large hydatiform moles (28).

The clinical usefulness of the vaginal route has however, recently been improved by the use of synthetic analogues with reduced gastro-intestinal stimulant activity. For both 15-methyl $PGF_{2\alpha}$ methyl ester and 16,16-dimethyl PGE_2 free acid or p-benzaldehyde semicarbazone ester promising results have been reported (4;29).

A total of 85 patients in the second trimester of pregnancy were given either 16-16-dimethyl PGE_2 (generally 1 mg every three hours) or 15-methyl $PGF_{2\alpha}$ methyl ester (1-2 mg every three hours in most cases) by the vaginal route. Both treatment schedules were equally effective resulting in an abortion rate of 97% within 30 hours. Only occasional patients aborted later than 24 hours after the initiation of therapy. Especially if 16,16-dimethyl PGE_2 was used the frequency of gastrointestinal side effects was low. The mean frequency of episodes of vomiting and diarrhea per patient was 0.7 and 0.3 respectively, which is comparable to the frequency found for the extra-amniotic route and lower than that following intra-amniotic administration of $PGF_{2\alpha}$ (29).

Karim obtained equally good results in 30 second trimester patients using suppositories containing 0.5 mg 16,16-dimethyl PGE_2 p-benzaldehyde semicarbazone ester given every four hours until abortion. All patients aborted in a mean time of 19.3 hours with minimum side effects (4).

The treatment by the vaginal route has recently been further simplified by the development of a long-acting suppository. One suppository containing 3.0-3.5 mg 15-methyl $PGF_{2\alpha}$ methyl ester in 2.2 or 2.5 g of the base Witepsol E-76 was given to 25 second trimester patients. Twenty-three of the patients aborted within 24 hours following one suppository (30).

If these promising results are verified in larger studies it is likely that the vaginal route will become the method of choice especially for the termination of early second trimester pregnancies.

INDUCTION OF LABOUR

In 1968, Karim et al (31) first reported on the successful use of $PGF_{2\alpha}$ for the induction of labour at term. A great number of reports have since then confirmed the value of $PGF_{2\alpha}$ and PGE_2 as oxytocic agents (for references see 32).

Intravenous Administration of $PGF_{2\alpha}$ or PGE_2

Many different dose schedules have been used. It is most common to increase dose levels at constant time intervals approximately every 30 minutes, until labour is established. As a rule, the infusion level at which labour progressed was kept constant until delivery if not uterine hyper stimulation, fetal bradycardia or dis-

turbing maternal side effects occurred. For $PGF_{2\alpha}$ generally an
initial dose of 2.5 µg/min has been administered. The dose was then
gradually increased up to 40 µg/min (33;34;35;). The corresponding
dose for PGE_2 has been 0.1 µg/min up to 4.0 µg/min respectively
(36;37;38).

Intravenous infusion of PGE_2 and $PGF_{2\alpha}$ as a rule result in
uterine contractions indistinguable from that of normal labour or
labour induced by oxytocin. If hypertonus and/or polysystole are
more common following prostaglandin than following oxytocin is a
matter of controversy. Because these types of contractile pattern
were also recorded during normal labour. Karim (36) considered them
as nonspecific for prostaglandins. However, Andersson et al (33)
who studied in a double blind fashion $PGF_{2\alpha}$, PGE_2 and oxytocin given
for induction of labour to 169 parous women found nine episodes of
uterine hypotonus all of which occurred in patients receiving
prostaglandin ($PGF_{2\alpha}$, seven episodes; PGE_2, two episodes) and the
author felt that the safety margin of $PGF_{2\alpha}$ in terms of uterine
response was not as wide as that for oxytocin.

TABLE V

Comparison of intravenous PGE_2 and oxytocin for induction
of labour in selected studies

Author and Reference	Number of Patients		Success Rate %	
	PGE_2	Oxytocin	PGE_2	Oxytocin
Karim (36)	100	100	96	56
Beazley et al (37)	146	146	73	73
Brown et al (38)	53	53	94	100

TABLE VI

Comparison of intravenous $PGF_{2\alpha}$ and oxytocin for induction
of labour in selected studies

Author and Reference	Number of patients		Success Rate %	
	$PGF_{2\alpha}$	Oxytocin	$PGF_{2\alpha}$	Oxytocin
Andersson et al (33)	91	51	81	84
Vakhariya et al (34)	50	50	94	96
Spellacy et al (35)	115	107	75	66

The overall success rate varied considerably beteen studies (Table V and VI). The most important variable is the state of the cervix. Andersson et al (33) found that both success rate and the mean induction delivery interval was closely related to the Bishop score. Vakhariya and Sherman (34) obtained a success rate of 88% in patients with an unripe cervix comparing to 100% in those with a ripe cervix. Spellacy et al (35) obtained 40% success in patients with a modified Bishop score of 0-3 compared to a 100% success rate in those with scores of 10-13. The majority of investigators agree, however, that PGE_2 and $PGF_{2\alpha}$ are equally effective as oxytocin for labour induction at term.

Maternal side effects are mainly vomiting and venous erythema at the site of infusion. With $PGF_{2\alpha}$ asthmatic attacks have been reported in occasional cases. With the doses of PGE_2 and $PGF_{2\alpha}$ used for induction of labour normally no effect on maternal cardiovascular function has been observed.

Oral Administration

Both PGE_2 and $PGF_{2\alpha}$ administered orally stimulate uterine contractions sufficiently to induce labour at or near term. Because this route may be more acceptable to the patient and requires less nursing attention than an intravenous infusion it has been extensively studied during recent years. It soon became obvious that oral administration of $PGF_{2\alpha}$ was associated with an unacceptable frequency of gastrointestinal side effects (44). The results with PGE_2 has, however, been more encouraging.

In most trials PGE_2 tablets are given in increasing doses, starting with 0.5 mg. The dose is then stepwise increased to 2.0 or 3.0 mg. The interval between the administration is generally one to two hours. Contractions are generally produced 15-30 minutes after the first dose of oral PGE_2 and gradually increased in frequency and intensity as in normal labour. It may be suspected that the dosage of orally adminstered oxytocics may be more difficult to adjust and therefore more often produces uterine hyperstimululation. Thiery found, however, that hypertonus was not more common with prostaglandins given orally than with intravenous oxytocin (45). The success rate has been reported to be around 85-95% depending on type of patient and criteria for success (39;40;41;42;43). Randomized comparisons between oral administration of PGE_2 and intravenous infusion of oxytocin indicate that both methods are equally effective. (Table VII).

With regard to side effects only those from the gastrointestinal tract are regularly mentioned. In most instances this type of side effect does not greatly affect the patient but the incidence of vomiting was higher following oral administration of PGE_2 than during intravenous infusion of the compound or during spontaneous labour.

TABLE VII

Oral administration of PGE_2 for induction of labour in selected studies

Author and Reference	No. of patients	Single Doses (mg)	Interval between Doses (hours)	Success Definition	Rate %
Karim et al (39)	764	0.5-2.0	2	Vaginal de-livery with-in 48 hours	90
Craft (40)	80	0.5-3.0	2	Satisfactory progress of labour	89
Filshie (41)	100	0.5-3.0	2	Vaginal de-livery with-in 48 hours	98
Thiery et al (42)	97	0.5-2.0	2	Full dilatat-ion within 24 hours	97
Elder et al (43)	70	0.5-2.0	2	In active la-bour within 12 hours	87

Thiery et al (46) have recently summarized the possible effects of prostaglandin induced labour on the fetus and concluded that only one risk factor consisting in the impairment of fetal oxygenation as a consequence of myometrial hyperstimulation has been identified. No specific adverse effects upon fetus or neonate were noted. Long term follow-up of children born following induction with intravenous $PGF_{2\alpha}$ showed normal psycho-motor development. Placental function was not modified.

It might be concluded that intravenous infusion of PGE_2 or $PGF_{2\alpha}$ and oral administration of PGE_2 appear as safe and as effective as intravenous oxytocin for the induction of labour at term. Oral prosta-glandin E_2 has the advantage of simplicity that is associated with an increased frequency of gastrointestinal side effects although the patients' acceptance is generally good.

REFERENCES

1. Borell, U., Bygdeman, M., Leader, A., Lundström, V. and Martin, J.N. Successful first trimester abortion following the use of 15(S)15-methyl prostaglandin $F_{2\alpha}$ methyl ester vaginal suppositories. Contraception 13:87-94, 1976.

2. Csapo, A.I. "Prostaglandin impact" for menstrual induction. Population Rept. Ser. G, No. 4, March p.p. 33-44, 1974.

3. Ragab, M.J. and Edelman, D.A. Early termination of pregnancy. A comparative study of intrauterine prostaglandin $F_{2\alpha}$ and vacuum aspiration. Prostaglandins 11:275-285, 1976.

4. Karim, S.M.M. Singapore experience with prostaglandin. Routine use and recent advances. In Obstetrics and Gynecological uses of Prostaglandins. (Ed. S.M.M. Karim) Eurasia Press, Singapore 1976, p.p. 127-154.

5. Bygdeman, M., Martin, J.N., Eneroth, P., Leader, A. and Lundström V. Outpatient postconceptional fertility control with vaginally administered 15(S)15-methyl $PGF_{2\alpha}$ methyl ester. Am. J. Obstet. Gynecol. 124:495-498, 1976.

6. Bygdeman, M., Martin, J.N., Leader, A., Lundström, V., Ramadan, M., Eneroth, P. and Green, K. Early pregnancy interruption by 15(S)15-methyl prostaglandin $F_{2\alpha}$ methyl ester. Obstet. Gynecol. 1976, in press.

7. Lundström, V., Bygdeman, M., Fotio, S. and Green, K. Abortion in early pregnancy by vaginal administration of 16,16-dimethyl PGE_2. To be published, 1976.

8. Edelman, D.A., Brenner, W.E. and Goldsmith, A. Menstrual regulation in four countries. IPPF Med. Bull. Vol 8, No. 6, Dec. 1974.

9. Toppozada, M., Bygdeman, M., Papageorgiou, C. and Wiqvist, M. Administration of 15-methyl prostaglandin $F_{2\alpha}$ as a preoperative means of cervical dilatation. Prostaglandins 4:371-379, 1973.

10. Brenner, W.E., Dingfelder, J.R., Staurovsky, L.G. and Hendrichs, C. Vaginally administered $PGF_{2\alpha}$ for cervical dilatation in nulliparas prior to suction curettage. Prostaglandins 4:829-836, 1973.

11. Wiqvist, N., Beguin, F., Bygdeman, M., Fernström, I. and Toppozada, M. Induction of abortion by extra-amniotic prostaglandin administration. Prostaglandins 1:37-53, 1972.

12. Embrey, M.P., Hillier, K. and Makendran, P. Termination of pregnancy by extra-amniotic prostaglandins and the synergistic effect of oxytocin. Adv. in the Biosciences 9:507-513, 1973.

13. Bygdeman, M., Beguin, F., Toppozada, M. and Wiqvist, N. Intrauterine administration of prostaglandin $F_{2\alpha}$ for induction of abortion. Adv. in the Biosciences 9:525-531, 1973.

14. MacKenzie, I.Z., Hillier, K. and Embrey, M.P. Single extra-amniotic injection of PGE_2 in viscous gel to induce midtrimester abortion. Brit. Med. J. 1:240-242, 1974.

15. Wiqvist, N., Bygdeman, M., Papageorgiou, C. and Toppozada, M. Intrauterine administration of prostaglandin by the extra-amniotic route. Prostaglandins 6:193-205, 1974.

16. Green, K., Bygdeman, M. and Wiqvist, N. Kinetic and metabolic studies of prostaglandin $F_{2\alpha}$ administered intra-amniotically for induction of abortion. Life Sciences 14:2285-2297, 1974.

17. Pace-Asciak, C., Wolfe, L.S., Gillet, R.G. and Kinch, R.A. Disappearence of prostaglandin $F_{2\alpha}$ from human amniotic fluid after intra-amniotic injection. Prostaglandins 1:469-477, 1972.

18. WHO Prostaglandin Task Force. Comparison of intra-amniotic prostaglandin $F_{2\alpha}$ and hypertonic saline for induction of second trimester abortion. Brit. Med. J. 1:1371-1376, 1976.

19. Karim, S.M.M., Sharma, S.D. and Filshie, G.M. Termination of second trimester pregnancy with intra-amniotic administration of prostaglandins E_2 and $F_{2\alpha}$. In he Prostaglandins-Clinical Application in Human Reproduction (Ed. E. Southern) Futura Press, Mount Kisco, N.Y. 1972, p.p. 403-416.

20. Brenner, W.E., Dingfelder, J.R., Hendricks, C.H. and Staurovsky,L. Induction of therapeutic abortion with a single dose of intra-amniotically administered prostaglandin $F_{2\alpha}$. Prostaglandins 4:485-498, 1973.

21. Wiqvist, N., Bygdeman, M. and Toppozada, M. Intra-amniotic prostaglandin administration - A challenge to the currently used methods for induction of midtrimester abortion. Contraception 8:113-131, 1973.

22. Brenner, W.E. The current status of prostaglandins as abortifacients. Am. J. Obstet. Gynecol. 123:306-328, 1975.

23. Karim, S.M.M. and Amy, J.J. Interruption of pregnancy. In Prostaglandins and Reproduction (Ed. S.M.M. Karim) MTP Press Ltd. Lancaster, p.p. 78-148, 1975.

24. Wiqvist, N., Martin, J.N., Bygdeman, M. and Green, K. Prostaglandin analogues and uterotonic potency: A comparative study of seven compounds. Prostaglandins 9:255-269, 1975.

25. Karim, S.M.M., Sivasamboo, R. and Ratnam, S.S. Abortifacient action of orally administered 16,16-dimethyl prostaglandin E_2 and its methyl ester. Prostaglandins 6:349-354, 1974.

26. Bolognese, R.J. and Corson, S.L. Prostaglandin E_2 vaginal suppository as an early second trimester abortifacient. Obstet. Gynecol. 43:104-108, 1974.

27. Beguin, F., Bygdeman, M., Toppozada, M. and Wiqvist, N. The response of the midpregnant human uterus to vaginal administration of prostaglandin suppositories. Prostaglandins 1:397-405, 1972.

28. Southern, E.M. Experience in the United States Clinical Trials of Prostaglandins in Obstetrics and Gynecology. In Obstetric and Gynecological uses of Prostaglandins. (Ed. S.M.M. Karim) Eurasia Press, Singapore 1976, p.p. 105-118.

29. Bygdeman, M. and Bergström, S. Prostaglandins as abortifacients. In Obstetric and Gynecological uses of Prostaglandins (Ed. S.M.M. Karim) Eurasia Press, Singapore 1976, p.p. 67-81.

30. Bygdeman, M., Ganguli, A., Kinoshita, K., Lundström, V., Green, K. and Bergström, S. Development of a vaginal suppository suitable for single administration for interruption of second trimester pregnancy. Contraception. In press.

31. Karim, S.M.M., Trussel, R.R., Patel, R.C. and Hillier, K. Response of pregnant human uterus to prostaglandin $F_{2\alpha}$ induction of labours. Brit. Med. J. 4:621-623, 1968.

32. Thiery, M. and Amy, J.J. Induction of labour with Prostaglandins. In Prostaglandins and Reproduction (Ed. S.M.M. Karim). MTP Press Ltd, Lancaster 1975, p.p. 149-229.

33. Andersson, G.G., Hobbins, J.C., Speroff, L. and Caldwell, B.V. Intravenous prostaglandins E_2 and $F_{2\alpha}$ and syntocinon for the induction of term labour. In The Prostaglandins Clinical Applications in Human Reproduction (Ed. E.M. Southern) Futura Press, Mount Kisco, N.Y. 1972, p.p. 85-94.

34. Vakhariya, V.R. and Sherman, A.I. Prostaglandin $F_{2\alpha}$ for induction of labour. Amer. J. Obstet. Gynecol. 113:212-222, 1972.

35. Spellacy, W.N., Gall, S.A., Shevach, A.B. and Holsinger, K.K. The induction of labor at term. Comparisons between prostaglandin $F_{2\alpha}$ and oxytocin infusions. Obstet. Gynecol. 41:14-21, 1973.

36. Karim, S.M.M., Patel, R.C., Sharma, S.D. and Trussel, R.R. Two years experience of labour J. Asian Fed. Obstet. Gynaecol. 2:1-6, 1971.

37. Beazley, J.M. and Gillespie, A. Double blind trial of prostaglandin E_2 and oxytocin in induction of labour. Lancet 1:152-155,1971.

38. Brown, A.A., Hamlett, J.D., Hibbard, M.B. and Howe, P.D. Induction of labour by amniotomy and intravenous infusion of oxytocic drugs - A comparison between prostaglandins and oxytocin. J. Obstet. Gynaecol. Brit. Cmwlth 80:111-115, 1973.

39. Karim, S.M.M. and Sharma, S.O. Oral administration of prostaglandin E_2 for the induction of acceleration of labour. In The Prostaglandins Clinical Applications in Human Reproduction. (Ed. E.M. Southern) Futura Press, Mount Kisco, N.Y. p.p. 207-217, 1972.

40. Craft, I.L. Oral prostaglandin E_2 and amniotomy for induction of labor. Adv. Biosciences 9:593-598, 1973.

41. Filshie, G.M. Labor induction with oral prostaglandin E_2. In The Prostaglandins Clinical Applications in Human Reproduction (Ed. E.M. Southern) Futura Press, Mount Kisco, N.Y. p.p. 223-226, 1972.

42. Thiery, M., Yo Le Gian, A., de Hemptinne, D., Derom, R., Martens, G., Vankets, J. and Amy, J.J. Induction of labour with prostaglandin E_2 tablets. J. Obstet. Gynaecol. Brit. Cmwlth. 81:303-306, 1974.

43. Elder, M.G. and Stone, M. Induction of labour by low amniotomy and oral administration of a solution compared to a tablet of prostaglandin E_2. Prostaglandins 6:427-432, 1974.

44. Barr, W. Induction of labour by prostaglandin E_2. In The Prostaglandins Clinical Applications in Human Reproduction. (Ed. E.M. Southern) Futura Press, Mount Kisco, N.Y. p.p. 219-222, 1972.

45. Thiery, M. Elective induction of labour at term with oxytocin
    and prostaglandins. Techniques and fetal and maternal effects.
    In Avortement et Parturition Provoqués (Eds. M.J. Bose,
    R. Palmer and C. Surean) Masson, Paris p.p. 267-287, 1974.

46. Thiery, M. and Amy, J.J. Recinatal effects of prostaglandin used
    for induction of labour. In Obstetric and Gynecological uses of
    prostaglandins (Ed. S.M.M. Karim) Eurasia Press, Singapore
    p.p. 31-54, 1976.

INTRODUCTION TO CIRCULATORY SYSTEM

F. Berti and C. Omini

Institute of Pharmacology and Pharmacognosy

School of Pharmacy, University of Milan

20129 Milan, Italy

The Prostaglandins are a family of naturally occurring acidic lipids which possess a variety of biological activities (1). The cardiovascular actions of
these compounds, originally described as vasodepressors,
have been widely investigated and the description of
their vasomotor activity is today highly complicated by
the differences not only among the various Prostaglandins
but also among animal species and often vascular beds in
the same animal (2).

A large number of reports illustrate that Prostaglandins of the E and A series are potent vasodepressors
when administered intravenously in normotensive animals.
These Prostaglandins inducing peripheral arteriolar dilation lead to a fall in peripheral resistance and to a
reflex increase in heart rate which in turn is responsible of the secondary increment in cardiac output (3,4).
When Prostaglandins of the E or A series are injected
intra-arterially an increase in blood flow in different
regional vascular beds of the coronary, femoral, mesenteric, renal vessels has been shown (5,6). The peripheral
arterial dilation induced by the above cited Prostaglandins does not appear to be connected with cholinergic or
adrenergic nerve endings and the real mechanism is still
unknown (7). However the fall in systemic pressure is
primarily due to an arteriolar vasodilation of the
splanchnic region since, following PGE_1 administration,

the time of maximal depression parallels with a significant rise in mesenteric arterial blood flow, while the
femoral or renal blood flow does not change (4,8). The
increase in heart rate brought about by PGE administration
is abolished by β-adrenergic blocking agents and therefore
this reflex phenomenon should be considered sympathetic
in nature (9).

The biochemical mechanism of action of Prostaglandins in dilating the peripheral arterioles is still an
open question, however Strong and Bohr (10) suggest a
decreased cell membrane stability secondary to a reduction in ionic calcium at the binding sites caused by
Prostaglandins. Kadar and Sunahara (11) on the contrary,
proposed a different mechanism of action based on the
Na^+ - K^+ dependent ATPase availability for PGE_1 activity
on vascular tissue. These Authors in fact demonstrated
that after pretreatment of the tissue with ouabain, PGE_1
loses its inhibiting activity on spontaneous contraction
of isolated canine mesenteric vein and artery.

In contrast to Prostaglandins of the E and A series,
$PGF_{2\alpha}$ has been shown to be a pressor agent in most species (rat and dog), with the exception of the cat and
rabbit, in which it is depressor. The experimental evidence suggests that the pressor activity of $PGF_{2\alpha}$ is a
consequence of venoconstriction (12). Mark et al. (13)
in this subject showed the capability of $PGF_{2\alpha}$ to induce
contraction of isolated colonic and mesenteric veins. As
far as cerebral blood flow is concerned there is a large
agreement on cerebral vasoconstrictor activity of $PGF_{2\alpha}$
(14,15), while the effect of Prostaglandin E on cerebral
circulation is still in dispute (16). The content of $PGF_{2\alpha}$
in cerebrospinal fluid is elevated during human cerebral
vasospasm indicating a possible role of this Prostaglandin in mediating such a syndrome (17).

Also the pulmonary circulation is affected in different way by the various Prostaglandins. Kadowitz et
al. (18,19) working on the canine pulmonary vascular
bed demonstrated that $PGF_{2\alpha}$ and PGB_2 were potent pulmonary pressor compounds. These Prostaglandins increase
pulmonary vascular resistance by constricting lobar veins
and small arteries. PGE_1 and PGA_1 on the contrary were
found to induce dilation of pulmonary vascular bed decreasing the resistance of veins and upstream vessels.
PGE_2 and PGA_2 nevertheless were found in similar experiments to increase lightly the resistance of the blood
flow in the lung. Secher and Andersen (20) measuring
central hemodynamics and regional lung function before

*and after infusion of $PGF_{2\alpha}$ in healthy women were able
to show a significant increase in both arterial and wedge
pressure and vasoconstriction of the arterial and venous
side of the pulmonary vascular bed.*

*The discovery that the kidney medulla possesses
Prostaglandins with potent antihypertensive and natri-
uretic activities proposed several roles for endogenously formed Prostaglandins in renal homeostasis control
(21). More recently McGiff (22) pointed out that kinins
increase Prostaglandin synthesis which in turn modifies
the renal effect of kinins suggesting a key role for
these coupled polipeptide-prostaglandin system on the
genesis on the hypertensive state. PGD_2 as a product of
renal Prostaglandin synthetase could also be very important on the modulation of vasoactive hormones and neurotransmitters in the kidney.*

*The cardiovascular activity of the new described
Prostaglandins C_2, D_2, G_2 and H_2 is complicated not only
by species differences but also by the variability of
the pressor responses which may be biphasic or triphasic
(23,24). Jones (25) described in detail a number of PGD
and PGE analogues on the cardiovascular system of the
sheep and evidentiated two distinct receptors in terms
of pressor and depressor activity. The endoperoxide
intermediates of Prostaglandin biosynthesis PGG_2 and
PGH_2, which were found to induce a more potent contraction of vascular tissue (26), may be of physiological
importance in spite of their short half-lives. In fact
if the synthesis of these intermediates could occur at
or near their site of action, as suggested for the primary prostaglandins, a more significant and fine regulation and control of hormonal activity in the cardiovascular area could be achieved.*

*Some of the pharmacological actions of Prostaglandins
may assume therapeutical meaning in many cardiovascular
diseases. The first prostaglandin of the PGA class to
be infused into patient with diastolic essential hypertension was PGA_2 isolated from the kidney as medullin
(27). A decrease in blood pressure associated with an
increase in cardiac output was observed. During the infusion of PGA_2 a marked diuresis occurred with-out side
effects at the gastrointestinal level. The availability
of PGA_1 and PGA_2, biosynthetically obtained (28), allowed
more systhematic investigations in order to evaluate the
anti-hypertensive properties of these compounds on their
relationships with the renal function and cardiovascular*

events (29). These studies confirmed the previous results
obtained with medullin and established that the anti-
hypertensive effect of PGA_1 and PGA_2 in patients with
essential hypertension is due to a peripheral vasodilation
leading to a fall in total resistance associated to a
secondary increase in heart rate and cardiac output (30).
However long-term therapy with these Prostaglandins is not
practical because of the route of administration (intravenously) and because of their short duration of action.
Some natural prostaglandins possess an indirect (31) anti-
arrhytmic property (32) and preliminary clinical trial
with $PGF_{2\alpha}$ in disorders of the cardiac rhythm is encouraging (33).

In conclusion prostaglandins have many cardiovascular
effects, they may play physiological roles as hormones
locally produced and may regulate kidney function and
blood flow in different organs and tissue, furthermore
a mechanism by which either renomedullary PGA_2 or PGE_2
might function as an intrarenal antihypertensive "hormone"
has been proposed (34).

REFERENCES

1. Bergström, S., Carlson, L.A., and Weeks, J.R. (1968).
   The Prostaglandins: A family of biologically active
   lipids. Pharmacol. Rev. 20, 11
2. Nakano, J. (1973). Cardiovascular actions. In: The
   Prostaglandins. Ed. Peter W. Ramwell, Plenum Press,
   pag. 239, vol. 1.
3. Nakano, J., and McCurdy, J.R. (1968). Hemodynamic
   effects of prostaglandins E_1, A_1 and $F_{2\alpha}$ in dogs.
   Proc. Soc. Exptl. Biol. Med. 128, 39.
4. Nakano, J., and McCurdy, J.R. (1967). Cardiovascular
   effects of prostaglandin E_1. J. Pharmacol. Exptl.
   Therap., 156, 538
5. Hauge, A., Lunde, P.K.M., and Waaler, B.A. (1967).
   Effects of prostaglandin E_1 and adrenaline on the
   pulmonary vascular resistance (PVR) in isolated rabbit
   lungs. Life Sci. 6, 573
6. Daugherty, R.M., Jr. (1971). Effects of iv and ia
   prostaglandin E_1 on dog forelimb skin and muscle blood
   flow. Am. J. Physiol., 220, 392
7. Smith, E.R., McMorrow, J.V., Covino, B.G., and Lee,
   J.B. (1968). Studies on the vasodilator action of
   prostaglandin E_1. In: Prostaglandin Symposium of the
   Worcester Foundation for Experimental Biology.
   Eds. P.W. Ramwell and J.E. Shaw, p. 259
   Interscience Publishers, New York.

8. Covino, B.G., Lee, J.B., and McMorrow, J.V. (1968). Circulatory effects of prostaglandins. Circulation 38, 60A (Suppl. 6).

9. Carlson, L.A., and Orö, L. (1966). Effect of prostaglandin E_1 on blood pressure and heart rate in the dog. Acta Physiol. Scand. 67, 89.

10. Strong, C.G., and Bohr, D.F. (1967). Effect of prostaglandins E_1, E_2, A_1 and $F_{1\alpha}$ on isolated vascular smooth muscle. Am. J. Physiol. 213, 725.

11. Kadar, D., and Sunahara, F.A. (1969). Inhibition of prostaglandin effects by ouabain in the canine vascular tissue. Can. J. Physiol. Pharmacol. 47, 871.

12. DuCharme, D.W., Weeks, J.R., and Montgomery, R.G. (1967). Studies on the mechanism of the hypertensive effect of prostaglandin $F_{2\alpha}$. J. Pharmacol. Exptl. Therap. 160, 1.

13. Mark, A.L., Schmid, P.G., Eckstein, J.W., and Wendling, M.G. (1971). Venous responses to prostaglandin $F_{2\alpha}$. Am. J. Physiol. 220, 222

14. Emerson, T.E., Radawski, D., Veenendaal, M., and Daugherty Jr., R.M. (1974). Effects of cerebral ventricular, systemic, and local administration of prostaglandin $F_{2\alpha}$ on canine cerebral hemodynamics. Prostaglandins 8, 521-530.

15. Welch, K.M.A., Spira, P.J., Knowles, L., and Lance, J.W. (1974). Effects of prostaglandins on the internal and external carotid blood flow in the monkey. Neurology 24, 705-710.

16. White, R.P. (1975). Role of prostaglandins in cerebrospinal tone. Prostaglandins 9, 405-407.

17. LaTorre, E., Patrono, C., Fortuna, A., and Grossi-Belloni, D. (1974). Role of prostaglandin $F_{2\alpha}$ in human cerebral vasospasm. J. Neurosurg. 41, 293-299.

18. Kadowitz, P.J., Joiner, P.D., and Hyman, A.L. (1975). Physiological and pharmacologic roles of prostaglandins. Annu. Rev. Pharmacol. 15, 285-306.

19. Kadowitz, P.J., Joiner, P.D., Greenberg, S., and Hyman, A. (1976). Comparison of the effects of Prostaglandins A, E, F, and B on the canine pulmonary vascular bed. In: Advances in Prostaglandin and Thromboxane Research. vol. 1, Eds. B. Samuelsson and R. Paoletti, Raven Press, New York, pag. 403-415.

20. Secher, N.J., and Andersen, L.H. (1976). Change in the central hemodynamic and the regional lung function in pregnant women during Prostaglandin $F_{2\alpha}$. In: Advances in Prostaglandin and Thromboxane Research. vol. 2, Eds. B. Samuelsson and R. Paoletti, Raven Press, New York, pag. 914.

21. McGiff, J.C., Crowshaw, K., and Itskovitz, H.D.
    (1974). Prostaglandins and renal function. Fed. Proc.
    $\underline{33}$, 39-47.

22. Terragno, N.A., Malik, K.V., Nasjletti, A., Terragno,
    D.A., and McGiff, J.C. (1976). Renal prostaglandins.
    In: Advances in Prostaglandin and Thromboxane
    Research, vol. 2, Eds. B. Samuelsson and R. Paoletti,
    Raven Press, New York, pag. 561-571.

23. Hedqvist, P., Strandberg, K., Hamberg, M., and
    Samuelsson, B. (1974). Some actions of prostaglandin
    endoperoxides on airway and vascular smooth muscle.
    Scand. J. Resp. Dis. (Suppl.) $\underline{88}$, 53

24. Jones, R.L., Kane, K.A., and Ungar, A. (1974).
    Cardiovascular actions of prostaglandin C in the cat
    and dog. Br. J. Pharmacol. $\underline{51}$, 157-160.

25. Jones, R.L. (1976). Cardiovascular actions of prosta-
    glandins D and E in the sheep: evidence for two
    distinct receptors. In: Advances in Prostaglandin and
    Thromboxane Research, vol. 1, Eds. B. Samuelsson and
    R. Paoletti, Raven Press, New York, pag. 221-230.

26. Tuvemo, G., Strandberg, K., Hamberg, M., and
    Samuelsson, B. (1974). Effects of prostaglandins E_1
    and E_2 and prostaglandin endoperoxides on the isolated
    human umbelical artery. Pediatr. Res. $\underline{8}$, 908

27. Lee, J.B. (1967). Chemical and physiological proper-
    ties of renal prostaglandins, the antihypertensive
    effects of medullin in essential hypertension. In:
    Prostaglandins. (Proc. Nobel Symp. 2), Eds. S.
    Bergström and B. Samuelsson, p. 197, Almgrist
    and Wiksell, Stockholm.

28. Daniels, E.G., and Pike, J.E. (1968). Isolation of
    prostaglandins. In: Prostaglandin Symposium of the
    Worcester Foundation for Experimental Biology.
    Eds. P.W. Ramwell and J.W. Shaw, p. 379
    Interscience Publishers, New York.

29. Westura, E.E., Kannegiesser, H., O'Toole, J.D., and
    Lee, J.B. (1970). Antihypertensive effects of prosta-
    glandin A_1 in essential hypertension. Circ. Res. $\underline{27}$,
    131   (Suppl. 1).

30. Carr, A.A. (1970). Hemodynamic and renal effects of
    a prostaglandin, PGA_1, in subjects with essential
    hypertension. Am. J. Med. Sci. $\underline{259}$, 21

31. Förster, W., Borbola, J., Papp, J.G., Schrör, K.,
    and Szekeres, L. (1974). Cardiac effects of prosta-
    glandins E_2 and $F_{2\alpha}$. Arch. Int. Pharmacodyn. Ther.
    $\underline{211}$, 133-140.

32. Mentz, P., and Förster, W. (1974). Über antiarrhyth-
    mische Wirkungen der Prostaglandine $F_{2\alpha}$, E_2 und A_2
    gegenuber $CaCl_2$ und Akonitinarrhythmien der Ratte.
    Acta Biol. Med. Ger. $\underline{32}$, 393-402.

33. Mann, D., Meyer, H.G., and Förster, W. (1973).
    Preliminary clinical experience with the anti-
    arrhythmic effect of $PGF_{2\alpha}$. Prostaglandins $\underline{3}$, 905-912.
34. Lee, J.B. (1973). Renal homeostasis and the Hyper-
    tensive state: A unifying Hypothesis. In: The Prosta-
    glandins. vol. 1, Ed. P.W. Ramwell, Plenum Press.
    pag. 133

POLYPEPTIDES: VASCULAR ACTIONS AS MODIFIED BY PROSTAGLANDINS

P. Y-K Wong, J. C. McGiff and A. Terragno

Departments of Pharmacology and Medicine

University of Tennessee, Memphis, Tennessee   38163

The general conclusion that prostaglandins are primarily local or tissue hormones (1), which exert their effects at or near
sites of synthesis, has particular significance for the vasculature. Thus, the capacity of blood vessels to synthesize prostaglandins intramurally (2) enables these local hormones to influence vascular tone and reactivity directly by affecting the vasoconstrictor actions of angiotensin and catechol amines (3) and by
modulating the release of norepinephrine from adrenergic nerves
(4). The proposal that prostaglandins participate in the regulation of vascular reactivity is supported by the following observations: 1) enhanced vascular reactivity to pressor stimuli occurs in organs with low basal rates of prostaglandin synthesis and
after inhibition of prostaglandin synthesis in organs with high
biosynthetic capacity (5); and 2) exogenous PGE_2 reversibly inhibits the vasoconstrictor activity of pressor stimuli (6).

The relationship between the vasoconstrictor action of a
pressor hormone in an organ and the prostaglandin biosynthetic
capacity of that organ has been studied in the uterus by Terragno
et al.(7). In late pregnancy in the chloralose-anesthetized dog,
intravenous infusion of angiotensin II, at a rate which elevated
mean aortic blood pressure by 20 mmHg, increased uterine blood
flow and concomitantly increased uterine venous efflux of a PGE
compound by two- to threefold (Figure 1). Under acute experimental conditions the concentration of prostaglandins of the E series
in uterine venous blood in late pregnancy is high, ca. 0.4 - 1.0
ng/ml, levels which are similar to those found in renal venous
blood in anesthetized-laparotomized dogs (8). In the non-gravid
state and in early pregnancy (before the fifteenth day), when
prostaglandins could not be detected in uterine venous blood, the

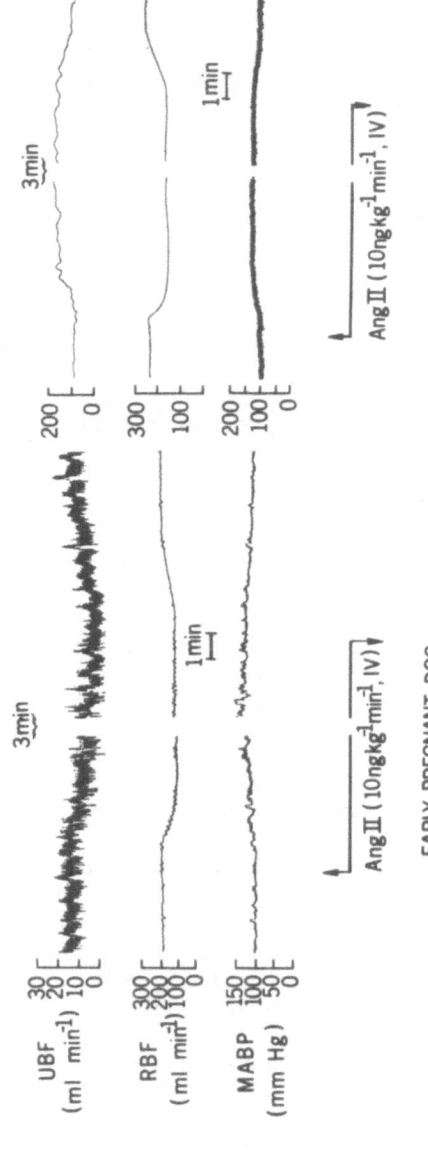

Fig. 1.   Effects of intravenous (IV) administration of angiotensin II in chloralose-anesthetized dogs, on uterine blood flow (UBF), renal blood flow (RBF) and mean aortic blood pressure (MABP) in early (left) and late (right) pregnancy.

canine uterine vascular bed constricted in response to angiotensin. As the assay method for PGE and PGF compounds permits detection of 0.02 ng/ml of blood (9), the uterine venous levels of prostaglandins in the non-pregnant animal and in early pregnancy are negligible. Further, administration of either angiotensin or bradykinin, each a potent stimulus to prostaglandin release (10,11), did not evoke efflux of prostaglandins from the uterus in early pregnancy. Thus, the capacity of the uterus to synthesize prostaglandins seems to protect this organ against the vasoconstrictor effect of pressor hormones; in this respect the uterine vasculature resembles that of the kidney (12). These studies may also contribute to an understanding of the development of tachyphylaxis to the vascular effects of pressor hormones. For those tissues capable of responding to vasoconstrictor hormones by increasing prostaglandin synthesis, tachyphylaxis to the pressor hormone derives in part from stimulation of prostaglandin production presumably within the vascular wall (2); one or more of the newly generated products of prostaglandin synthetase then opposes the pressor action of the polypeptide. Although within the kidney PGE_2 appears to subserve this modulatory antipressor function (13), in extrarenal tissues other products of prostaglandin synthetase such as the intermediates in the biosynthetic pathways, the endoperoxides, may act as modulators (14). Moreover, there are other products of renal prostaglandin synthetase, e.g., PGD_2, which could be as important as PGE_2 to the regulation of the renal effects of one or more vasoactive hormones and neurotransmitters.

VASCULAR SYNTHESIS OF PROSTAGLANDINS

As prostaglandins synthesized in the walls of the major resistance blood vessels modify the vasoconstrictor responses to hormonal and neural stimuli and, thereby, peripheral resistance, we sought to define the prostaglandin biosynthetic capabilities of blood vessels. In this study (2) bovine mesenteric blood vessels were selected because: 1) the demonstration of extrarenal vascular synthesis of prostaglandins is germane to the proposal that one or more products of prostaglandin synthetase participate in the regulation of vascular reactivity (13); 2) the splanchnic circulation, of which the mesenteric vasculature is a large component, contributes importantly to peripheral resistance (15); and 3) as relatively small amounts of prostaglandins are released, we selected a species which would provide abundant vascular tissue. We included observations on veins as well as arteries in view of the importance of venous tone in regulating cardiac output (15). For example, the earliest hemodynamic change in renovascular hypertension, increased cardiac output (16), may result from primary changes in venomotor tone.

The mean rate of synthesis of prostaglandins by mesenteric

arteries was 216 ng/g wet weight after 1 hr incubation. A comparable rate was achieved by mesenteric veins: 186 ng/g after 1 hr. Similar rates of synthesis were found in main (inside diameter {ID} 3-5 mm) and small (ID 0.5 - 1.5 mm) arteries and veins. Twice as much PGE was produced as PGF by both arteries and veins under control conditions, although when stimulated, mesenteric arteries and veins responded differently. Thus, PGE was released from arteries whereas PGF was released from veins by the kinin. The implication of this finding will be considered later. The addition of the prostaglandin synthetase inhibitor meclofenamate (1.0 mM) decrease the rate of synthesis by 90% in both arteries and veins. The prostaglandin biosynthetic capacity of bovine mesenteric blood vessels is high, as indicated by greater than 20% conversion of added substrate, arachidonic acid, to PGE_2 and $PGF_{2\alpha}$. Comparable biosynthetic rates occur only in the seminal vesicles, renal medulla, lung, and urinary bladder (17). A much lower capacity has been reported for the aorta; about 1% to 2% substrate was converted to a PGE compound (18). We did not recover prostaglandins from the walls of the mesenteric blood vessels. This finding supports the concept that prostaglandins are not stored in tissues, i.e., once synthesized they are released into the extracellular fluid. The demonstration of prostaglandin synthesis within the walls of blood vessels complements previous studies showing the presence of renin (19) as well as adrenergic fibers (20) in vascular tissue. Thus, the major elements of pressor and antipressor systems are located in blood vessels where the resultant of their opposing activities, as influenced by sodium, potassium, and calcium ions, determines vascular reactivity. Further, the demonstration of prostaglandin synthesis within the walls of blood vessels, including the major resistance vessels, raises alternative mechanisms whereby products of prostaglandin synthesis affect peripheral resistance without achieving the status of a circulating hormone. In this regard the evidence which supports a role for PGA_2 as a circulating hormone is weak, being based primarily on radioimmunoassay of questionable specificity (21). Further, there is no biochemical evidence which indicates that PGA_2 can be synthesized (22).

PROSTAGLANDINS: PROHYPERTENSIVE IN THE RAT?

There is an important difference in species response observed by Malik and McGiff (23) which has raised major questions concerning the suitability of the rat as a model for investigating the role of vasodepressor systems in regulating blood pressure. Prostaglandins of the E series were shown to augment the renal vasoconstrictor responses of the isolated perfused kidney of the rat to nerve stimulation in concentrations which did not affect vascular tone (100 pg/ml) and to constrict the renal vasculature in higher concentrations. Under identical conditions in the isolated perfused kidney of the rabbit, PGE compounds inhibited adrenergic vasoconstriction and dilated blood vessels. Concentrations of PGE_2 in the perfusing medium as low as 20 pg/ml frequently reduced the vasoconstrictor response to nerve stimulation of the rabbit kidney. At this concentration it is appropriate to invoke the basal state and to consider the physiological relevance of these observations to the regulation of the renal circulation. The anomalous response of the isolated kidney of the rat to PGE_2 takes on greater significance in view of two additional observations. First, *in vivo* PGE_2 can be shown to constrict the renal vasculature of the rat albeit in very high doses. In this regard it should be noted that the renal vascular bed of the rat is peculiarly resistant to PGE_2 when compared to other species as dog and rabbit. However, as noted previously in terms of a proposed role for PGE_2 as a modulator of adrenergic neurotransmitter release, the effects of prostaglandins as modulators may be demonstrated at concentrations one-two hundred and fiftieth (1/250) those which have a direct vascular action. It is this activity as modulators, whereby small amounts of prostaglandins influence the action of hormones for periods in excess of their known biological activity, that compels our attention, particularly in terms of their interactions with vasoactive hormones. The second observation which forces consideration of the significance of the paradoxical vascular response to PGE_2 in the rat isolated kidney is that the principal product of prostaglandin synthetase has the same effect as administered PGE_2. Thus, enhanced synthesis of renal prostaglandins evoked by arachidonic acid administration in the isolated kidney resulted in vasoconstriction and inhibition of prostaglandin synthesis with indomethacin produced vasodilation (23). It should be recalled that PGE_2 has been identified as the principal product of prostaglandin synthesis in both the rat and rabbit kidneys.

The study of Malik and McGiff (23) urges consideration that renal prostaglandins contribute to the development of hypertension in the rat as an increase in renal vascular resistance by itself might initiate hypertension (24). Armstrong et al. have explored this possibility in the New Zealand strain of the genetically hypertensive (GH) rat (25). Comparison of the vasoconstrictor responses to norepinephrine revealed that the renal vascular bed of GH rats was about 50% more sensitive to norepinephrine than that of normotensive rats and was more susceptible to indomethacin-induced attenuation of the constrictor action of norepinephrine. PGE_2 in doses (100–300 pg/ml) which were without a direct effect on renal blood vessels restored to pre-indomethacin levels the vasoconstrictor responses to low doses of norepinephrine. Thus, increased renal vascular reactivity to pressor hormones in the GH rat could result partly from elevated levels of prostaglandins. If increased levels of prostaglandins occur, either increased synthesis or decreased degradation of prostaglandins must be

responsible. Production of prostaglandins by homogenates of kidneys of GH rats was, however, not different from normal as indicated by measurements of conversion of labeled arachidonic acid to
prostaglandins. However, important differences were found between
the kidneys of GH and normotensive rats in their abilities to degrade prostaglandins to their 15-keto metabolites, the first and
most important step in the metabolism of prostaglandins.

As modulators, prostaglandins may amplify or attenuate the
vascular actions of hormones or neurotransmitters in concentrations
which do not affect blood vessels directly. There is an additional
and less well-known corollary to the range of interactions between
modulators and hormones, and one which has bearing on the concept
that hypertension may develop in the face of normal activity of
either the adrenergic nervous system or the renin-angiotensin system. That is, that an abnormality of the modulator, by itself,
could result in elevated blood pressure; this may be the case in
the GH rat. Thus, the primary or initiating event would be increased levels of PGE_2, or a similar product of prostaglandin synthetase, which results from a deficiency of the major catabolizing
enzyme in the face of normal rates of prostaglandin synthesis. It
is likely that this disturbance need only occur in the kidney for
the development of hypertension.

There is one additional complicating factor that must be ac-
knowledged, although information on its operation in blood vessels
is meager. That is, increased activity of prostaglandin synthetase results in a cascade of vasoactive substances including cyclic endoperoxides and their metabolic products (26), as well as
other prostaglandins such as PGD_2 (27), all of which have vascular
activity. In some tissues, endoperoxides may be the most important products of prostaglandin synthetase (28). Thus, it is possible that one or more of the aforementioned products of prostaglandin synthetase in vascular walls will prove to be of greater
importance than the primary prostaglandins in mediating changes in
vascular reactivity and in modulating the vascular actions of hormones.

PROSTAGLANDINS AS MODULATORS AND MEDIATORS OF KININS

In contrast to attenuation of the vascular actions of pressor
hormones, the rat excepted, prostaglandins contribute to the vascular effects of kinins (29). As noted earlier, bradykinin dif-
ferentially increased synthesis of prostaglandins in bovine mesenteric blood vessels, i.e., enhanced release of a PGE compound from
arteries and of a PGF compound from veins. This finding provides
an explanation for the variable effects of bradykinin on blood
vessels, e.g., bradykinin dilates arteries and may constrict veins
(30). Thus, in those organs in which bradykinin increased synthesis

of PGE, e.g., the canine kidney (11) and uterus (7) and rat skeletal muscle (31), the released PGE may reinforce the vasodilator
action of the kinin. Moreover, the venoconstrictor action of
bradykinin may depend on the capacity of the vein to increase synthesis of $PGF_{2\alpha}$, a known venoconstrictor (2), in response to kinin.
In keeping with this interpretation, contraction of the bovine
mesenteric vein evoked by bradykinin was selectively abolished by
indomethacin. However, the release of PGF from veins by bradykinin
cannot be simply due to enhanced delivery of substrate to prostaglandin synthetase consequent to activation of a phospholipase,
the postulated effect of the kinin on production of prostaglandins
(32), for such activity should not change the ratio of PGE:PGF
(2:1 before bradykinin) in the released prostaglandins. That
bradykinin altered these ratios in a widely different manner in
veins (0.3:1) and arteries (5.7:1) also argues against a major effect of the kinin on the primary degradative enzyme, prostaglandin
15-hydroxydehydrogenase, as the K_m values of PGE_2 and $PGF_{2\alpha}$ with
respect to this enzyme are similar (33). Therefore, an additional
action of bradykinin was sought; i.e., one which might be exerted
through a mechanism that affects the ratio of PGE to PGF released
by a tissue. Such a mechanism has been described by Leslie and
Levine (34); viz, the stereospecific reduction of the 9-keto group
of PGE_2 by PGE 9-ketoreductase results in the formation of $PGF_{2\alpha}$.
This enzyme which is present in many tissues may be an important
prostaglandin regulatory mechanism whereby the functional consequences of changes in rates of prostaglandin synthesis are governed by regulating the ratio of PGE to PGF within a tissue.

We used a high speed supernatant fraction of bovine mesenteric
blood vessels which contains the greatest activity of this enzyme.
Incubation of radioactive PGE_2 with this fraction at 37°C in the
presence of a NADPH generating system resulted in time-dependent
conversion of PGE_2 to $PGF_{2\alpha}$. Bradykinin (0.01 mM) increased the
activity of PGE 9-ketoreductase by two- to threefold as measured
by conversion of PGE_2 to $PGF_{2\alpha}$ in the cytoplasmic fraction obtained
from veins (Figure 2). We also demonstrated PGE 9-ketoreductase
activity in the high speed supernatant fractions obtained from
arteries, as well as veins, which appears to invalidate the hypothesis that bradykinin released $PGF_{2\alpha}$ from veins primarily through
its effect on this enzyme. A possible explanation for this apparent discrepancy is that the venous and arterial enzymes may be
different. Indeed, evidence that we may be dealing with isoenzymes
was provided by the pH optimum and the shape of the pH profiles of
the PGE 9-ketoreductase in veins and arteries; the pH optima were
7.5 and 7.2 for the venous and arterial enzymes, respectively. In
support of this interpretation was the demonstration by Lee and
Levine (35) that the most purified PGE 9-ketoreductase activity,
thus far obtained, appeared to contain two forms of the enzyme.
An alternative explanation which remains to be explored for the
differential release by bradykinin of PGF from veins and PGE from

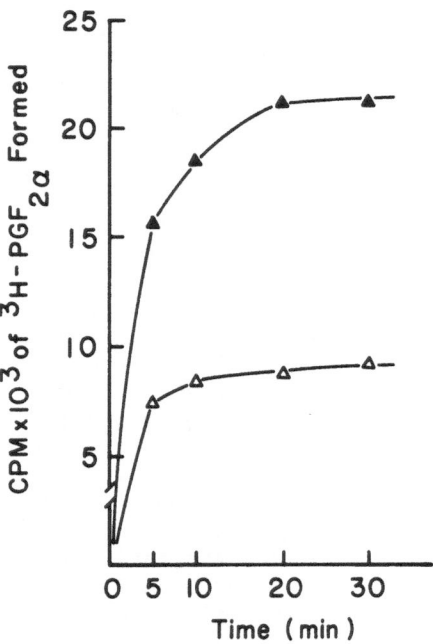

Fig. 2.   Time course study of PGE 9-ketoreductase activity in
bovine mesenteric veins (Δ) and the effect of bradykinin (▲).

arteries is that kinins also affect those enzymes which regulate
breakdown of the endoperoxide intermediates to PGE_2 and $PGF_{2\alpha}$.
However, evidence for enzymic regulation of $PGF_{2\alpha}$ formation from
endoperoxides is lacking (36,37).

    There is an additional consideration, however, which raises
the question whether the PGE 9-ketoreductase can be directly affected by kinins. Thus, an intermediate step through which bradykinin
stimulates PGE 9-ketoreductase is probable in view of three considerations: 1) PGE 9-ketoreductase activity is associated with a
cytoplasmic fraction and, therefore, would not likely be accessible
to the direct action of agents which penetrate the cell poorly; 2)
bradykinin, a nonapeptide, has limited access to intracellular
sites (38); 3) bradykinin-induced contractions of blood vessels
were shown to be associated with increased levels of cyclic GMP
(39,40). Changes in the cellular levels of this cyclic nucleotide
have been suggested to serve as a common mechanism to elevate vascular tone in response to diverse stimuli (41,42). We have obtained evidence that those vascular actions of bradykinin mediated
by PGF may occur through a mechanism involving cyclic GMP. The effects of various concentrations of cyclic GMP on the activity of

PGE 9-ketoreductase in mesenteric arteries are shown in Figure 3. At a concentration of $5 \times 10^{-6}M$ cyclic GMP maximally stimulated PGE 9-ketoreductase activity; a rapid decline occurred as the concentration of cyclic GMP was further increased. The apparent Michaelis constants (apparent K_m) and maximum reaction rates (V_{max}) were determined using PGE_2 as substrate in the presence of either bradykinin (0.01 mM) or cyclic GMP ($5 \times 10^{-6}M$) by using the Lineweaver-Burk double reciprocal plot; the apparent K_m and V_{max} were obtained by linear least-squares analysis. Table 1 shows the effect of bradykinin and cyclic GMP on these kinetic parameters. Cyclic GMP mimicked the effect of bradykinin on PGE 9-ketoreductase activity; i.e., each affected the conversion of PGE_2 to $PGF_{2\alpha}$ by increasing the rate of PGE_2 reduction by PGE 9-ketoreductase without affecting the apparent K_m of the enzyme. Cyclic AMP ($10^{-6}M$) did not affect PGE 9-ketoreductase activity.

The functional significance of the finding that cyclic GMP affects the activity of PGE 9-ketoreductase is that this mechanism may represent an intermediate in some of the vascular actions of bradykinin. Clyman et al recently have shown that bradykinin increased the accumulation of cyclic GMP in human umbilical arteries, without affecting the level of cyclic AMP (42). The venoconstrictor action of the kinin was also associated with increased levels of cyclic GMP (40). Thus, bradykinin-induced constriction of isolated bovine mesenteric arteries may be mediated through accumulation of cyclic GMP, the latter increasing the activity of PGE 9-ketoreductase and, thereby, promoting formation of PGF. Whether bradykinin has a direct effect on guanylate cyclase or cyclic GMP specific phosphodiesterase, either of which may be responsible for the accumulation of cyclic GMP, still remains to be established. We propose that modulation of prostaglandin activity through determining the ratio of PGE to PGF is effected by a cyclic GMP-dependent system. This system, by determining the ratio of PGE to PGF within the vascular wall, could contribute to the regulation of vascular tone; e.g., increased formation of PGF would facilitate, whereas elevated levels of PGE would inhibit hormonally and neurally induced vasoconstriction (23). This hypothesis embodies two earlier proposals: 1) that increased levels of cyclic GMP are associated with constriction of blood vessels (41), and 2) that the biological activities of prostaglandins of the F series are related to the guanylate cyclase system as those of the E series are related to the adenylate cyclase system (43).

The coupling of the kallikrein-kinin and prostaglandin systems may be unique: prostaglandins mediate some of the actions of kinins and modulate others; in addition, generation of kinins affects the level and type of prostaglandins. Thus, not only production of prostaglandins but the functional consequences of their enhanced production, as determined by the ratio of PGE_2 to $PGF_{2\alpha}$, may be regulated in part by kinins.

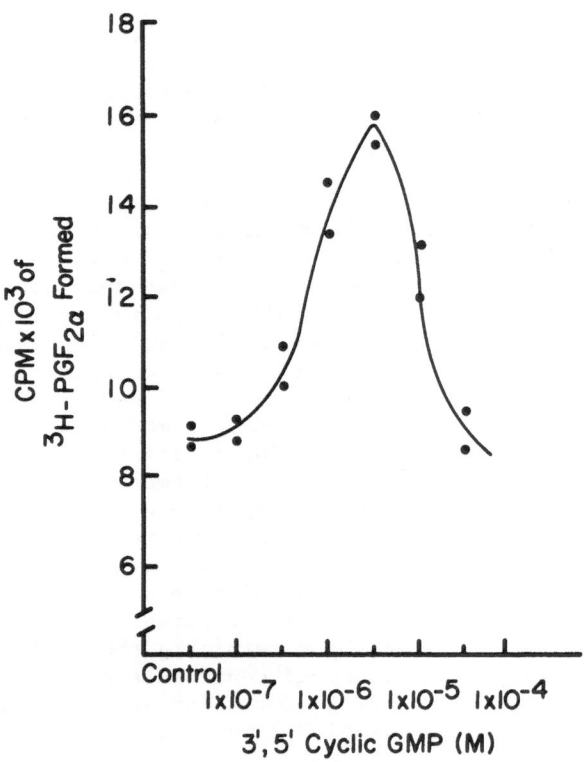

Fig. 3.   Relationship between the concentrations of cyclic GMP
and activity of PGE 9-ketoreductase in a bovine mesenteric artery.

Table 1.   Effects of bradykinin and 3', 5' cyclic GMP on PGE
9-ketoreductase activity in bovine mesenteric arteries

TREATMENT	APPARENT K_m(M)	V_{max} (nmole/min)
Control	5.4×10^{-4}	0.90
Bradykinin (0.01 mM)	5.4×10^{-4}	1.14
Control	5.2×10^{-4}	0.95
3', 5' cyclic GMP (5×10^{-6}M)	5.2×10^{-4}	1.25

CONCLUSIONS

Prostaglandins are local hormones which have their effects at
or near the site of synthesis. Blood vessels have variable capacities to synthesize prostaglandins; their release intramurally affects vascular tone and reactivity to neural and hormonal stimuli.
Interactions of prostaglandins and vasoactive hormones are of particular importance within the kidney because of the pivotal role of
renal mechanisms in the regulation of blood pressure.

ACKNOWLEDGMENTS

We thank Ms. Jo Lariviere for typing the manuscript. This
work was supported by grants from the National Heart and Lung Institute, HL-18845, from the American Heart Association and from the
Memphis and Tennessee Heart Associations.

1. Vane, J.R.: Brit. J. Pharmacol. 35:209-242, 1969.

2. Terragno, D.A., Crowshaw, K., Terragno, N.A. and McGiff, J.C.:
   Circ. Res. 36 and 37 (suppl I):I-76-I-80, 1975.

3. McGiff, J.C., Crowshaw, K., Terragno, N.A. and Lonigro, A.J.:
   Nature (Lond) 227:1255-1257, 1970.

4. Hedqvist, P.: Acta. Physiol. Scand. (suppl) 345:1-40, 1970.

5. Aiken, J.W. and Vane, J.R.: J. Pharmacol. Exp. Ther. 184:678-
   687, 1973.

6. Lonigro, A.J., Terragno, N.A., Malik, K.U. and McGiff, J.C.:
   Prostaglandins 3:595-606, 1973.

7. Terragno, N.A., Terragno, D.A., Pacholczyk, D. and McGiff, J.C.:
   Nature (Lond) 249:57-58, 1974.

8. Lonigro, A.J., Itskovitz, H.D., Crowshaw, K. and McGiff, J.C.:
   Circ. Res. 32:712-717, 1973.

9. McGiff, J.C., Crowshaw, K., Terragno, N.A., Malik, K.U. and
   Lonigro, A.J.: Clin. Sci. 42:223-233, 1972.

10. McGiff, J.C., Crowshaw, K., Terragno, N.A. and Lonigro, A.J.:
    Circ. Res. 26 and 27 (suppl I):I-121-I-130, 1970.

11. McGiff, J.C., Terragno, N.A., Malik, K.U. and Lonigro, A.J.:
    Circ. Res. 31:36-43, 1972.

12. Vane, J.R. and McGiff, J.C.: Circ. Res. 36 and 37 (suppl I):
    I-68-I-75, 1975.

13. McGiff, J.C. and Itskovitz, H.D.: Circ. Res. 33:479-488, 1973.

14. Hamberg, M., Svensson, J. and Samuelsson, B.: Proc. Nat. Acad.
    Sci. (Wash.) 71:3824-3828, 1974.

15. Mellander, S. and Johansson, B.: Pharmacol. Rev. 20:117-196,
    1968.

16. Ledingham, J.M. and Cohen, R.D.: Lancet 2:979-981, 1963.

17. Christ, E.J. and Van Dorp, D.A.: Advances in the Biosciences,
    International Conference on Prostaglandins, Vol 9, edited by
    Bergstrom, S., Pergamon Press, New York, 1972, pp. 35-38.

18. Hollander, W., Kramsch, D.M., Franzblau, C., Paddock, J. and
    Colombo, M.A.: Circ. Res. 34 and 35 (suppl I):I-131-I-141, 1974.

19. Gould, A.B., Skeggs, L.T., Jr. and Kahn, J.R.: J. Exp. Med. 119:389-399, 1964.

20. Burnstock, G., Gannon, B. and Iwayama, T.: Circ. Res. 26 and 27 (suppl II):II-5-II-23, 1970.

21. Zusman, R.M., Spector, D., Caldwell, B.V., Speroff, L., Schneider, G. and Mulrow, P.J.: J. Clin. Invest. 52:1093-1098, 1973.

22. Hamberg, M.: FEBS Letter 5:127-130, 1969.

23. Malik, K.U. and McGiff, J.C.: Circ. Res. 36:599-609, 1975.

24. Tobian, L.: Fed. Proc. 33:138-142, 1974.

25. Armstrong, J.M., Blackwell, G.J., Flower, R.J., McGiff, J.C., Mullane, K.M. and Vane, J.R.: Nature 260:582-586, 1976.

26. Hamberg, M. and Samuelsson, B.: Proc. Nat. Acad. Sci. 71: 3400-3404, 1974.

27. Blackwell, G.J., Flower, R.J. and Vane, J.R.: Biochim. Biophys. Acta 398:178-190, 1975.

28. Needleman, P., Minkes, M. and Raz, A.: Science 193:163-165, 1976.

29. McGiff, J.C., Itskovitz, H.D., Terragno, A. and Wong, P.Y-K: Fed. Proc. 35:175-180, 1976.

30. Bobbin, R.P. and Guth, P.S.: J. Pharmacol. Exp. Ther. 160: 11-21, 1968.

31. Messina, E.J., Weiner, R. and Kaley, G.: Circ. Res. 37:430-437, 1975.

32. Hong, S.L. and Levine, L.: J. Biol. Chem. (In Press)

33. Nakano, J., Änggard, E. and Samuelsson, B.: Eur. J. Biochem. 11:386, 1969.

34. Leslie, C.A. and Levine, L.: Biochem. Biophys. Res. Com. 52: 717-724, 1973.

35. Lee, S.C. and Levine, L.: J. Biol. Chem. 250:4549, 1975.

36. Pace-Asciak, C. and Nashat, M.: Biochim. Biophys. Acta 388: 243, 1975.

37. Chan, J.A., Nagasawa, M., Takeguchi, C. and Sih, C.J.: Biochemistry 14:2987, 1975.

38. Barakè, J., Park, W.K. and Regoli, D.: Can. J. Physiol. Pharmacol. 53:345-353, 1974.

39. Clyman, R.I., Sandler, J.A., Manganiello, V.C. and Vaughan, M.: J. Clin. Invest. 55:1020-1025, 1975.

40. Dunham, E.W., Haddox, M.K. and Goldberg, N.D.: Proc. Nat. Acad. Sci., U.S.A. 71:815-819, 1974.

41. Amer, M.S., Gomoll, A.W., Perach, J.L., Ferguson, H.C. and McKinney, G.R.: Proc. Nat. Acad. Sci., U.S.A. 71:4930-4934, 1974.

42. Clyman, R.I., Blacksin, A.S., Manganiello, V.C. and Vaughan, M.: Proc. Nat. Acad. Sci., U.S.A. 72:3883-3887, 1975.

43. Kuehl, F.A., Jr., Circilo, V.J., Ham, E.A. and Humes, J.L.: Advances in the Biosciences, International Conference on Prostaglandins, Vol 9, edited by Bergstrom, S. and Bernhard, S., Pergamon Press, Vieweg, Braunschweig, 1973, pp. 155-172.

PROSTAGLANDINS AND VASCULAR WALL

Ryszard J. Gryglewski

Department of Pharmacology, Copernicus
Academy of Medicine in Cracow
31-531 Cracow, 16 Grzegórzecka, Poland

INTRODUCTION

Prostaglandins (PGs) are potent vasoactive agents. Prostaglandins of the F series (PGFs) constrict mainly capacitance vessels, whereas prostaglandins of the E series (PGEs) dilate arterioles, metaarterioles, pre-capillaries and venules (Greenberg and Sparks, 1969) in most organs except for nasal mucosa (Nakano, 1973). There are also species differences in vascular reactivity to PGEs. PGE_2 constricts renal blood vessels of rats and potentiates vascular sensitivity to pressor hormones in renal (Malik and McGiff, 1975a; Mullane et al., 1976), mesenteric (Malik and McGiff, 1975b; Horrobin et al., 1974) and coronary (Horrobin et al., 1974) regions. Also isolated rat aorta, carotid and femoral arteries keep their tone owing to a continuous generation of PGEs and are relaxed by PG synthetase inhibitors (Rioux and Regoli, 1975). Therefore speaking about vasodilator properties of PGEs we have to exclude rats. In most other species, including man, PGEs relax vascular smooth muscle, counteract vasoconstriction induced by catecholamines, angiotensin and vasopressin (see Nakano, 1973) and are supposed to inhibit the release of adrenergic transmitter (Hedqvist, 1970, 1972). In contrast to a brief hypotensive effect following an intravenous injection of PGEs, their local vascular effects are long lasting and cannot be abolished by atropine, propranolol, methysergide or antihistaminics (Nakano, 1973). Ferreira (1972) has shown that erythema induced by an intradermal injection of PGE_1 in man lasts up to 10

265

hours. The proposed role of endogenous PGEs in mediation
or modulation of reactive and functional hyperemia, in
regulation of regional blood flow and in development
of vascular inflammatory responses has been reviewed by
Staszewska-Barczak and Vane (1975), McGiff (1975) and
Malik and McGiff (1976).

It is unlikely that circulating PGEs can affect
arterial tone since they are effectively destroyed in
lungs (Ferreira and Vane, 1967). The regulatory role of
PGEs should be sought in their local action. If so, PGEs
either should be born in vascular wall and act in loco
nascendi or could be generated in closely adjacent
tissues and act at a short distance on blood vessels,
as it might be a case in kidney.

PROSTAGLANDIN BIOSYNTHESIS BY VASCULAR WALLS

There are few evidence for generation of PGs by
vascular wall. Homogenates of rabbit aorta convert only
1% of the radioactive substrate to PGE_1, whereas in
rabbit kidney medulla this conversion is 41% (Christ
and van Dorp, 1972). A similar low capacity of aorta to
convert the substrate to PGEs has been reported by Hollander et al. (1974). Incubated human umbilical arteries produce an immunoreactive $PGF_{2\alpha}$ -like material in
an amount of 15 ng/g wet weight/2 hours (Tuvemo and
Wide, 1973). Cultured endothelial cells of human umbilical veins produce PGEs (detected by radioimmunoassay)
when stimulated by angiotensin II (Gimborne and Alexander, 1975). Recently Terragno et al. (1975) have reported that incubated slices of bovine mesenteric arteries
and veins release to the medium a material, which has
been biologically assayed as PG-like substances. The
rate of generation of this material was 200 ng/g wet
weight/ 1 hour. Arachidonic acid (1 µg/ml) does not
affect the rate of synthesis of PG-like substances,
however, meclofenamic acid (1 mM) decreases the rate of
synthesis by 90%. On the other hand radiochromatograms
of PGs biosynthetized by blood vessels from $1-^{14}C$ -
arachidonic acid demonstrated the appearance of ^{14}C-
labelled compounds indistinguishable from PGE_2 and $PGF_{2\alpha}$
with overall yield of 20% conversion of the substrate.

INDIRECT EVIDENCE FOR PROSTAGLANDIN RELEASE
BY VASCULAR WALL

PGEs have been suggested to be local mediators of
stimuli evoking vasodilatation (Horton, 1969). In 1972

Hedqvist has written: " Although PG release from vascular tissue has not been demonstrated so far ..... it is
tempting to assume that endogenous and locally released
PGE_1 and/or PGE_2 may operate to modulate the functional
state of contraction of vascular smooth muscle, thereby
acting as a link in regulation of regional blood flow
and pressure". Aiken (1974) has shown that the ability
of indomethacin and meclofenamic acid to reverse angiotensin tachyphylaxis in the isolated strips of rabbit
coeliac and mesenteric arteries is associated with a
high sensitivity of these arteries to the relaxing
effects of PGE_1 and PGE_2. Aiken (1974) has suggested
that during angiotensin-induced vasoconstriction arterial strips generate PGEs which contribute to tachyphylaxis. PG synthetase inhibitors potentiate the contractile responses to low frequency of stimulation in the
strips of rabbit portal vein (Greenberg, 1974) and in
the perfused rabbit arteries (Hadházy and Nádor, 1976),
indicating the existance of a negative feedback mechanism in vascular walls which is governed by PGEs. We
have shown (Gryglewski and Ocetkiewicz, 1974) that acute
tolerance to noradrenaline infusions in cats is associated with the release of a PGE-like substance into mixed
venous blood. This effect is abolished by α-adrenolytic
drugs, but then there is no pressor response to noradrenaline and by PG synthetase inhibitors, but then the
pressor response is enhanced. We have also observed that
following intravenous injections of E.coli endotoxin
(Korbut et al., 1975) or rat blood (Gryglewski et al.,
1975) into anaesthetized cats there appears (in parallel
to severe hypotension) a PGE-like material in mixed
blood. Both effects can be abolished by indomethacin.
These data might point at vascular wall as a site for
PGE generation. A variety of stimuli release PG-like
substances from perfused lungs, kidney, spleen, uterus,
hind legs, heart and from many other perfused organs
(for review see Staszewska-Barczak and Vane, 1975; Malik and McGiff, 1976). PG synthetase inhibitors block
this release, however, it is difficult to evaluate to
what extend (if at all) vascular wall contributes to
the output of PGs from these perfused organs.

            RELEASE OF PROSTAGLANDIN-LIKE SUBSTANCES
            BY NORADRENALINE-INDUCED VASOCONSTRICTION

        It is difficult to separate the vascular network
from surrounding tissues. We have used two vascular preparations which contain little of adjacent tissues.
These were isolated perfused rabbit´s ear (Gryglewski

and Korbut, 1975) and isolated perfused rabbit mesenteric vessels (Grodzińska et al., 1976). Noradrenaline
was used to stimulate the release of PG-like substances
for two reasons. Firstly, the noradrenaline-induced vasoconstriction may be abolished by α-adrenergic blockade but not by β-adrenergic blockade. We have found
that only α-adrenergic blockade inhibits the release
of PG-like substances from the perfused vascular preparations and thereby solely vasoconstriction but none of
metabolic events in adjacent tissues is responsible for
the releasing effect of noradrenaline. Secondly, owing
to our modification (Gryglewski et al., 1975; Gryglewski
and Korbut, 1976) of the technique of Vane (1964) it is
possible to remove noradrenaline from the effluent before its contact with the bioassay organs and thus a vasoconstrictor does not interfere with the bioassay of PG-
like substances.

Originally we have used a rat stomach strip and
a rat colon as the bioassay tissues. This bank of superfused organs was later supplemented by strips of rabbit
aorta, rabbit vena cava, rabbit mesenteric artery, rabbit coeliac artery (Bunting et al., 1976) and a chick
rectum. PG-like substances which are released by an infusion of noradrenaline (1 - 3 µg/ml) from the vascular
preparations have the following properties: they do not
contract rabbit aorta, rabbit vena cava and rat colon,
they relax rabbit mesenteric and coeliac arteries and
they contract rat stomach strip and chick rectum. The
closest behaviour of the assay tissues to that described
above can be obtained when PGE_2 is infused over the assay
organs. Using contractions of rat stomach for quantitation of PG-like substances released, we have found that
the maximal concentration of these substances in the
effluent from rabbit ear vessels is 5.1 ± 0.7 ng PGE_2
equivalents/ml (9 experiments) and from rabbit mesenteric vessels 3.2 ± 0.7 ng PGE_2 equivalents/ml (20 experiments) (mean $\pm$ S.E.). However, the released PG-like
substances seem to have a weaker relaxing activity on
strips of rabbit mesenteric and coeliac arteries and
stronger contracting activity on a chick rectum than
PGE_2 infused at a dose that matches the contraction of
rat stomach produced by PG-like substances. The noradrenaline-induced release of PG-like substances from blood
vessels is inhibited by indomethacin and mefenamate
(1 - 3 µg/ml) as well as by dexamethasone (2 µg/ml),
whereas it is augmented by arachidonic acid (0.2 µg/ml).
The inhibition of generation of PG-like substances is
associated with the disappearance of acute tolerance
to noradrenaline.

LIMITATIONS OF BIOASSAY

Bioassay technique of Vane (1964) with later modifications (Gilmore et al., 1968; Piper and Vane,1969) is still unbeatable for detection of small amounts of PG-like substances in a dynamic biological experiment. This technique has, however, its limitations. One cannot positively identify a PG-like substance as PGE_1, PGE_2, $PGF_{2\alpha}$ etc. By increasing a number of assay organs which are superfused in cascade it is possible to reach a better approximation, however, even then it might be neccessary to use a biological description for PG-like substances appearing in the effluent, e.g. a rabbit aorta contracting substance (Piper and Vane, 1969) or chick rectum contracting substances (Gryglewski and Vane, 1970; Lewis and Piper, 1975). This procedure is obligatory when biological activities of these substances do not correspond to the activities of the known PGs. Eventually these "biologically active PG-like substances" are identified as chemical entities. For instance, a rabbit aorta contracting substance from lungs seems to be a mixture of thromboxane A_2, PGG_2 and PGH_2 (Svensson et al., 1975). Such findings stimulate the development of new bioassay techniques which differentiate the components of a mixture (Bunting et al., 1976). The uncertainty concerning the chemical character of PG-like substances released from tissues is not the unique feature of the bioassay technique. Radiochromatographic techniques suffer from the same difficulty. $6(9)$-Oxy-$PGF_{2\alpha}$ from rat stomach (Pace-Asciak, 1972) and thromboxane A_2 from blood platelets (Hamberg et al., 1975) cochromatograph with PGE_2, and only because of the sophisiticated physicochemical techniques employed, and the ingeniousness of the authors it has been possible to separate and identify the above substances. Until an endogenous PG-like substance is not isolated and chemically identified, it is much safer to stick to an awkward but precise biological description than to classify it as a chemical entity.

CONCLUSION

There is a number of evidence that vascular walls generate prostaglandin E-like substances which counteract the excessive vasoconstriction. The chemical nature of these substances remains to be established.

ACKNOWLEDGEMENTS

I gratefully acknowledge the generous grant of equipment from the Trustees of The Wellcome Trust, London, Great Britain.

REFERENCES

Aiken, J.W., 1974, Effects of prostaglandin synthesis inhibitors on angiotensin tachyphylaxis on the isolated coeliac and mesenteric arteries of the rabbit. Pol. J. Pharmac. Pharmacol. 26: 217-227.

Bunting, S., Moncada, S., and Vane, J.R., 1976, The effects of prostaglandin endoperoxides and thromboxane A_2 on strips of rabbit coeliac artery and certain other smooth muscle preparations. Proc. Br. Pharmacol. Soc. 1 - 2 April, p. 48.

Christ, E.J., and van Dorp, D.A., 1972, Comparative aspects of prostaglandin biosynthesis in animal tissues. Biochem. Biophys. Acta 270: 537-545.

Ferreira, S.H., 1972, Prostaglandins, aspirin-like drugs and analgesia. Nature New Biology 240: 200-203.

Ferreira, S.H., and Vane, J.R., 1967, Prostaglandins: their disappearance from and release into circulation. Nature (London), 216: 868-873.

Gilmore, N., Vane, J.R., and Wyllie, J.H., 1968, Prostaglandins released by the spleen. Nature 218: 1135-1140.

Gimborne, M.A., Jr., and Alexander, R.W., 1975, Angiotensin II stimulation of prostaglandin production in cultured human vascular endothelium. Science 189: 219-220.

Greenberg, R., 1974, The effects of indomethacin and eicosa-5,8,11,14-tetraynoic acid on the response of the rabbit portal vein to electrical stimulation. Br. J. Pharmac. 52: 61-68.

Greenberg, R.A., and Sparks, H.V., 1969, Prostaglandins and consecutive vascular segments of the canine hindlimb. Amer. J. Physiol. 216: 567-571.

Grodzińska, L., Panczenko- B., and Gryglewski, R.J., 1976, Release of prostaglandin-like material from perfused mesenteric blood vessels of rabbits. J. Pharm. Pharmac. 28: 40-43.

Gryglewski, R.J., Grodzińska, L., Korbut, R., Ocetkiewicz, A., and Panczenko, B., 1975, Regulatory role of prostaglandins in the vascular system. Materia Med. Pol. 7: 314-321.

Gryglewski, R.J., and Korbut, R., 1975, Prostaglandin feedback mechanism limits vasoconstrictor action of norepinephrine in perfused rabbit ear. Experientia 31: 89-90.

Gryglewski, R.J., and Korbut, R., 1976, Bioassay of histamine in the presence of prostaglandins. Br. J. Pharmac. 56: 39.

Gryglewski, R.J., and Ocetkiewicz, A., 1974, A release of prostaglandins may be responsible for acute tolerance to norepinephrine infusions. Prostaglandins 8: 31-42.

Gryglewski, R.J., and Vane, J.R., 1970, The inactivation of noradrenaline and isoprenaline in dogs. Br. J. Pharmac. 39: 573-584.

Hadházy, P., and Nádor, T., 1976, Effects of indomethacin and PGE_1 on the vasoconstrictor responses of the rabbit ear artery to nerve stimulation. Prostaglandins 11: 241-250.

Hamberg, M., Svensson, J., and Samuelsson, B., 1975, Thromboxanes: A new group of biologically active compounds derived from prostaglandin endoperoxides. Proc. Nat. Acad. Sci. USA 72: 2994-2998.

Hedqvist, P., 1970, Studies on the effect of prostaglandins E_1 and E_2 on the sympathetic neuromuscular transmission in some animal tissues. Acta physiol. scand. (suppl.), 345: 1-40.

Hedqvist, P., 1972, Prostaglandin-induced inhibition of vascular tone and reactivity in the cat's hindleg in vivo. Europ. J. Pharmacol. 17: 157-162.

Hollander, W., Kramsch, D.M., Franzblau, C., Paddock, J., and Colombo, M.A., 1974, Supression of atheromatous fibrous plaque formation by antiproliferative and anti-inflammatory drugs, in: Hypertension XXII: Peptides, Lipids, Electrolytes and Hypertension, (J.C. Hunt, ed.), pp. 131-141, Circulation Res. 34: and 35 (Suppl.I).

Horrobin, D.F., Manko, M.S., Karmali, R., Nassae, B.A., and Davis, P.A., 1974, Aspirin, indomethacin, catecholamine and prostaglandin interaction on rat arteriales and rabbit hearts. Nature (London) 250: 425-426.

Horton, E.W., 1969, Hypotheses on physiological roles of prostaglandins. Physiol. Rev. 49: 122-141.

Korbut, R., Ocetkiewicz, A., and Gryglewski, R.J., 1975, Release of a prostaglandin E-like substance into mixed venous blood during endotoxin hypotension in cats. Pol. J. Pharmacol. 27: 439-443.

Lewis, G.P., and Piper, P.J., 1975, Inhibition of release of prostaglandins as an explanation of some of the actions of anti-inflammatory corticosteroids. Nature 254: 308-311.

Malik, K.U., McGiff, J.C., 1975a, Modulation by prostaglandins of adrenergic transmission in the isolated perfused rabbit and rat kidney. Circ. Res. 36: 599-609.

Malik, K.U., and McGiff, J.C., 1975b, Modulation by prostaglandin E₁ of adrenergic transmission in isolated perfused rabbit and rat mesenteric arteries. Fed. Proc. 34: 763.

Malik, K.U., and McGiff, J.C., 1976, Cardiovascular actions of prostaglandins, in: Recent Advances in Prostaglandin Research, Vol. 3, (S.M.M. Karim, ed.), MTP, Lancaster.

McGiff, J.C., 1975, Prostaglandins as regulators of blood pressure. Hosp. Pract. 10: 101-112.

Mullane, K.M., Armstrong, J.M., McGiff, J.C., 1976, Potentiation by prostaglandins of send vascular sensitivity to pressor hormones in normotensive and genetically hypertensive rats of the New Zealand strain, in: Advances in Prostaglandin and Thromboxane Research, Vol. 2, (B.Samuelsson and R. Paoletti eds.), p. 954, Raven Press, New York.

Nakano, J., 1973, Cardiovascular action, in: The Prostaglandins, Vol. 1 (P.W. Ramwell, ed.), p. 239, Plenum Press, New York-London.

Pace-Asciak, C., 1972, Prostaglandin synthetase activity in the rat stomach fundus. Activation by l-norepinephrine and related compounds. Biochim. Biophys. Acta 280: 161-171.

Piper, P.J., and Vane, J.R., 1969, Release of additional factors in anaphylaxis and its antagonism by anti-inflammatory drugs. Nature 233: 29-35.

Rioux, F., and Regoli, D., 1975, In vitro production of prostaglandins by isolated aorta strips of normotensive and hypertensive rats. Can. J. Physiol. Pharmacol. 53: 673-677.

Staszewska-Barczak, J., and Vane, J.R., 1975, The role of prostaglandins in the local control of circulation. Clin. Exp. Pharmacol. Physiol. Suppl.2, 71-78.

Svensson, J., Hamberg, M., and Samuelsson, B., 1975, Prostaglandin endoperoxides IX. Characterization of rabbit aorta contracting substance (RCS) from guinea pig lungs and human platelets. Acta physiol. scand. 94: 222-228.

Terragno, D.A., Crowshaw, K., Terragno, N.A., and McGiff, J.C., 1975, Prostaglandin synthesis by bovine mesenteric arteries and veins. Circulation Res. 36 and 37, suppl. I. 76-80.

Tuvemo, T., and Wide, L., 1973, Prostaglandin release
    from the human umbilical artery _in vitro_.
    Prostaglandins 4: 689-694.

THE ROLE OF PROSTAGLANDINS IN GASTROINTESTINAL TONE AND
MOTILITY

Alan Bennett

Department of Surgery, King's College Hospital

Medical School, London SE5 8RX, UK

Background Information

Prostaglandin-like material occurs in the alimentary
tract of all species studied and is released from the
tissue in vitro (e.g. frog intestine (1), rat stomach
(2,3), human stomach (4) and in other tissues examined
subsequently). Prostaglandin E_2 (PGE_2) has been formally
identified in the gastrointestinal tract of the shark
(Triakis scyllia)(5). It might therefore be expected
that PGs contribute to the activity of isolated muscles.
However, the effect would depend on the PGs released and
the muscle layer concerned. In general, longitudinal
muscle of the gut is contracted by PGE and F compounds,
but the circular muscle is usually inhibited by PGE
compounds and contracted by PGF compounds (6,7).

Bovine sphincter pupillae (8) was the first tissue
used which indicated the importance of PGs in maintaining
muscle tone. Certain PG antagonists reduce gastrointestinal muscle tone, and it was thought that they might
do so by inhibiting PG action (9). PGs were considered
to maintain tone of rat gastric fundus (10), and the
longitudinal muscle of rabbit isolated jejunum (11) and
guinea-pig ileum (12,13). More recent work along these
lines is described below.

Recent Studies on PG Release

Release of PGE_2-like material occurred at rest and

275

during electrical stimulation of guinea-pig isolated
ileum (13). We have found that finely cut pieces of
guinea-pig ileum release PG-like material during
incubation and inactivate added PGE_2 in Krebs solution at
37ºC (unpublished). Release also occurs from finely cut
human gastrointestinal mucosa, the amounts being greater
from gastric tissue than from ileum or colon; human
mucosal tissue metabolises the released material and
added PGE_2 (unpublished).

Segments of rabbit isolated jejunum released PGE_2-
and $PGF_{2\alpha}$-like material (14). This was not due to
mechanical changes in muscle length, although it is known
that stretching rat gastric fundus releases PG-like
material (2). The amounts obtained from the jejunum
declined during the first two hours and then increased
steadily, possibly indicating release due to damage. In
support of this view, greater output occurred after
refrigeration for 48 hours. Damage by pinching released
PG-like material from rat gastrointestinal tissues (15),
and a similar effect occurred with human gastric mucosa
(unpublished). It may also be relevant that capsaicin (a
constituent of Cayenne pepper) or ethanol (substances
which cause gastric mucosal irritation) stimulated PG
synthesis by bull seminal vesicles (16).

Recent Studies on Muscle Tone and Reactivity

The longitudinal muscle of guinea-pig colon (unlike
the ileal muscle) possesses tone, and relatively low doses
of aspirin (20 or 100µg/ml) or indomethacin (1 or 2µg/ml)
caused relaxation (17). The amplitude of spontaneous contractions increased, and submaximal contractions to acetylcholine (ACh) and PGE_2 were enhanced, probably because of
the fall in tone. However, contractions to electrical
stimulation were increased to an even greater extent so
that a lowering of tone does not seem the only factor
responsible.

The circular muscle of guinea-pig colon behaved
differently, and it is important to remember that PGE
compounds relax this tissue. Aspirin 50-100µg/ml raised
the tone and increased or initiated spontaneous activity.
An even greater stimulation occurred in ileal circular
muscle, and since adding PGE_2 antagonised the effect, PGE
release in the circular muscle seems to keep the tissue
relaxed (17). Somewhat similar results have been obtained
in human isolated stomach and ileum (18). Indomethacin

$2\mu g/ml$ relaxed longitudinal muscle to an extent which
correlated with the level of initial tone. Contractions
to ACh were increased, presumably because of the reduced
tone. Unlike guinea-pig colon, indomethacin also reduced
the tone of ileal circular muscle, but as in the guinea-
pig tissue, it initiated or increased rhythmic activity
and usually enhanced ACh-induced contractions. PGE_2
reversed these increases. In the human isolated internal
anal sphincter, indomethacin $0.1-1\mu g/ml$ blocked relaxations
to ACh but not to electrical field stimulation, so that
PG might mediate the response to ACh (19). The evidence
that PGs contribute to the control of tone and responses
in isolated gastrointestinal muscle is therefore substantial, but it is not clear to what extent the effects
observed are due to PG released from tissue damaged during
preparation and subsequent maintenance in vitro. However,
there is some evidence that PGs play a similar role in vivo.
Administration of indomethacin rectally to human subjects
increased the tone of the lower oesophageal sphincter (20).
In this tissue, as in most other circular muscles, PGE_2
is inhibitory and $PGF_{2\alpha}$ is excitatory. As already discussed, human gastric mucosa contains PGE_2-like material
but the PG content of the muscle of the lower oesophageal
sphincter has not been studied. However, the data is
consistent with the possibility that released PGE_2 in vivo
helps keep this circular muscle relaxed. The authors
suggest that PGE released from the inflamed lower
oesophagus might relax the sphincter and increase gastro-
oesophageal reflux (20). In anaesthetised guinea-pigs,
5,8,11,14-eicosatetraynoic acid (TYA) applied to the
serosal surface of the ileum reduced or prevented spontaneous contractions, but there was no effect with TYA
injected into the bloodstream (21).

In human subjects, $PGF_{2\alpha}$ infused iv increased the
frequency of antral contractions (22), but in contrast to
what might be expected from the in vitro data, it reduced
the intestinal activity as judged from intraluminal pressure recordings (23). Perhaps this arose because $PGF_{2\alpha}$
caused maintained contraction of the circular muscle
(this would probably cause only a transient increase of
intraluminal pressure), and superimposed waves of increased
pressure due to $PGF_{2\alpha}$ were small compared with basal
activity because the muscle was already shortened.

Diarrhoea with PGE compounds is likely to be due
substantially to increased intestinal secretion. However,
reduced colonic pressure activity is associated with
diarrhoea (24) and if PGE compounds inhibit human colonic

circular muscle <u>in vivo</u> as they do <u>in vitro</u>, this might
contribute to the effect; mass movements in the colon
might be aided by reducing the resistance to propulsion.
Alternatively, PGE might tend to increase intestinal
propulsion by increasing cholinergic contraction of the
circular muscle.

Prostaglandins and Nerve Activity

There is substantial evidence in the peripheral
nervous system that PGs inhibit the release of noradrenaline (25,26). Little is known about this aspect in
the gut, but there have been several studies on intestinal
cholinergic nerves.

Morphine, PG antagonists of PG synthesis inhibitors
reduced neurogenic contractions of guinea-pig isolated
ileum, and since low concentrations of PGE_1 or E_2
reversed the reduction, the authors suggested that PGE
couples cholinergic nerve terminal excitation with ACh
release ·(27). However, there are various flaws in the
experimental design and reasoning (28).

Indomethacin 20µg/ml greatly reduced contractions of
guinea-pig isolated ileum to angiotensin II, a substance
which acts partly by stimulating cholinergic nerves (29).
This depression was mainly reversed by low concentrations
of PGE_2, and the authors thought that released PG contributes to some aspects of the contraction. Since
indomethacin had less effect on the response to ACh than
to angiotensin, it seems that part of the effect was
neuronal.

Indomethacin 10µg/ml prevented both spontaneous and
electrically induced release of PGE (13), and since 1Hz
and 10Hz elicited contractions of similar amplitude,
whereas output of PGE_2 was greater at the higher frequency,
the release was considered to be due to nerve stimulation
and not to muscle contraction. However, they obtained
only transient reduction of contractions to field stimulation at 0.2Hz with 10µg/ml indomethacin. A concentration of 1µg/ml indomethacin seems effective in
inhibiting spontaneous generation of tone in guinea-pig
ileum (13), but more than 10 times the amount is needed
to inhibit evoked contractions. Perhaps this is due to
synthetases with different sensitivities in the tissue,
or to poor penetration of indomethacin to neuronal sites.
These aspects are discussed in reference 28.

Indomethacin 1µM reduced responses of guinea-pig isolated ileum to electrical stimulation and angiotensin, but not to ACh (30). The effect was restored or, with electrical stimulation, increased by a low concentration of PGE_2. Guanethidine or α-methyl-p-tyrosine (which block adrenergic neurones and deplete stores of noradrenaline) prevented the reduction of response to nerve stimulation. The authors considered that indomethacin acts in the untreated ileum by inhibiting PG synthesis and thereby allowing noradrenaline release to increase. Other investigators (28), however, found that 1µM indomethacin did not alter contractions at 0.1Hz, and depression of responses occurred only above 10µM. The discrepancy was because of differences in bathing fluid; similar results were obtained (28) with 1µM indomethacin using modified Krebs solution (as in reference 30) in place of Krebs solution. This seems to have important implications for the action of indomethacin, and it would be interesting to see if this affects selectivity, if other PG synthetase inhibitors are affected similarly, and if ionic changes in vivo affect the therapeutic response to indomethacin.

Indomethacin 40µg/ml (in Krebs solution) prevented nicotine-induced contractions of the longitudinal muscle of guinea-pig ileum, and greatly reduced those to electrical stimulation at 0.1-4Hz (28). Submaximal contractions to ACh, histamine and electrical stimulation at 8 and 16Hz were reduced to a somewhat smaller extent. PGE_2 partly restored all these responses. Lower concentrations of indomethacin (1-4µg/ml) reduced contractions to nicotine and PGE_2, but had no significant effect on those to ACh; aspirin 10-200µg/ml had little effect on responses of the longitudinal muscle of guinea-pig isolated ileum to ACh, histamine, nicotine or PGE_2 (31).

The experiments indicate that PGs might be involved in ACh release, but this only partly explains their potentiating effect. Low concentrations of PGs also increased the response to substances such as ACh and histamine which act directly on the muscle. Furthermore, PG release seems important for the maintenance of contractions of guinea-pig colonic muscle to histamine (10). Nevertheless, some of the results described must be interpreted with care, particularly those employing high concentrations of indomethacin.

Our unpublished findings show that PGE- and often PGF-like material can be extracted from human ileal muscle,

and that the effects of indomethacin (2-10μg/ml) on the
multiphasic responses to electrical stimulation of
intrinsic nerves are often complex. Because indomethacin
lowered the tone in the longitudinal or circular muscle
layers, nerve mediated relaxations were reduced. The drug
affected contractions differently in the two muscle layers.
In the longitudinal muscle the contractions during
stimulation were enhanced (presumably because of the
decreased tone) whereas the after-contractions were
reduced. This is similar to the effect on after-contraction in the longitudinal muscle (taenia) of guinea-pig
caecum (32), although these authors used high concentrations of indomethacin. By contrast in the circular
muscle, indomethacin depressed the contraction which
occurred during stimulation but enhanced the after-contraction. Thus PG might be involved in the contraction
which occurs during stimulation of the circular muscle,
and in the after-contraction in the longitudinal (but not
circular) muscle.

Lastly, with regard to nerves, PGs do not seem
important for non-adrenergic relaxations of the longitudinal muscle of guinea-pig ileum or caecum (30) or in
either muscle layer of human ileum (unpublished).

Effect of Prostaglandins on Peristalsis

The effect of PGs on peristalsis is complicated
because in some instances the PGE and F compounds exhibit
different activities, and because of regional differences
in the effect on the gut. All the studies to date have
been done on guinea-pig isolated intestine. Serosally
applied PGE_1 or E_2 (0.1-1μg/ml) inhibited ileal
peristalsis (33). This was later mainly confirmed (with
PGE_1)(34) but low concentrations (10-50ng/ml) first
stimulated peristalsis and then caused inhibition after
washout; only inhibition occurred with higher concentrations (0.5-1μg/ml). Subsequently another group found
that PGE_1 or $F_{2\alpha}$ 0.1-0.2μg/ml slightly increased ileal
peristalsis studied by the Trendelenberg method (35).
We have obtained increased ileal peristaltic activity with
serosally applied $PGF_{1\alpha}$ or $F_{2\alpha}$, but the PGs applied
mucosally had no significant effect, presumably because
they did not penetrate to the muscle and did not affect
the mucosal nerves (unpublished).

In the colon PGE_1 10^{-7} or PGE_2 10^{-7} to 10^{-6}M relaxed
strips of circular muscle and contracted strips of

longitudinal muscle. Nevertheless, PGE_1 $10^{-8}M$ slightly stimulated aboral propulsion, although concentrations above $10^{-7}M$ showed mainly initial inhibition of movement followed by one strong peak of stimulation. $PGF_{2\alpha}$ $10^{-7}M$ to $10^{-6}M$ contracted strips from both layers and $10^{-6}M$ stimulated aboral propulsion (36). Our unpublished data show that even though PGE_1 or E_2 relaxed circular muscle strips of guinea-pig colon, propulsion and circular muscle peristaltic activity were increased by serosally applied PGE or F compounds in the colon following initial contraction of the longitudinal muscle.

The last piece of evidence for the involvement of PGs in peristalsis is that aspirin 20-100µg/ml or indomethacin 1-4µg/ml applied serosally reduced peristaltic activity in guinea-pig isolated ileum and colon. The inhibition of ileal peristalsis by aspirin was antagonised by adding PGE_2 to the serosal surface. The effect of indomethacin was removed by $PGF_{2\alpha}$ or ACh; PGE_2 was not tried. Inhibition of colonic peristalsis by aspirin was antagonised by PGE_2 but rarely by ACh; PGE_1 or E_2 counteracted the effect of indomethacin. Mucosal application of aspirin had little effect on either ileum or colon, but indomethacin caused some inhibition (31).

It therefore seems likely that PGs are involved in peristalsis in the isolated intestine, although an effect of aspirin-like drugs on a non-PG-synthetase pathway cannot be excluded. We do not know how much of the PG release is due to tissue damage in vitro, and to what extent PGs are involved normally in peristalsis. As in most tissues, PGs seem unlikely to be essential for activity but might have a modulating effect. PG synthetase inhibitors reduce but do not stop peristalsis in vitro, and treatment with aspirin-like drugs is not usually associated with constipation in human subjects. However, we do not know how much therapeutic doses of these drugs inhibit PG synthesis in the human gut, or whether they have a tendency to produce constipation which is masked by a compensatory change in bowel activity.

References

1. Vogt, W., Suzuki, T. and Babilli, S. (1966). Prostaglandins in SRS-C and in darmstoff preparation from frog intestinal dialysates. Mem. Soc. Endocrinol. 14, 137-142.

2. Bennett, A., Friedmann, C.A. and Vane, J.R. (1967).
    Release of prostaglandin E_1 from the rat stomach.
    Nature (Lond) <u>216</u>, 873-876.

3. Coceani, F., Pace-Asciak, C., Volta, F. and Wolfe,L.S.
    (1967). Effect of nerve stimulation on prostaglandins
    formation and release from the rat stomach. Am. J.
    Physiol. <u>213</u>, 1056-1064.

4. Bennett, A., Murray, J.G. and Wyllie, J.H. (1968).
    Occurrence of prostaglandin E_2 in the human stomach,
    and a study of its effects on human isolated gastric
    muscle. Br. J. Pharmacol. Chemother. <u>32</u>, 339-349.

5. Ogata, H. and Nomura, T. (1975). Isolation and
    identification of prostaglandin E_2 from the gastro-
    intestinal tract of shark triakis scyllia. Biochim.
    Biophys. Acta. <u>388</u>, 84-91.

6. Bennett, A. and Fleshler, B. (1970). Prostaglandins
    and the gastrointestinal tract. Gastroenterology,
    <u>59</u>, 790-800.

7. Bennett, A. (1976). Prostaglandins and the alimentary
    tract. In: Prostaglandins: physiological, pharmaco-
    logical and pathological aspects. Ed. S.M.M. Karim,
    MTP Press Ltd., Lancaster, England, pp 247-276.

8. Posner, J. (1970). The release of prostaglandin E_2
    from the bovine iris. Br. J. Pharmacol. <u>40</u>, 163P-
    164P.

9. Bennett, A. and Posner, J. (1971). Studies on
    prostaglandin antagonists. Br. J. Pharmacol. <u>42</u>,
    584-594.

10. Eckenfels, A. and Vane, J.R. (1972). Prostaglandins,
    oxygen tension and smooth muscle tone. Br. J.
    Pharmacol. <u>45</u>, 451-462.

11. Ferreira, S.H., Herman, A. and Vane, J.R. (1972).
    Prostaglandin generation maintains the smooth muscle
    tone of the rabbit isolated jejunum. Br. J.
    Pharmacol. <u>44</u>, 328P-329P.

12. Davison, P., Ramwell, P.W. and Willis, A.L. (1972).
    Inhibition of intestinal tone and prostaglandin
    synthesis by 5,8,11,14-tetraynoic acid. Br. J.
    Pharmacol. <u>46</u>, 547P-548P.

13. Botting, J.H. and Salzmann, R. (1974). The effect
    of indomethacin on the release of prostaglandin E_2
    and acetylcholine from guinea-pig isolated ileum
    at rest and during field stimulation. Br. J.
    Pharmacol. <u>50,</u> 119-124.

14. Ferreira, S.H., Herman, A.G. and Vane, J.R. (1976).
    Prostaglandin production by rabbit isolated jejunum
    and its relationship to the inherent tone of the
    preparation. Br. J. Pharmac. <u>56</u>, 469-477.

15. Collier, H.O.J. (1974). Prostaglandin synthetase
    inhibitors and the gut. In: "Prostaglandin
    synthetase inhibitors". H.J. Robinson and J.R. Vane,
    Raven Press, New York, p121-133.

16. Collier, H.O.J., McDonald-Gibson, W.J. and Saeed, S.A.
    (1975). Stimulation of prostaglandin biosynthesis
    by capsaicin, ethanol, and tyramine. Lancet, <u>1</u>,702.

17. Bennett, A., Eley, K.G. and Stockley, H.L. (1975).
    The effects of prostaglandins on guinea-pig isolated
    intestine and their possible contribution to muscle
    activity and tone. Br. J. Pharmacol. <u>54</u>, 197-204.

18. Stockley, H.L. and Bennett, A. (1976). Modulation of
    activity by prostaglandins in human gastrointestinal
    muscle. Proceedings of the Fifth International
    Symposium on gastrointestinal motility. Typoff-Press,
    Herentals, Belgium.

19. D'Mello, A., Burleigh, D.E. and Parks, A.G. (1975).
    A non-adrenergic inhibitory mechanism in the human
    internal anal sphincter probably involving the release
    of prostaglandins. Abstracts, Sixth International
    Congress of Pharmacology, Helsinki, p158.

20. Dilawari, J.B., Newman, A., Poleo, J. and Misiewicz,
    J.J. (1975). Response of the human cardiac
    sphincter to circulating prostaglandins $F_{2\alpha}$ and E_2
    and to anti-inflammatory drugs. Gut, <u>16</u>, 137-143.

21. Willis, A.L., Davison, P. and Ramwell, P.W. (1974).
    Inhibition of intestinal tone, motility and
    prostaglandin biosynthesis by 5,8,11,14-eicosatetray-
    noic acid (TYA). Prostaglandins <u>5</u>, 355-368.

22. Newman, A., De Moraes-Filho, J.P.P., Philippakos, D.
    and Misiewicz, J.J. (1975). The effect of intravenous

infusions of prostaglandins E_2 and $F_{2\alpha}$ on human gastric function. Gut, <u>16</u>, 272-276.

23. Cummings, J.H., Newman, A., Misiewicz, J.J., Milton-Thompson, G.J. and Billings, J.A. (1973). Effect of intravenous prostaglandin $F_{2\alpha}$ on small intestinal function in man. Nature (Lond). <u>243</u>, 169-171.

24. Connell, A.M. (1962). The motility of the pelvic colon. Part II. Paradoxical motility in diarrhoea and constipation. Gut, <u>3</u>, 342-348.

25. Hedqvist, P. (1971). Prostaglandin E compounds and sympathetic neuromuscular transmission. Ann. N.Y. Acad. Sci. <u>180</u>, 410-415.

26. Hedqvist, P. (1976). Effects of prostaglandins on the autonomic neurotransmission. In: Prostaglandins: physiological, pharmacological and pathological aspects. Ed. S.M.M. Karim, MTP Press Ltd., Lancaster, England, pp 37-61.

27. Ehrenpreis, S., Greenberg, J. and Belman, S. (1973). Prostaglandins reverse inhibition of electrically-induced contractions of guinea pig ileum by morphine, indomethacin and acetylsalicylic acid. Nature (New Biol), <u>245</u>, 280-282.

28. Bennett, A., Eley, K.G. and Stockley, H.L. (1975). Modulation by prostaglandins of contractions in guinea-pig ileum. Prostaglandins, <u>9</u>, 377-384.

29. Chong, E.K.S. and Downing, O.A. (1974). Reversal by prostaglandin E_2 of the inhibitory effect of indomethacin on contractions of guinea-pig ileum induced by angiotensin. J.Pharm. Pharmacol. <u>26</u>,729-730.

30. Kadlec, O., Masek, K. and Šeferna, I. (1974). A modulating role of prostaglandins in contractions of the guinea-pig ileum. Br. J. Pharmacol. <u>51</u>, 565-570.

31. Bennett, A., Eley, K.G. and Stockley, H.L. (1976). Inhibition of peristalsis in guinea-pig ileum and colon by drugs which block prostaglandin synthesis. Br. J. Pharmac. <u>57</u>, 335-340.

32. Burnstock, G., Cocks, T., Paddle, B. and Staszewska-Barczac, J. (1975). Evidence that

prostaglandin is responsible for the 'rebound contraction' following stimulation of non-adrenergic non-cholinergic ('purinergic') inhibitory nerves. Eur. J. Pharmacol. $\underline{31}$, 360-362.

33. Bennett, A., Eley, K.G. and Scholes, G.B. (1968). Effect of prostaglandins E_1 and E_2 on intestinal motility in the guinea-pig and rat. Br. J. Pharmacol. $\underline{34}$, 639-647.

34. Radmanovič, B.Z. (1972). Effect of prostaglandin E_1 on the peristaltic activity of the guinea-pig isolated ileum. Arch. Int. Pharmacodyn. Ther. $\underline{200}$, 396-404.

35. Takai, M., Matsuyama, S. and Yagasaki, O. (1974). Prostaglandin release during extension of the small intestine of the guinea pig. (In Japanese). Jap. J. Smooth Muscle Res. $\underline{10}$, 187-189.

36. Ishizawa, M. and Miyazaki, E. (1973). Effect of prostaglandin on the movement of isolated guinea pig intestine. (In Japanese). Jap. J. Smooth Muscle Res. $\underline{9}$, 235-237.

EFFECT OF PROSTAGLANDINS ON GASTROINTESTINAL FUNCTIONS

André Robert

Department of Experimental Biology
The Upjohn Company
Kalamazoo, Michigan 49001 U.S.A.

INTRODUCTION

Several prostaglandins (PG) of the E and A types were found to affect several functions of the gastrointestinal tract, and to exert therapeutic effect on a variety of lesions of the stomach, the duodenum and the small intestine in animals. They have already been used successfully on a limited basis, in patients with peptic ulcer. PG are known to be produced within the gastrointestinal tract, and, because of their very short half life, are believed to exert their activity primarily where they are produced (1,2,3,4,5,6,7,8). The reader is referred to reviews that have been published on the various effects of PG on the gastrointestinal tract (9,10,11,12,13). We will discuss certain properties of PG related to gastric secretion, ulcer formation, diarrhea, and a property discovered recently called "cytoprotection."

GASTRIC SECRETION

Animal Studies

Natural prostaglandins. PGE_1, PGE_2 and PGA_1 inhibited gastric secretion stimulated by a variety of agents. In particular, secretion induced in dogs by histamine (14,15,16,17,18), pentagastrin (15,17), food (15,16,17), 2-deoxy-D-glucose (15,17), reserpine (17), carbachol (17), and CCK octapeptide (19) was blocked by parenteral administration of certain natural PG of either the E or A type (Fig. 1). Basal secretion in rats was also inhibited by the same kinds of PG given parenterally (18,20,21).

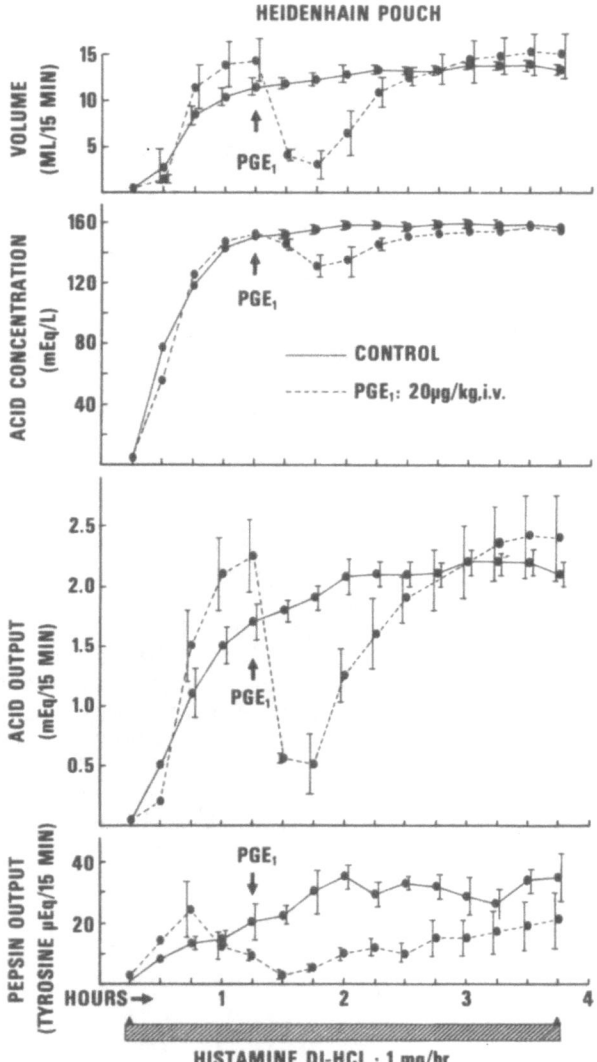

Figure 1. Effect of PGE$_1$ given intravenously at a single dose
(20 µg/kg), on gastric secretion stimulated with histamine dihydrochloride (1 mg/hr), in Heidenhain pouch
dogs. Each curve represents the average of two dogs
(control curve: 11 studies; PGE$_1$ curve: 7 studies).
Vertical bars: ±S.E. of mean. (From J. Physiol. <u>218</u>:
369, 1971).

When given orally at high doses, PGE_2 inhibited gastric secretion of pylorus ligated rats (18,22), although in dogs PGE_1 and PGE_2 were ineffective by this route (23). In rats, perfusion of PGE_1 through the stomach inhibited acid secretion measured in the effluent when the animals were stimulated with pentagastrin (24,25,26,27), histamine (25,26,27) or vagal excitation (25,26). PGE inhibited gastric acid secretion in vitro when applied on the frog gastric mucosa stimulated with histamine (28), gastrin (28) or pentagastrin (29). A similar antisecretory effect was observed in vitro with the dog isolated gastric mucosa stimulated with histamine, pentagastrin or vagal excitation (26).

   Prostaglandin analogs. In view of the relative low activity of these agents when given orally, analogs were prepared with the hope that the antisecretory property of PG could be exerted after oral administration. In particular 15-methyl PGE_2 (30,31) and 16,16-dimethyl PGE_2 (32) were synthesized and were found to inhibit gastric secretion when given orally or parenterally to dogs and rats, to be much more potent than the parent compound, and to exert activity for a much longer duration (18,33,34,35,36, 37) (Fig. 2). Gastric secretion was inhibited regardless of the gastric stimulant [histamine (18,33,34,35,36), pentagastrin (35,36,37), urecholine (35,36), 2-deoxy-D-glucose (35,36), betanechol (37), bombesin (37) or caerulein (37)]. Similar results were obtained in cats (38) and rats (18,39,40). Other analogs of PGE_1 or PGE_2 were also used in animals and found to inhibit gastric secretion (41,42,43,44).

   In addition to systemic and in vitro effects, 15(R)-15-methyl PGE_2, 15(S)-15-methyl PGE_2 methyl ester and 16,16-dimethyl PGE_2 exerted a local antisecretory effect when applied directly onto the gastric mucosa of a Heidenhain pouch. This was demonstrated in dogs in which two Heidenhain pouches were made. Gastric acid secretion from both pouches was stimulated by an intravenous infusion of histamine dihydrochloride (1 mg/hr). When 16,16-dimethyl PGE_2 was administered into one pouch, at doses ranging from 5 to 50 μg per pouch, only that pouch was inhibited, and the inhibition was total for up to 6 hr, depending on the dose. At higher doses, the other pouch was also inhibited, a fact indicating that some of the PG was absorbed from one pouch, carried through the blood stream to the other pouch where it exerted a systemic antisecretory activity (45). However, the local effect was much more potent than the systemic effect. The inhibition of secretion from the second pouch, although total when the dose of PG was high enough, lasted for a much shorter duration than the inhibition from the first pouch. Subsequent studies confirmed this local action of 16,16-dimethyl PGE_2 in dogs (46,47). It is likely that in humans when a methyl PG is administered orally, inhibition of gastric secretion is achieved partly by the local contact of the PG with the gastric mucosa,

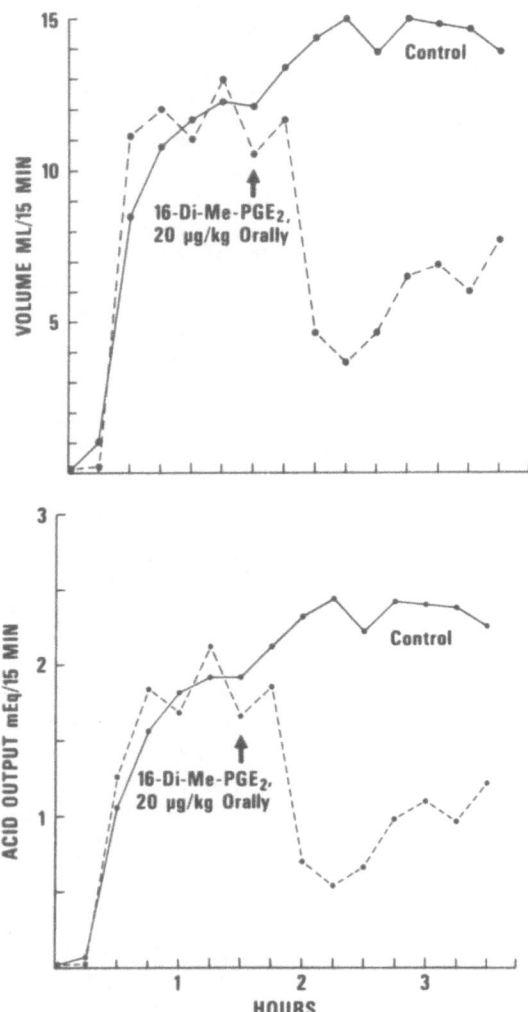

Figure 2. Effect of 16,16-dimethyl PGE_2, methyl ester, given
orally on volume and acid output from a gastric pouch.
Secretion was stimulated by intravenous infusion of
histamine.

and secondarily by systemic absorption and return of the PG to the stomach via the blood stream. This would be an advantage since doses could be adjusted to be high enough to exert a local action but too low to exert significant systemic effect on other organs.

The fact that 15(R)-15-methyl PGE_2 methyl ester inhibited gastric secretion (57,63,66) was somewhat unexpected since the 15(R) configuration usually shows much reduced biological activity. Studies performed in dogs showed that this PG is almost ineffective as a gastric antisecretory agent when given intravenously, but shows marked activity after oral administration. The explanation for this apparent anomaly was found in experiments performed in dogs. It was found that the low pH of stimulated gastric juice transforms approximately half of the 15(R) PG into the correspondent 15(S) PG through epimerization. It was concluded that the newly formed 15(S)-15-methyl PGE_2 methyl ester was responsible for most of the antisecretory activity (48).

Interestingly, PGE_2 and 16,16-dimethyl PGE_2, administered intravaginally to pouch dogs stimulated with histamine, inhibited gastric secretion (49). This result shows that the PG were readily absorbed from the vaginal mucosa. In this respect, the methyl esters of each of these two PG were 4-5 times more potent than the corresponding free acids. The enhanced activity of the methyl esters may be due mainly to their faster absorption through the vaginal mucosa rather than greater intrinsic activity.

Human Studies

Natural prostaglandins. Several studies were performed in humans, both in healthy volunteers and in patients with gastric or duodenal ulcers. Several natural PG inhibited gastric acid secretion when given intravenously to humans. This was the case for PGA_1 [histamine stimulation (50,51)], PGE_1 [pentagastrin (52), tetragastrin (53), basal secretion (54)], PGE_2 [pentagastrin (55), tetragastrin (53), basal secretion (53)]. As in the case of the dog, neither PGE_1 (56), PGE_2 (57), nor PGA_1 (58) affected gastric secretion when given orally, whereas PGA_2 (58) caused a short lived reduction. $PGF_2\alpha$ given intravenously at high doses caused a transient inhibition of gastric secretion stimulated with pentagastrin (52). PGE_1 given orally at doses of 2.5 mg/subject did not inhibit pentagastrin-induced secretion, but induced diarrhea 2-4 hr after treatment (56).

Prostaglandin analogs. 15(S)-15-Methyl PGE_2 methyl ester (57,60,61,62) and 15(R)-15-methyl PGE_2 (63) or its methyl ester (57,64,65,66), given orally to healthy subjects stimulated with pentagastrin, reduced volume of secretion, acid concentration and

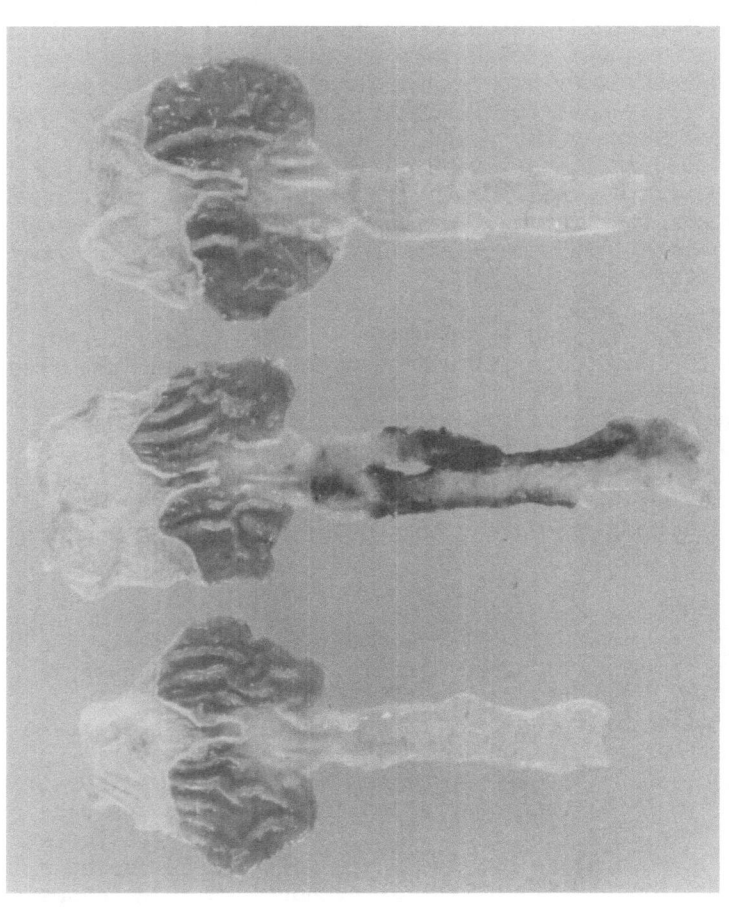

Figure 3. Prevention of duodenal ulcers by 16,16-dimethyl PGE₂. LEFT: stomach and duodenum of control rat. MIDDLE: multiple, bleeding duodenal ulcers (one is perforated) produced in a rat receiving histamine dihydrochloride (0.1 mg/kg-min) plus carbachol (0.35 μg/kg-min) subcutaneously for 24 hr. RIGHT: same treatment as in the middle, plus 16,16-dimethyl PGE₂ 0.2 mg/kg, subcutaneously twice, at 12 hr intervals. Complete protection.

acid output for several hours. The ED50 (dose inhibiting acid output by 50%) was approximately 1 μg/kg. Similarly, 15(R)-15-methyl PGE_2 inhibited pentagastrin-induced gastric acid secretion in duodenal ulcer patients (67). 16,16-Dimethyl PGE_2 was also antisecretory in healthy volunteers when administered either orally (60,62,68,69,70,71,72,73) or intravenously (68,69). These methyl PG [15(R)-15-methyl PGE_2 methyl ester (74), 15(S)-15-methyl PGE_2 methyl ester (74), and 16,16-dimethyl PGE_2 (71,74)] were particularly potent in inhibiting food-induced gastric secretion both in healthy volunteers and duodenal ulcer patients. Only when administered at high doses, were abdominal cramps and loose stools reported. Such doses were higher than those required to reduce gastric acid secretion by 50-70%.

ANTIULCER EFFECT

Animal Studies

Because of the antisecretory effect of several PG, it was thought that they could be used to either prevent or cure gastro-duodenal ulcers. Several PG were therefore tested in various animal models of experimental ulcers.

Gastric Ulcers. PGE_1, PGE_2 and some of the methyl analogs of PGE_2, given either subcutaneously or orally to rats, inhibited the following gastric ulcers: (a) Shay ulcers (produced by pylorus ligation) (2,15,18,20,33,41), (b) stress ulcers [produced by restraint (139) or exertion (41)], (c) steroid-induced ulcers (produced by subcutaneous injections of prednisolone) (18,20), (d) ulcers produced by administration of either serotonin (75), reserpine (22), or bile (76), (e) ulcers produced by oral or intraperitoneal administration of several nonsteroidal anti-inflammatory compounds (NOSAC) such as aspirin (77), indomethacin (40,78,79), and flurbiprofen (79).

Duodenal Ulcers. Either PGE_2 or 16,16-dimethyl PGE_2, given either orally or subcutaneously, inhibited in a dose dependent manner duodenal ulcers produced in rats by (a) a single subcutaneous injection of histamine (80), (b) a constant subcutaneous infusion for 24 or 48 hours of histamine plus carbachol (Fig. 3); histamine plus pentagastrin; or pentagastrin plus carbachol (18,33,45,48,81, 82,83), (c) a single injection of cysteamine (81,84), (d) injections for two days of propionitrile (81,85), (e) administration of several NOSAC (indomethacin, ibuprofen, flurbiprofen) to bile duct ligated animals (86). In other species, PGE_2 or certain analogs of PGE_1 and PGE_2 prevented the formation of duodenal ulcers produced by pentagastrin [guinea pigs (22,41), cats (38)], and histamine [guinea pigs (22,41)]. The main differences between the natural PG and the methyl analogs were that the latter were much more potent, and could inhibit gastric and duodenal ulcers

more effectively when given orally than in the case of the parent
compound.

Human Studies

15(R)-15-Methyl PGE$_2$ (or its methyl ester) was administered
to gastric ulcer patients at a dose of 150 µg/subject, 4 times a
day for 2 weeks. The ulcer, measured endoscopically before and
after treatment, was found to be reduced in size in most patients,
and to have completely disappeared in several subjects, whereas
in the controls, receiving only a placebo, most of the ulcers
were active and of a larger size (63,87,89). In a double blind
study gastric and duodenal ulcer pain following a single oral
treatment with 15(R)-15-methyl PGE$_2$ methyl ester (150 µg/subject)
was of shorter duration than in placebo-treated patients (88).
During a two week treatment of gastric ulcer patients with 15(R)-
15-methyl PGE$_2$ (150 µg/subject four times a day), pain disappeared
in more patients receiving the PG than in those receiving placebo
(89). Finally, a single oral treatment with 15(R)-15-methyl PGE$_2$
methyl ester was reported to increase gastric mucus formation in
patients with either gastric or duodenal ulcer (90). Similar
improvement (symptomatic and from measurement of ulcer size) was
reported after a three week treatment with either 15(R) or 15(S)-
15-methyl PGE$_2$ methyl ester (1 to 1.5 µg/kg, three times a day)
(91).

MECHANISM OF ACTION

The mechanism by which certain PG inhibit gastric secretion
is not known. It is to be noted that the inhibition was demonstrated against all known gastric stimulants, an observation
showing that these PG do not discriminate among stimulants, as
some other inhibitors do. PG seem to block the activity of
parietal cells, regardless of the secretogogue used. The content
in cyclic AMP of gastric mucosa cells was found to be increased
by PGE$_2$ in guinea pigs (92,93) and dogs (94). Similarly, in the
dog gastric mucosa, 16,16-dimethyl PGE$_2$ activated adenylyl cyclase
(95), the enzyme transforming ATP into cyclic AMP. Whether
cyclic AMP is the second messenger responsible for initiating
acid secretion is still debated. Some reported that administration
of cyclic AMP or its dibutyryl derivative stimulates acid secretion
in rats in vivo (25,26,96) and in the frog gastric mucosa in
vitro (97,98), whereas others found that it is inactive in rats
(99), and is even inhibitory in dogs (100). At present, the best
evidence indicates that, at least in dogs, cyclic AMP does not
initiate gastric acid secretion (101,102).

Although certain PG of the E type decrease gastric mucosal
blood flow [dogs (46,103,104,105) and rats (106,107,108, 109,110)]
this effect is not responsible for its antisecretory activity,

but rather is the result of it. In studies comparing the effect of PGE_1 with that of norepinephrine, it was shown that PGE_1 first decreases the metabolic activity of parietal cells, which in turn results in a reduced mucosal blood flow. Norepinephrine works in an opposite manner. Norepinephrine exerts primarily a vasoconstricting effect on the vessels of the gastric mucosa, and in turn this ischemia inhibits gastric secretion. (103).

It was also well demonstrated that PG do not exert an anticholinergic effect (28,29,111,112). Therefore, the inhibition of gastric secretion is not due to an inhibition of vagal activity.

It is of interest that PG are antisecretory whether they are applied directly to the surface of the gastric mucosa or whether they reach the parietal cells from the blood stream. Such local action was shown for natural PG of the E type in rats (18,22,24,25, 26,27) and frogs (28,29), and for 16,16-dimethyl PGE_2 in a dog gastric pouch (45,47,113). These observations suggest that PG can penetrate the gastric mucosa and exert a direct inhibitory effect on parietal cells.

It was also demonstrated that antisecretory PG do not decrease the acid content of gastric juice by a phenomenon of acid back diffusion through a breaking of the mucosal barrier. Although 16,16-dimethyl PGE_2, introduced at an excessive dose (300 µg) into a denervated gastric pouch, altered the mucosal barrier (114), the same PG showed opposite effects in an isolated dog gastric mucosa preparation (95,138). Not only did 16,16-dimethyl PGE_2 tighten the mucosal barrier (increase in potential difference and of net sodium flux), but it reversed the deleterious effects of indomethacin on the gastric mucosal barrier. In another study in humans (115), PGE_2 administered orally raised the potential difference of the gastric mucosa (relative to blood), and reversed the damaging effects of both aspirin and indomethacin also given orally. This result indicates that PGE_2 protects the mucosal barrier in humans.

Regardless of the mechanism of action, the antisecretory and antiulcer studies performed in animals and humans permit to conclude that certain PG, especially those active orally, may eventually be used in the treatment of peptic ulcer.

ENTEROPOOLING

It is well known that most prostaglandins, when administered at high doses, tend to induce diarrhea, both in animals and humans. Although PG are known to stimulate contraction of the longitudinal smooth muscle of the intestine, this property does not explain the development of diarrhea. Increased motility can accelerate the transit of intestinal contents, but does not

liquefy these contents, a prerequisite for diarrhea to occur.
The diarrheogenic effect of large doses of PG was investigated
and found to result from the accumulation of large amounts of
fluid in the small intestine. The large amounts of fluid that
fill the small intestine are then carried to the large intestine
and expelled as diarrhea. This phenomenon of fluid accumulation
in the small intestine after PG administration was observed in
dogs (116,117), rats (118,119), and humans (11,59,120,121,122,123,
124,125,126). This property was called "enteropooling," which is
defined: the accumulation of fluid into the small intestine due
to the passage of fluid and electrolytes from the blood into the
intestinal lumen, plus a certain degree of inhibition of absorption
of water and electrolytes already present in the intestinal lumen
(119). Enteropooling seems to be mediated by stimulation of
cyclic AMP formation in the intestinal mucosa (127). Indeed,
cyclic AMP itself was shown to be enteropooling in rabbits (128).
Moreover, cholera toxin, which also stimulates adenylyl cyclase
activity (127) and cyclic AMP formation (129) in the small intestine
is well known to induce formation of enormous amounts of intestinal
fluid.

This property of PG to favor accumulation of intestinal
fluid was used as the basis of an enteropooling assay which can
determine quantitatively the diarrheogenic property of PG (119).
It consists of administering a PG, either orally or subcutaneously,
to rats which have been fasted for 24 hr. Thirty minutes after
treatment, the animals are killed and the entire small intestine,
from the pylorus to the terminal ileum, is dissected out. The
intestinal contents are then emptied into a graduated test tube
by milking the whole length of the small intestine with fingers.
The volume of fluid thus collected is measured to the nearest 0.1
ml, and is in direct relation to the diarrheogenic potential of a
PG. At maximum doses, a volume of 3.5 to 4 ml of fluid can be
collected from the small intestine of a PG-treated animal.
Therefore, the ED50 has been defined as a dose producing a volume
of 1.75 ml in 30 minutes.

The degree of enteropooling is predictive of diarrhea, which
usually occurs one hour after administration of a diarrheogenic
PG. The presence of diarrhea, however, is dependent upon the
degree of enteropooling. For a small to moderate dose of PG,
although enteropooling can be observed and measured, the fluid is
usually reabsorbed so that no diarrhea occurs. With a higher
dose, however, the extent of reabsorption is insufficient for all
the fluid that has entered the large intestine, and diarrhea
occurs. Finally, as in humans, diarrhea is more difficult to
produce with a PG or with a laxative, in a fed rather than in a
fasted animal, probably because the enteropooling fluid is mixed
with the solid products of food digestion, so that the contents

of the small intestine consist of a thick paste rather than a liquid. As a result, what is expelled is solid or semisolid, rather than frank diarrhea.

CYTOPROTECTION

Intestinal Lesions

Several nonsteroidal antiinflammatory compounds (NOSAC) are known to produce gastrointestinal side effects in a certain percentage of people treated for inflammatory diseases. In animals, such as the rat, high doses of these compounds, such as indomethacin, produce within 2-3 days a severe, fatal syndrome of multiple intestinal lesions characterized by ulcerations of the small intestine, from the upper jejunum to the lower ileum (130). These ulcers perforate and produce peritonitis with abundant exudate. The animals usually die within 3 to 4 days. These NOSAC are known to inhibit the formation of PG in the body by blocking the activity of PG synthetase (131). It was thought that these gastrointestinal side effects in humans, and the severe intestinal lesions in animals produced by NOSAC could be due to a deficiency in PG produced by these very same compounds. To test the hypothesis, several PG were administered to rats treated with indomethacin given at doses capable of producing the fatal intestinal syndrome. These studies showed that administration of PG of different types prevented completely the development of the syndrome (79,132,133,134) (Fig. 4). Actually a single treatment, either orally or subcutaneously with a PG such as 16,16-dimethyl PGE$_2$ given near the time of indomethacin prevented the syndrome (134). These results suggest that indeed the intestinal lesions are not due to indomethacin per se, but to the PG deficiency produced by indomethacin. The situation is analogous to that of several endocrine diseases in which a hormone deficiency can be corrected by administration of that same hormone (such as cortisone for Addison's disease, insulin for diabetes, thyroxine for myxedema).

This property of PG to protect the intestinal epithelium from injury was called "cytoprotection" (by Dr. E. D. Jacobson, University of Texas, Houston, Texas, after examining our data), since it consists of protecting the cells of the intestinal epithelium from the damaging effects of agents such as indomethacin. It is not known what are the immediate damaging agents that cause the intestinal lesions. However, it appears that in the absence of PG in the cells of the intestinal epithelium, noxious agents present in the intestinal lumen, such as microorganisms, bile acids, food debris, chemical irritants, can damage the cells to the point of causing necrosis and penetration of the full thickness of the intestinal wall, with eventual perforation and peritonitis. The role of microorganisms in this syndrome was demonstrated by

Figure 4. Prevention of indomethacin-induced intestinal lesions by 15-keto PGF$_2\alpha$. TOP: normal rat intestinal loop. LEFT: small intestine from a rat treated with 20 mg/kg of indomethacin, given once orally. After 3 days, the small intestine is transformed into an adhesive mass as a result of peritonitis. RIGHT: indomethacin as above, plus 15-keto PGF$_2\alpha$ infused at the rate of 40 µg/kg-min for 3 days. Complete protection. This is an example of intestinal cytoprotection.

studies in germfree rats. In such animals, indomethacin did not produce the intestinal lesions (135). Similarly, administration of antibiotics prevented the intestinal syndrome from developing (130). Whether the protection in germfree rats and in animals treated with antibiotics is due to the absence of microorganisms per se, or to the absence of secondary bile acids normally formed by the intestinal flora from primary bile acids, is unknown.

Gastric Lesions

The cytoprotective property of PG was then tested for the stomach. Gastric lesions were produced with either aspirin, indomethacin, or ethanol. These lesions consist of multiple bleeding ulcers of the glandular portion of the stomach (the corpus), and are produced particularly well in fasted rats. When PG of several types were administered immediately prior to the administration of these damaging agents, the lesions were prevented in a dose dependent manner (79,136). In the case of aspirin and indomethacin, agents other than PG, such as anticholinergics and antacids, also protected the mucosa. The reason seems to be that in the case of these two irritants, acid is essential for the gastric lesions to form. This was further demonstrated in the case of aspirin, where the lesions were prevented by an anticholinergic, but developed fully when, in addition to an anticholinergic, hydrochloric acid 0.1 $\underline{N}$ was also administered orally (137). In the case of alcohol, however, an anticholinergic agent (methscopolamine bromide) administered at doses that completely abolished gastric acid secretion, failed to prevent the gastric lesions. They were as severe and as numerous as when alcohol alone was given (Fig. 5). Moreover, PG that do not inhibit gastric acid secretion, such as 16,16-dimethyl PGA_2 and 15-methyl $PGF_{2\beta}$, prevented the alcohol-induced lesion. Finally, 16,16-dimethyl PGE_2 administered at a very low dose either orally (0.5 µg/kg) or subcutaneously (1.5 µg/kg) prevented the alcohol-induced lesions (Fig. 5). Such doses are approximately 100 times lower than the threshold dose inhibiting gastric secretion in rats. It is quite clear, therefore, that a number of PG are cytoprotective for the stomach as well as for the small intestine.

Mechanism of Cytoprotection

The mechanism of cytoprotection by PG is unknown. It was recently reported that after local application of indomethacin to the gastric mucosa of an isolated dog preparation, the sodium pump (actively transporting sodium from the lumen to the serosal side) was severely altered. When either 16,16-dimethyl PGE_2 or dibutyryl cyclic AMP was added to the preparation, the sodium pump was returned to normal (95,138). The interpretation was that the gastric sodium pump is essential for the integrity of

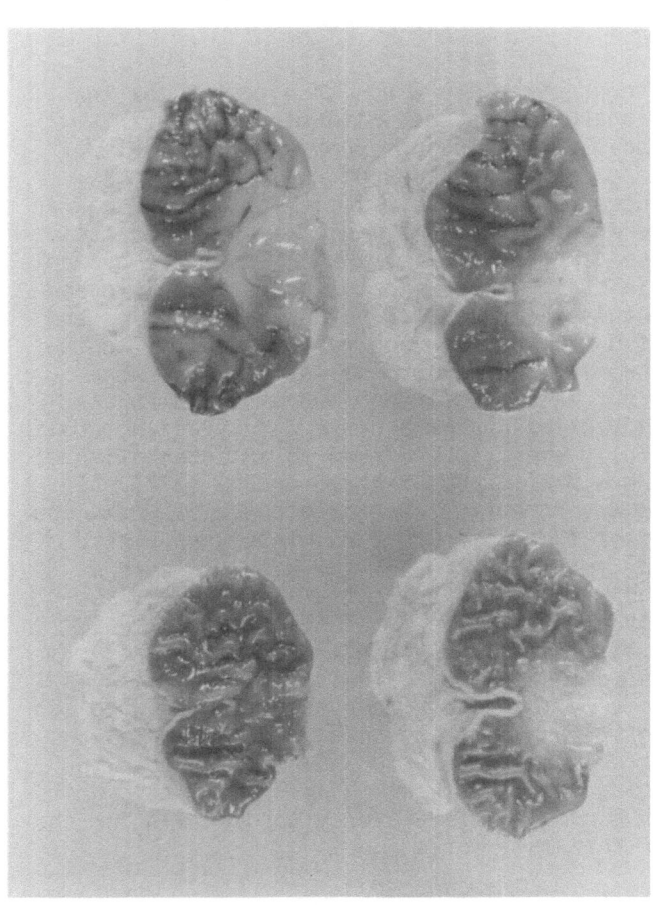

Figure 5. Prevention of alcohol-induced gastric lesions by 16,16-dimethyl PGE₂. TOP LEFT: control rat stomach. TOP RIGHT: stomach from a rat that received 1 ml of 80% ethanol orally, 5 hr earlier. Severe, numerous and bleeding ulcers in the corpus (glandular portion). BOTTOM LEFT: stomach from a rat that received same amount of ethanol, plus two oral administrations of 16,16-dimethyl PGE₂ 0.5 μg/kg (30 min before and 2-1/2 hr after ethanol). Complete protection. This is an example of gastric cytoprotection. BOTTOM RIGHT: stomach from a rat that received same amount of ethanol, plus two oral administrations of methscopolamine bromide 10 mg/kg (dose inhibiting gastric acid secretion completely). Note that the gastric ulcers are severe and numerous as with ethanol alone.

the cells of the gastric epithelium, and that it is vulnerable to damaging agents such as indomethacin. The sodium pump is thus protected by certain PG which, by activating adenylyl cyclase and therefore increasing the intracellular cyclic AMP content, stimulate the active transport of sodium. This appears to be a valid hypothesis, which needs to be tested also for the intestinal epithelium.

Clinical Implications

The implications of these results are that certain PG could be used in a variety of gastrointestinal ailments, either spontaneous such as inflammatory bowel diseases, peptic ulcer, hemorrhagic gastritis, or drug-induced lesions, such as damage produced by NOSAC. It is also conceivable that the antiulcer effect (gastric and duodenal) of certain PG is not due solely to their gastric antisecretory property, but also to their cytoprotective effect. Those two properties would potentiate each other in providing increased efficacy in the case of peptic ulcer.

REFERENCES

1. Bartels, J, Junze, H, Vogt, W and Willie G. Prostaglandin: liberation from and formation in perfused frog intestine. Naunyn-Schmiedebergs Arch Pharmakol 226:199-207 (1970).

2. Bennett, A, Friedmann, CA and Vane JR. Release of prostaglandin El from the rat stomach. Nature (Lond) 216:873-876 (1967).

3. Bennett, A, Murray, JG and Wyllie, JH. Occurrence of prostaglandin E2 in the human stomach, and a study of its effects on human isolated gastric muscle. Br J Pharmacol Chemother 32:339-349 (1968).

4. Coceani, F, Pace-Asciak, C, Volta, F and Wolfe, LS. Effect of nerve stimulation on prostaglandin formation and release from the rat stomach. Am J Physiol 213:1056-1064 (1967).

5. Miyazaki, Y. Isolation of prostaglandin E-like substance from the mucous membrane layer of large intestine of pig. Sapporo Med J 34:141-154 (1968); (Japanese).

6. Miyazaki, Y. Occurrence of prostaglandin El in the mucous membrane layer of swine large intestine. Sapporo Med J 34:321-334 (1968); (Japanese).

7. Pace-Asciak, C, Morawska, K, Coceani, F and Wolfe, LS. The

biosynthesis of prostaglandins E2 and F2α in homogenates of the rat stomach. Prostaglandin Symp of Worcester Found for Exp Biol, PW Ramwell and JE Shaw, Interscience, NY: 371-378 (1968).

8. Wolfe, LS, Coceani, F and Pace-Asciak, C. The relationship between nerve stimulation and the formation and release of prostaglandins. Pharmacologist 9:171-172 (1967).

9. Bennett, A. Effects of prostaglandins on the gastrointestinal tract. In: The Prostaglandins: Progress in Research, SMM Karim, MTP Med and Tech Publ Co, Oxford:205-221 (1972).

10. Main, IHM. Prostaglandins and the gastrointestinal tract. In: The Prostaglandins. Pharmacological and Therapeutic Advances, MF Cuthbert, Heinemann, Lond:287-323 (1973).

11. Matuchansky, C and Bernier, JJ. Prostaglandines et appareil Digestif. Biol Gastroéntérol (Paris) 6:251-268 (1973).

12. Waller, SL. Prostaglandins and the gastrointestinal tract. Gut 14:402-417 (1973).

13. Wilson, DE. Prostaglandins. Their actions on the gastro-intestinal tract. Arch Intern Med 133:112-118 (1974).

14. Nezamis, JE, Robert A, Stowe, DF. Inhibition by prostaglandin E1 of gastric secretion in the dog. J Physiol (Lond) 218:369-383 (1971).

15. Robert, A. Antisecretory property of prostaglandins. Prostaglandin Symp of Worcester Found for Exp Biol, PW Ramwell and JE Shaw, Interscience, NY:47-54 (1968).

16. Robert, A, Nezamis, JE, Phillips, JP. Inhibition of gastric secretion by prostaglandins. Am J Dig Dis 12:1073-1076 (1967).

17. Robert, A, Phillips, JP, Nezamis, JE. Inhibition by prosta-glandin E1 of gastric secretion in the dog. Gastroenterology 54:1263 (1968).

18. Robert, A, Schultz, JR, Nezamis, JE and Lancaster, C. Gastric antisecretory and antiulcer properties of PGE2, 15-methyl PGE2, and 16,16-dimethyl PGE2. Intravenous, oral and intrajejunal administration. Gastroenterology 70:359 (1976).

19. Kaminski, DL, Ruwart, M, Willman, VL. The effect of prosta-glandin A1 and E1 on canine hepatic bile flow. J Surg Res 18:391-397 (1975).

20. Robert, A, Nezamis, JE, Phillips, JP. Effect of prostaglandin E1 on gastric secretion and ulcer formation in the rat. Gastroenterology 55:481-487 (1968).

21. Main, IHM. Effects of prostaglandin E2 (PGE2) on the output of histamine and acid in rat gastric secretion induced by pentagastrin or histamine. Br J Pharmacol 36:214P-215P (1969).

22. Lee, YH, Cheng, WD, Bianchi, RG, Mollison, K, Hansen J. Effects of oral administration of PGE2 on gastric secretion and experimental peptic ulcerations. Prostaglandins 3:29-45 (1973).

23. Robert, A. (Unpublished).

24. Banerjee, AK, Phillips, J, Winning, WW. E-Type prostaglandins and gastric acid secretion in the rat. Nature (New Biol) 238:177-179 (1972).

25. Shaw, JE, Ramwell, PW. Inhibition of gastric secretion in rats by prostaglandin E1. Prostaglandin Symp of Worcester Found for Exp Biol, PW Ramwell and JE Shaw, Interscience, NY:55-66 (1968).

26. Shaw, JE, Urquhart, J. Parameters of the control of acid secretion in the isolated blood perfused stomach. J Physiol (Lond) 226:107P-108P (1972).

27. Whittle, BJR. Studies on the mode of action of cyclic 3'5'-AMP and prostaglandin E2 on rat gastric acid secretion and mucosal blood flow. Br J Pharmacol 46:546P-547P (1972).

28. Way, L, Durbin, RP. Inhibition of gastric acid secretion in vitro by prostaglandin E1. Nature (Lond) 221:874-875 (1969).

29. Nakaji, NT, Charters, AC, Guillemin, RCL, Orloff, MJ. Inhibition of gastric secretion by somatostatin. Gastro-enterology 70:989 (1976).

30. Yankee, EW, Axen, U, Bundy, GL. Total synthesis of 15-methylprostaglandins. J Am Chem Soc 96:5865-5876 (1974).

31. Yankee, EW, Bundy, GL. (15S)-15-Methylprostaglandins. J Am Chem Soc 94:3651-3652 (1972).

32. Magerlein, BJ, DuCharme, DW, Magee, WE, Miller, WL, Robert, A, Weeks, JR. Synthesis and biological properties of 16-alkylprostaglandins. Prostaglandins 4:143-145 (1973).

33. Robert, A, Lancaster, C, Nezamis, JE, Badalamenti, JN. A
    gastric antisecretory and antiulcer prostaglandin with oral
    and long-acting activity. Gastroenterology 64:790 (1973).

34. Robert, A, Magerlein, BJ. 15-Methyl PGE2 and 16,16dimethyl
    PGE2: potent inhibitors of gastric secretion. Adv Biosci
    9:247-253 (1973).

35. Mihas, A, Gibson, R, Hirschowitz, BI. Inhibition of gastric
    secretion in dogs by a synthetic prostaglandin (PGE2). Fed
    Proc 34:442 (1975).

36. Mihas, AA, Gibson, RG, Hirschowitz, BI. Inhibition of
    gastric secretion in the dog by 16,16-dimethyl prostaglandin
    E2. Am J Physiol 230:351356 (1976).

37. Impicciatore, M, Bertaccini, G, Usardi, MM. Effect of a new
    synthetic prostaglandin on acid gastric secretion in different
    laboratory animals. In: Advances in Prostaglandin and
    Thromboxane Research. Raven Press, NY, p. 945 (1976).

38. Konturek, SJ, Radecki, T, Demitrescu, T, Kwiecien, N, Pucher,
    A, Robert, A. Effect of synthetic 15-methyl analog of
    prostaglandin E2 on gastric secretion and peptic ulcer
    formation. J Lab Clin Med 84:716-725 (1974).

39. Carter, DC, Ganesan, PA, Bhana, D, Karim, SMM. The effect
    of locally administered prostaglandin 15(R)15 methyl-E2
    methyl ester on gastric ulcer formation in the Shay rat
    preparation. Prostaglandins 5:455-463 (1974).

40. Whittle, BJR. Gastric antisecretory and antiulcer activity
    of prostaglandin E2 methyl analogs. In: Advances in Prosta-
    glandin and Thromboxane Research. Raven Press, NY, p. 948
    (1976).

41. Lee, YH, Bianchi, RG. The antisecretory and antiulcer
    activity of a prostaglandin analog, SC-24655, in experimental
    animals. Abstr 5th Int Congr Pharmacol San Franc 136 (1972).

42. Lippmann, W. Inhibition of gastric acid secretion by a
    potent synthetic prostaglandin. J Pharm Pharmacol 22:65-67
    (1970).

43. Lippmann, W. Inhibition of gastric acid secretion in the
    rat by synthetic prostaglandin analogues. Ann NY Acad Sci
    180:332-335 (1971).

44. Lippmann, W, Seethaler, K. Oral antiulcer activity of a
    synthetic prostaglandin analogue (9-oxoprostanoic acid: AY-

22,469). Experientia 29:993-995 (1973).

45. Robert, A. Prostaglandins and the Digestive System. In: Prostaglandines 1973. Inserm, Paris:297-315 (1973).

46. Cheung, LY, Lowry, SF. Effect of prostaglandin E2 (PGE2) and 16,16-dimethyl PGE2 (DMPGE2) on gastric acid secretion and blood flow. Gastroenterology 70:870 (1976).

47. Andersson, S, Nylander, B. Effects of some newer prostaglandin E2 derivatives on gastrointestinal functions in the dog and their probable mode of action. Acta Hepatol-Gastroenterologica 23:79 (1976).

48. Robert, A, Yankee, EW. Gastric antisecretory effect of 15(R)-15-methyl PGE2, methyl ester and of 15(S)-15-methyl PGE2, methyl ester. Proc Soc Exp Biol Med 148:1155-1158 (1975).

49. Robert, A, Lancaster, C, Nezamis, JE. Inhibition of gastric secretion after intravaginal administration of prostaglandins. In: Advances in Prostaglandin and Thromboxane Research. Raven Press, NY, p. 946 (1976).

50. Wilson, DE, Phillips, C, Levine, RA. Inhibition of gastric secretion in man by prostaglandin A1 (PGA1). Gastroenterology 58:1007 (1970).

51. Wilson, DE, Phillips, C, Levine, RA. Inhibition of gastric secretion in man by prostaglandin A1. Gastroenterology 61:201-206 (1971).

52. Classen, M, Koch, H, Bickhardt, J, Topf, G, Demling, L. The effect of prostaglandin E1 on the pentagastrin-stimulated gastric secretion in man. Digestion 4:333-344 (1971).

53. Wada, T, Ishizawa, M. Effects of prostaglandin on the function of the gastric secretion. Jap J Clin Med 28:2465-2468 (1970); (Japanese).

54. Classen, M, Koch, H, Deyhle, P, Weidenhiller, S, Demling, L. Wirkung von Prostaglandin E1 auf die basale Magensekretion des Menschen. Klin Wschr 48:876-878 (1970).

55. Newman, A, DeMoraes-Fil, JPP, Philippakos, D, Misiewicz, JJ. The effect of intravenous infusions of prostaglandins E2 and F2α on human gastric function. Gut 16:272-276 (1975).

56. Horton, EW, Main, IHM, Thompson, CJ, Wright, PM. Effect of orally administered prostaglandin E1 on gastric secretion and gastrointestinal motility in man. Gut 9:655-658 (1968).

57. Karim, SMM, Carter, DC, Bhana, D, Ganesan, PA. Effect of
    orally administered prostaglandin E and its 15-methyl analogues
    on gastric secretion. Br Med J 1:143-146 (1973).

58. Bhana, D, Karim, SMM, Carter, DC, Ganesan, PA. The effect of
    orally administered prostaglandins A, A2 and 15 epi-A2 on
    human gastric acid secretion. Prostaglandins 3:307-316
    (1973).

59. Cummings, JH, Newman, A, Misiewicz, JJ, Milton-Thompson, GJ
    and Billings, JA. Effect of intravenous prostaglandin F2α on
    small intestinal function in man. Nature (Lond) 243: 169-171
    (1973).

60. Nylander, B, Andersson, S. Gastric secretory inhibition
    induced by three methyl analogs of prostaglandin E2 admin-
    istered intragastrically to man. Scand J Gastroenterol
    9:751-758 (1974).

61. Nylander, B, Robert, A, Andersson, S. Gastric secretory
    inhibition by certain methyl analogs of prostaglandin E2
    following intestinal administration in man. Scand J Gastro-
    enterol 9:759-762 (1974).

62. Robert, A, Nylander, B, Andersson, S. Marked inhibition of
    gastric secretion by two prostaglandin analogs given orally
    to man. Life Sci 14:533-538 (1974).

63. Karim, SMM, Fung, WP. Effect of 15(R)-15-methyl prostaglandin
    E2 on gastric secretion and a preliminary study on the healing
    of gastric ulcers in man. Int Res Commun Syst Med Sci 3:348
    (1975).

64. Amy, JJ, Jackson, DM, Ganesan, PA, Karim, SMM. Prostaglandin
    15(R)-15-methyl-E2 methyl ester for suppression of gastric
    acidity in gravida at term. Br Med J 4:208-211 (1973).

65. Carter, DC, Karim, SMM, Bhana, D, Ganesan, PA. Inhibition of
    human gastric secretion by prostaglandin. Br J Surg 60:828-
    831 (1973).

66. Karim, SMM, Carter, DC, Bhana, D, Ganesan, PA. Effect of
    orally and intravenously administered prostaglandin 15(R)-15-
    methyl E2 on gastric secretion in man. Adv Biosci 9:255-264
    (1973).

67. Chen, FWK, Teck, HS, Karim, SMM. The effect of 15(R)-15-
    methyl prostaglandin E2 on gastric acid secretion in duodenal
    ulcer patients. Prostaglandins (in press).

68. Karim, SMM, Carter, DC, Bhana, D, Ganesan, PA. Inhibition of basal and pentagastrin induced gastric acid secretion in man with prostaglandin 16:16-dimethyl E2 methyl ester. Int Res Commun Syst (73-3)8-3-2 (1973).

69. Karim, SMM, Carter, DC, Bhana, D, Ganesan, PA. The effect of orally and intravenously administered prostaglandin 16:16 dimethyl E2 methyl ester on human gastric acid secretion. Prostaglandins 4:71-83 (1973).

70. Karim, SMM, Fung, WP. Effects of some naturally occurring prostaglandins and synthetic analogs on gastric secretion and ulcer healing in man. In: Advances in Prostaglandin and Thromboxane Research, Raven Press, NY, p. 529 (1976).

71. Ippoliti, AF, Isenberg, JI, Maxwell, VJ, Walsh, JH. The effect of 16,16-dimethyl PGE2 on meal-stimulated gastric acid secretion and serum gastrin in duodenal ulcer patients. Gastroenterology 70:488-491 (1976).

72. Wilson, DE, Quertermus, J, Raiser, M, Curran, J, Robert, A. Inhibition of stimulated gastric secretion by an orally administered prostaglandin capsule: a study in normal men. Ann Int Med 84:688-691 (1976).

73. Wilson, DE, Winnan, G, Quertermus, J, Tao, P. Effects of an orally administered prostaglandin analogue (16,16-dimethyl prostaglandin E2) on human gastric secretion. Gastroenterology 69:607-611 (1975).

74. Konturek, SJ, Kwiecien, N, Swierczek, J, Oleksy, J, Sito, E, Robert A. Comparison of methylated prostaglandin E2 analogs given orally in the inhibition of gastric responses to pentagastrin and peptone meal in man. Gastroenterology 70:683-687 (1975).

75. Ferguson, WW, Edmonds, AW, Starling, JR, Wangensteen, SL. Protective effect of prostaglandin E1 (PGE1) on lysosomal enzyme release in serotonin-induced gastric ulceration. Ann Surg 177:648-654 (1973).

76. Mann, NS. Prevention of bile induced acute erosive gastritis by prostaglandin E2, Maalox and cholestyramine. Gastro-enterology 68:946 (1975).

77. Robert, A. Antisecretory prostaglandins in the treatment of gastric hypersecretion and peptic ulcer. In: Progress in Gastroenterology, Volume III. G. B. Jersy Glass, Ed. 1977 (in press).

78. Daturi, S, Franceschini, J, Mandelli, V, Mizzutti, B, Usardi, MM. A proposed role for PGE2 in the genesis of stress-induced gastric ulcers. Br J Pharmacol 52:464P (1974).

79. Robert, A. Antisecretory, antiulcer, cytoprotective and diarrheogenic properties of prostaglandins. In: Advances in Prostaglandin and Thromboxane Research. Raven Press, NY, p. 507 (1975).

80. Robert, A, Standish, WL. Production of duodenal ulcers, in rats, with one injection of histamine. Fed Proc 32:322 (1973).

81. Robert, A. Prevention by prostaglandins of duodenal ulcers produced experimentally in rats. Fifth World Congress of Gastroenterology, Mexico City, October 13-18, p. 123 (1974).

82. Robert, A. Duodenal ulcers in the rat: Production and prevention. In: Peptic Ulcer, CJ Pfeiffer, Munksgaard, Copenh:21-33 (1971).

83. Robert, A, Stowe, DF, Nezamis, JE. Prevention of duodenal ulcers by administration of prostaglandin E2 (PGE2). Scand J Gastroenterol 6:303-305 (1971).

84. Robert, A, Nezamis, JE, Lancaster, C, Badalamenti, JN. Cysteamine-induced duodenal ulcers: a new model to test antiulcer agents. Digestion 11:199214 (1975).

85. Robert, A, Nezamis, JE, Lancaster, C. Duodenal ulcers produced in rats by propionitrile: factors inhibiting and aggravating such ulcers. Toxicol Appl Pharmacol 31:201-207 (1975).

86. Robert, A., Nezamis, JE, Lancaster, C. Duodenal ulcers produced by nonsteroidal antiinflammatory compounds plus bile duct ligation. Fed Proc 34:442 (1975).

87. Fung, WP, Karim, SMM, Tye, CY. Effect of 15(R)-15-methyl-prostaglandin E2 methyl ester on healing of gastric ulcers. Controlled Endoscopic Study. Lancet 2:10-12 (1974).

88. Fung, WP, Karim, SMM, Tye, CY. Double-blind trial of 15(R)-15-methyl prostaglandin E2 methyl ester in the relief of peptic ulcer pain. Ann Acad Med 3:375 (1974).

89. Fung, WP, Karim, SMM. Effect of 15(R)-15-methyl prostaglandin E2 on the healing of gastric ulcers. Med J Aust 2:127 (1976).

90. Fung, WP, Lee, SK, Karim, SMM. Effect of prostaglandin 15(R)-15-methyl-E2-methyl ester on the gastric mucosa in patients with peptic ulceration - an endoscopic and his- tological study. Prostaglandins 5:465-472 (1974).

91. Gibinski, K, Rybicka, J, Mikos, E, Nowak, A. Gastroduodenal ulcer healing by oral M-prostaglandins. Tenth Internat. Congr Gastroent Budapest, June 23-29, p 9 (1976).

92. Perrier, CV, Laster, L. Adenyl cyclase activity of guinea pig gastric mucosa: stimulation by histamine and prostaglandins J Clin Invest 49:73A (1970).

93. Wollin, A, Code, CF, Dousa, TP. Evidence for separate histamine and prostaglandin sensitive adenylate cyclases (AC) in guinea pig gastric mucosa (GM). Clin Res 22:606A (1974).

94. Dozois, RR, Wollin, A. Prostaglandine E2, adenylate cyclase et secretion gastrique chez le chien. Biol Gastro-enterol 8:122 (1975).

95. Chaudhury, TK, Jacobson, ED. Basis for gastric mucosal cytoprotection by prostaglandins. Fed Proc 35:393 (1976).

96. Main, IHM, Whittle, BJR. Investigation of the vasodilator and antisecretory role of prostaglandins in the rat gastric mucosa by use of nonsteroidal antiinflammatory drugs. Br J Pharmacol 53:217-224 (1975).

97. Harris, JB, Nigon, K, Alonso, D. Adenosine-3',5'-monophosphate: intracellular mediator for methylxanthine stimulation of gastric secretion. Gastroenterology 57:377 (1969).

98. Harris, JP, Alonso, D. Stimulation of the gastric mucosa by adenosine-3,5'-monophosphate. Fed Proc 24:1368 (1965).

99. Taft, RC, Sessions, JT. Inhibition of gastric acid secretion by dibutyryl cyclic adenosine 3',5'-monophosphate (DB.cAMP). Clin Res 20:43 (1972).

100. Levine, RA, Wilson, DE. The role of cyclic AMP in gastric secretion. Ann NY Acad Sci 185:363-375 (1971).

101. Mao, CC, Shanbour, LL, Hodgins, DS, Jacobson, ED. Cyclic adenosine-3',5'-monophosphate (cyclic AMP) and secretion in the canine stomach. Gastroenterology 63:427 (1972).

102. Thompson, WJ, Rosenfeld, GC, Jacobson, ED. Adenylyl cyclase and gastric acid secretion. Fed Proc (in press) (1976).

103. Jacobson, ED. Comparison of prostaglandin E1 and norepine-
     phrine on the gastric mucosal circulation. Proc Soc Exp
     Biol Med 133:516-519 (1970).

104. Wilson, DE, Levine, RA. Decreased canine gastric mucosal
     blood flow induced by prostaglandin E1: a mechanism for its
     inhibitory effect on gastric secretion. Gastroenterology
     56:1268 (1969).

105. Wilson, DE, Levine, RA. The effect of prostaglandin E1 on
     canine gastric acid secretion and gastric mucosal blood
     flow. Am J Dig Dis 17:527-532 (1972).

106. Banerjee, AK, Christmas, AJ, Hall, CE. Effects of H2-
     receptor antagonists, prostaglandins E2 and 16:16-dimethyl-
     E2-methyl ester on gastric acid secretion, mucosal blood
     flow and ulceration. Abstr 6th Int Congr Pharmacol:120
     (1975).

107. Main, IHM, Whittle, BJR. Effects of prostaglandins on the E
     and A series on rat gastric mucosal blood flow as determined
     by 14C-aniline clearance. Abstr 5th Int Congr Pharmacol San
     Franc 145 (1972).

108. Main, IHM, Whittle, BJR. Effects of prostaglandin E2 on rat
     gastric mucosal blood flow, as determined by 14C-aniline
     clearance. Br J Pharmacol 44:331P-332P (1972).

109. Main, IHM, Whittle, BJR. The relationship between rat
     gastric mucosal blood flow and acid secretion during oral or
     intravenous administration of prostaglandins and dibutyryl
     cyclic AMP. Adv Biosci 9:271-275 (1973).

110. Main, IHM, Whittle, BJR. The effects of E and A prosta-
     glandins on gastric mucosal blood flow and acid secretion in
     the rat. Br J Pharmacol 49:428-436 (1973).

111. Bergström, S, Eliasson, R, vonEuler, US, Sjovall, J. Some
     biological effects of two crystalline prostaglandin factors.
     Acta Physiol Scand 45:133-144 (1959).

112. Waitzman, MB, King, CD. Prostaglandin influences on intra-
     ocular pressure and pupil size. Am J Physiol 212:329-334
     (1967).

113. Andersson, S, Nylander, B. Local inhibitory action of
     16,16-dimethyl PGE2 on gastric acid secretion in the dog.
     In: Advances in Prostaglandin and Thromboxane Research.
     Raven Press, NY, p. 943 (1976).

114. O'Brien, PE, Carter, DC. Effect of gastric secretory in-
     hibitors on the gastric mucosal barrier. Gut 16:437-442
     (1975).

115. Cohen, MM, Pollett, J. Prevention of human gastric mucosal
     damage by prostaglandin E2. Surg Forum, October 1976 (in
     press).

116. Greenough, WB, Pierce, NF, Al-Awqati, Q, Carpenter, CCJ.
     Stimulation of gut electrolyte secretion by prostaglandins,
     theophylline and cholera exotoxin. J Clin Invest 48:32A-33A
     (1969).

117. Pierce, NF, Carpenter, CCJ, Elliot, HL, Greenough, WB.
     Effects of prostaglandins, theophylline, and cholera exotoxin
     upon transmucosal water and electrolyte movement in the
     canine jejunum. Gastroenterology 60:22-32 (1971).

118. Robert, A, Nezamis, JE, Hanchar, AJ, Lancaster, C, Klepper,
     MS. The enteropooling assay to test for diarrhea due to
     prostaglandins. Fed Proc 35:457 (1976).

119. Robert, A, Nezamis, JE, Lancaster, C, Hanchar, AJ, Klepper,
     MS. Enteropooling assay: a test for diarrhea produced by
     prostaglandins. Prostaglandins 11:809-828 (1976).

120. Cummings, JH, Milton-Thompson, GJ, Billings, J, Newman, A,
     Misiewicz, JJ. Studies on the site of production of diarrhea
     induced by prostaglandins. Clin Sci Mol Med 46:15P (1974).

121. Matuchansky, C, Bernier, JJ. Effects of prostaglandin E1 on
     net and unidirectional movements of water and electrolytes
     across jejunal mucosa in man. Gut 12:854-855 (1971).

122. Matuchansky, C, Bernier, JJ. Effect of prostaglandin E1 on
     glucose, water, and electrolyte absorption in the human
     jejunum. Gastroenterology 64:1111-1118 (1973).

123. Matuchansky, C, Bernier, JJ. Effects of prostaglandin E1 on
     jejunal absorption in man. Digestion 9:86-87 (1973).

124. Matuchansky, C, Mary, JY, Bernier, JJ. Effets de la prosta-
     glandine E1 sur l'absorption du glucose et les mouvements trans-
     intestinaux de l'eau et des electrolytes dans le jejunum
     humain. Biol Gastroenterol (Paris) 5:636C (1972).

125. Newman, A, Milton-Thompson, G, Cummings, JH, Billings, JA,
     Misiewicz, JJ. Differential response of the human small and
     large intestine to prostaglandin F2α. Gastroenterology
     66:A-100/754 (1974).

126. Matuchansky, C, Mary, JY, Bernier, JJ. Further studies on prostaglandin E1-induced jejunal secretion of water and electrolytes in man, with special reference to the influence of ethacrynic acid, furosemide and aspirin. Gastroenterology 71:274 (1976).

127. Kimberg, DV, Field, M, Johnson, J, Henderson, A, Gershon, E. Stimulation of intestinal mucosal adenyl cyclase by cholera enterotoxin and prostaglandins. J Clin Invest 50:1218-1230 (1971).

128. Al-Awqati, Q, Field, M, Greenough, WB. Reversal of cyclic AMP-mediated intestinal secretion by ethacrynic acid. J Clin Invest 53:687 (1974).

129. Schafer, DE, Lust, WD, Sircar, B. Elevated concentration of adenosine 3',5'-cyclic monophosphate in intestinal mucosa after treatment with cholera toxin. Proc Nat Acad Sci USA 67:851 (1970).

130. Kent, TH, Cardelli, RM, Stamler, FW. Small intestinal ulcers and intestinal flora in rats given indomethacin. Am J Path 54:237 (1969).

131. Vane, JR. Inhibition of prostaglandin synthesis as a mechanism of action for aspirin-like drugs. Nature (New Biol) 231:232-235 (1971).

132. Robert, A. An intestinal disease in the rat probably caused by a prostaglandin deficiency. Gastroenterology 66:A-111/765 (1974).

133. Robert, A. Effects of prostaglandins on the stomach and the intestine. Prostaglandins 6:523-532 (1974).

134. Robert, A. An intestinal disease produced experimentally by a prostaglandin deficiency. Gastroenterology 69:1045-1047 (1975).

135. Robert, A, Asano, T. Resistance of germfree rats to intestinal lesions produced by indomethacin. 10th Int Congr Gastroent Budapest June 23-29 (1976).

136. Robert, A. The role of prostaglandins in the etiology and treatment of gastrointestinal diseases. 6th Int Congr Pharmacol, Helsinki, July, p. 161 (1975).

137. Brodie, DA, Chase, BJ. Role of gastric acid in aspirin-induced gastric irritation in the rat. Gastroenterology 53:604 (1967).

138. Jacobson, ED, Chaudhury, TK, Thompson, WJ. Mechanism of gastric mucosal cytoprotection by prostaglandins. Gastro-enterology 70:897 (1976).

139. Kawarada, Y, Lambek, J, Matsumoto, T. Pathophysiology of stress ulcer and its prevention. II. Prostaglandin E1 and microcirculatory responses in stress ulcer. Am. J. Surg. 129:217-222 (1975).

ANTISECRETORY PROSTAGLANDINS AND GASTRIC MUCOSAL EROSIONS

IN THE RAT

B. J. R. Whittle

Department of Pharmacology, Institute of Basic Medical

Sciences, Royal College of Surgeons, London, WC2A 3PN

INTRODUCTION

Antisecretory prostaglandins inhibit the formation of gastric
ulcers and erosions induced by a variety of experimental techniques
including pyloric-ligation, steroid administration or stress
(Robert et al, 1968;Usardi et al, 1974). In the present paper,
the relationship between inhibition of of gastric acid secretion
and the prevention of gastric mucosal erosions, induced acutely
by indomethacin or bile salts, has been investigated in the rat.
In addition, since changes in gastric mucosal blood flow and in
the resistance of the mucosa to acid back-diffusion may be
possible mechanisms underlying such erosion formation, the effects
of prostaglandins on these parameters have been studied.

METHODS

Assessment of Mucosal Erosions

Wistar rats (200-250 g) starved for 18 h but allowed water,
were killed at various time intervals following drug treatment.
The stomachs were removed, opened along the lesser curvature and
rinsed under a stream of water. They were then pinned flat on a
cork board and studied with the aid of the binocular microscope
(x 10). Erosions, which formed only in the glandular mucosa, were
counted, and each one given a severity rating on a 1-3 scale as
previously described (Main and Whittle, 1975a). The total, divided by a factor of 10, was designated the 'erosion index' for that
stomach. In further experiments where the rat gastric lumen was
perfused, erosions were assessed 3 h after starting the perfusion.

Acid back-diffusion

The gastric lumen of the urethane-anaesthetised rat was
perfused (0.1 or 0.2 ml/min) with acid saline (pH 1) and the loss
of acid across the mucosa from the lumen determined by titration
(autoburette, Radiometer). The potential difference (PD) across
the mucosa, which is related to hydrogen- and sodium-ion flux
and gives an indication of the integrity of the mucosal barrier
(Chvasta and Cooke, 1972), was measured with calomel electrodes.

Gastric mucosal blood flow

Mucosal blood flow was determined by the ^{14}C-aniline clearance technique (Main and Whittle, 1973a). This technique, like
aminopyrine clearance used to measure mucosal blood flow in dogs,
depends on the clearance of the unionized lipophilic weak base
from the mucosal blood into the acidic gastric lumen, where the
base is ionized and hence trapped. The ratio of gastric output
to blood concentration of aniline (i.e. clearance) has been shown
to give an estimate of mucosal blood flow.

RESULTS

Formation of gastric erosions by indomethacin

Administration of indomethacin (5-30 mg/kg, s.c.) gave a
dose-dependent increase in the incidence and severity of mucosal
erosions over the 6 h period of measurement. A submaximal dose
of indomethacin (15 or 20 mg/kg, s.c.) was chosen for subsequent
experiments.

Inhibition of gastric erosions by antisecretory prostaglandins

The (15S)-15-methyl analogue of prostaglandin E_2 (2.5 µg/kg,
s.c.) was administered immediately prior to indomethacin (15 µg/
kg, s.c.) and erosion formation in groups of rats was assessed
at various time intervals. With this dose of the PGE_2 methyl
analogue there was a marked inhibition of erosions at 2.5 h
(P<0.001) but this inhibitory activity was much reduced when
assessed after 5 h (Fig. 1), presumably reflecting the relatively
short half-life of the prostaglandin. In subsequent experiments,
the potency of prostaglandins as inhibitors of indomethacin-induced
erosions was determined 3 h after administration. The dose-dependent reduction of the gastric erosion index by prostaglandin E_2
and its analogues is shown in Table 1, the dose causing 50% reduction (ID_{50}) in the control erosion index was 220, 0.5 and 0.5 µg/
kg, s.c. for PGE_2, (15S)-15 methyl E_2 and 16, 16 dimethyl E_2,
respectively.

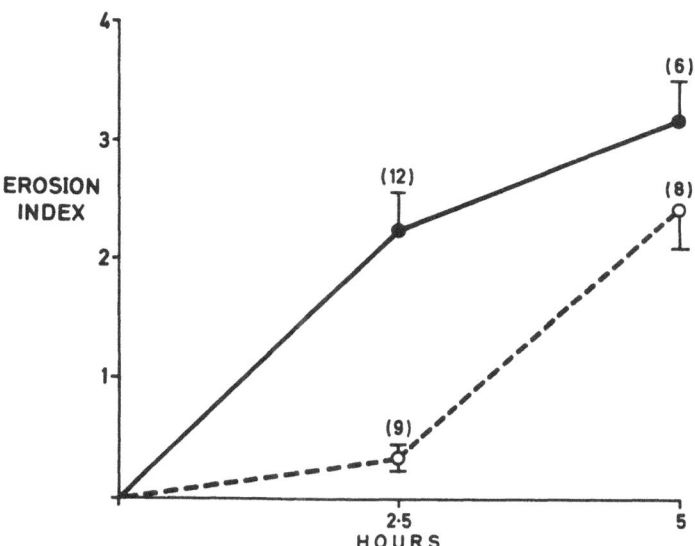

Figure 1. Inhibition of indomethacin (15 mg/kg, s.c.)-induced gastric mucosal erosions (•——•) by (15S)-15-methyl prostaglandin E_2, 2.5 μg/kg, s.c. (o——o). Results are shown as the mean$^+$s.e. mean, where (n) is the number of values in each group.

Table 1. Inhibition of indomethacin (15 mg/kg, s.c.)-induced gastric erosions by subcutaneous administration of antisecretory prostaglandins, determined 3 h after administration. Results are expressed as mean$^+$s.e.mean, (n) being number of values in each group.

	μg/kg	% inhibition	n
Prostaglandin E_2	62.5	33^+_-11	11
	125	45^+_-5	12
	250	50^+_-11	12
(15S)—methyl E_2	0.625	56^+_-11	12
	1.25	67^+_-9	12
	2.5	77^+_-7	12
16, 16 dimethyl E_2	0.625	62^+_-10	11
	2.5	90^+_-4	12

Effects of taurocholate perfusion

Acid-saline perfusion (100 mM HCl, pH 1) for 3 h in the urethane-anaesthetised rat did not damage the mucosa (Table 2). Simultaneous perfusion of the surface active bile salt, sodium taurocholate (2 mM) significantly ($P < 0.01$) increased acid loss across the mucosa from the gastric lumen (Table 2), lowered the PD (by 10.4 ± 1.5 mV; mean-s.e.mean, n=4) and increased mucosal blood flow ($P < 0.001$). These effects were dependent on the concentration of the taurocholate and of the acid saline and there was a direct correlation between the rise in mucosal blood flow and the increase in acid back-diffusion. A low erosion-index was observed after 3 h perfusion of taurocholate and acid saline (Table 2).

Effects of indomethacin

Indomethacin (20 mg/kg, i.v.), injected during acid perfusion, had no consistent effect on acid-loss or PD, but significantly ($P < 0.01$) decreased mucosal blood flow, and a low erosion score was observed after 3 h (Table 2). However, during concomitant taurocholate (2 mM) perfusion, indomethacin significantly ($P < 0.01$) reduced the elevated mucosal blood flow seen with acid and taurocholate alone, without any marked additional effect on acid-loss, and led to a high incidence of mucosal erosions (Table 2).

Effects of prostaglandin analogues

Administration of (15S)-15 methyl E_2 (5 μg kg^{-1}h^{-1}, s.c.) reduced the acid back-diffusion during combined taurocholate and indomethacin administration (by $20 \pm 5\%$, n=8), in a dose causing $53 \pm 12\%$, n=4 ($P < 0.002$) inhibition of the erosion formation. This dose of prostaglandin analogue also increased PD (6.7 ± 0.9 mV, n=3) and MBF (by $66 \pm 13\%$, n=3).

Table 2. Changes in acid-loss, mucosal blood flow and erosion formation during administration of acid saline (100 mM HCl), taurocholate (2 mM) and indomethacin (20 mg/kg). Results expressed as the mean-s.e.mean of four experiments.

	Acid loss (μ-equiv./min)	MBF (% of control)	Erosion index
HCl	0.9 ± 0.1	100 ± 2.5	0
HCl + Indo	0.9 ± 0.2	81 ± 1.4	1.1 ± 0.8
HCl + Tauro	3.1 ± 0.3	345 ± 18	0.5 ± 0.1
HCl + Indo + Tauro	3.3 ± 0.4	251 ± 21	6.3 ± 0.1

DISCUSSION

In the present study, the relationship between the inhibition
of gastric secretion and the prevention of indomethacin-induced
mucosal erosions by several prostaglandins has been investigated.
The (15S)-15-methyl and 16, 16-dimethyl analogues of PGE_2, which
inhibit gastric acid secretion in the rat (Main and Whittle, 1975b)
were extremely potent inhibitors of indomethacin-induced erosions,
being approximately 400 times as active as the parent prostaglandin.

The acid concentration in the gastric lumen is important in
the development of these erosions and it is known that other
antisecretory agents unrelated to prostaglandins also inhibit
indomethacin erosions (Lee et al, 1971). Thus, in order to study
closely the relationship between acid concentration and the inhibition or erosion formation, the gastric lumen of the urethane-
anaesthetised rat was perfused with acid-saline. Any changes in
the low basal secretion would be expected to have little effect
on the total acid concentration in the gastric lumen, although it
is not known whether such a perfusion procedure is sufficient to
mask any local changes in acid concentration within the mucosa.
In this preparation, which excludes duodenal contamination, a low
incidence of erosions following systemic indomethacin administration was observed, which was greatly increased by addition of the
surface-active bile salt, sodium taurocholate, to the gastric
perfusate. The mechanisms underlying this potentiation of erosion
formation may be related to the observed changes in local blood
flow and in the mucosal "barrier" to acid diffusion. The present
results suggest that whereas a reduction in mucosal blood flow,
as observed with parenteral indomethacin, or an increased acid
back-diffusion, as seen with taurocholate, can lead to a low
incidence of erosions, a combination of both produces extensive
mucosal damage.

The fall in mucosal blood flow induced by the aspirin-like
drug indomethacin, in doses sufficient to inhibit mucosal prostaglandin formation (Main and Whittle, 1975a), could suggest a role
for endogenous prostaglandins in the local regulation of the
gastric microcirculation; it has been found that primary prostaglandins of the E and A series can increase resting mucosal blood
flow when perfused locally or administered systemically (Main and
Whittle, 1973b). The fall in mucosal blood flow following indomethacin would be expected to lead to areas of ischaemia within
the mucosa and it is likely that these areas are the sites of the
subsequent erosion formation.

The mechanism by which the mucosal hyperaemia in response to
the surface-active agent, taurocholate, is brought about is not
known, but may be a direct response of the mucosal microvascu-
lature to the increased acid back-diffusion or may be in response

to the local release of mediators, possibly prostaglandins. This mucosal hyperaemia was directly correlated with the acid back-diffusion and hence may represent a protective mechanism preventing an accumulation of acid within the mucosa, since few erosions were subsequently observed. This is supported by the finding that when indomethacin was administered simultaneously and the hyperaemic response to taurocholate was reduced, a very high incidence or erosion formation ensued. It is of interest that several aspirin-like drugs are known to cause acid back-diffusion following intragastric administration. Thus, the potency of such drugs in causing gastric erosions when orally administered may be related to both changes in acid back-diffusion and the concurrent effects on the gastric mucosal microcirculation.

Although the inhibition of gastric acid secretion is likely to be the primary mechanism by which these prostaglandins can inhibit gastric mucosal erosion formation in the intact rat, the finding that prostaglandins can prevent erosion formation in the presence of exogenous perfused acid suggests that prostaglandins may exert other protective mechanisms on the mucosa. Such mechanisms may involve the observed actions of the prostaglandin analogues on local blood flow and on the permeability of the mucosa. It is possible that similar effects may account for the protective effect of prostaglandins observed by Robert (1974) on intestinal lesion formation following indomethacin administration.

The finding that the prostaglandin analogues could prevent the acute toxicity of indomethacin raised the question whether the anti-inflammatory actions of indomethacin would also be inhibited by these analogues (McCall et al, 1976). It is known that local administration of E-prostaglandins can increase carrageenan-induced rat paw oedema (Moncada et al, 1973). In the present work, the methyl analogues of PGE_2 likewise increased paw volume when injected locally into carrageenan-treated paws of rats pretreated with indomethacin, but had only comparable activity with the parent prostaglandin (Fig. 2). Furthermore, systemic administration of the analogues, in doses far greater than those required to inhibit the indomethacin-induced gastric erosions, had no effect on paw volume, indicating that the anti-inflammatory actions of indomethacin were not inhibited. This raises the possibility that concurrent administration of low doses of a prostaglandin analogue could prevent the gastric irritancy associated with the use of non-steroid anti-inflammatory agents, without suppressing their therapeutic actions.

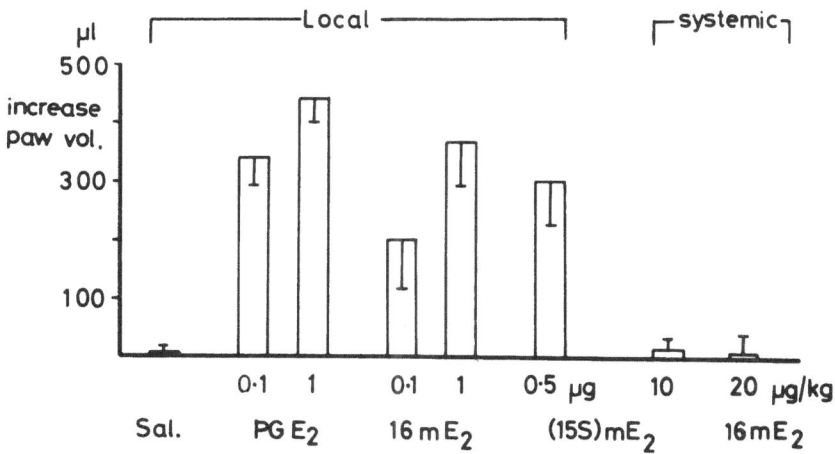

Figure 2. Change in paw volume following prostaglandin administration in rats treated with carrageenan (0.1 ml, 2% locally) and indomethacin (20 mg/kg, s.c., 1 h before carrageenan). Prostaglandins administered 2.25 h after carrageenan, either locally to the paw or systemically, and volume change determined 1.5 h later. Results are expressed as mean±s.e.mean.

Acknowledgements

This work was supported by the Medical Research Council. Prostaglandins were kindly supplied by the Upjohn Company, Kalamazoo.

References

CHVASTA, T. E. & COOKE, A. R. (1972).
    The effect of several ulcerogenic drugs on the canine gastric mucosal barrier. J. Lab. clin. Med., 79, 302-315.

LEE, Y. H., MOLLISON, K. W. & CHENG, W. D. (1971).
    The effects of antiulcer agents on indomethacin-induced
    gastric ulceration in the rat. Arch. Int. Pharmacodyn.,
    191, 370-377.

MAIN, I. H. M. & WHITTLE, B. J. R. (1973a).
    Gastric mucosal blood flow during pentagastrin- and histamine-
    stimulated acid secretion in the rat. Br. J. Pharmac., 49,
    534-542.

MAIN, I. H. M. & WHITTLE, B. J. R. (1973b).
    The effects of E and A prostaglandins on gastric mucosal
    blood flow and acid secretion in the rat. Br. J. Pharmac.,
    49, 428-436.

MAIN, I. H. M. & WHITTLE, B. J. R. (1975a)
    Investigation of the vasodilator and antisecretory role of
    prostaglandins in the rat gastric mucosa by use of non-ster-
    oidal anti-inflammatory drugs. Br. J. Pharmac., 53, 217-224.

MAIN, I. H. M. & WHITTLE, B. J. R. (1975b).
    Potency and selectivity of methyl analogues of prostaglandin
    E_2 on rat gastrointestinal function. Br. J. Pharmac., 54,
    309-317.

McCALL, E., WHITTLE, B. J. R. & YOULTEN, L. J. F. (1976).
    Effects of prostaglandin E_2 methyl analogues on the anti-
    inflammatory and gastric erosive activity of indomethacin.
    J. Pharm. Pharmac., 28, 588-589.

MONCADA, S., FERREIRA, S. H. & VANE, J. R. (1973).
    Prostaglandins, aspirin-like drugs and the oedema of inflam-
    mation. Nature, New Biol., 246, 217-219.

ROBERT, A. (1974).
    Effects of prostaglandins on the stomach and intestine.
    Prostaglandins, 6, 523-532.

ROBERT, A., NEZAMIS, J. F. & PHILLIPS, J. P. (1968).
    Effect of prostaglandin E_1 on gastric secretion and ulcer
    formation in the rat. Gastroenterology, 55, 481-487.

USARDI, M. M., FRANCESCHINI, J., MANDELLI, V., DATURI, S. &
MIZZOTTI, B. (1974).
    'Prostaglandins VIII': A proposed role for PGE_2 in the
    genesis of stress-induced gastric ulcers. Prostaglandins,
    8, 43-51.

THE ROLE OF PROSTAGLANDINS AND CYCLIC NUCLEOTIDES

IN INFLAMMATION

J.P. GIROUD[+] -G.P. VELO[++] -D.A. WILLOUGHBY[+++]

[+]Département de Pharmacologie - ERA 629 - C.N.R.S.
Faculté de Médecine Cochin Port-Royal - 27, rue du
Faubourg Saint-Jacques - 75014 PARIS (France)
[++]Istituto di Farmacologia - Centro Ospedalie-
ro Clinicizzato di Borgo Roma, Università di Padova-
37100 VERONA (Italy)
[+++]Department of Experimental Pathology - St
Bartholomew's Hospital Medical College - LONDON,
EC1A 7BE (England)

INTRODUCTION

One of the main problems surrounding the study of
the PG's in the inflammatory response has been the
lack of suitable models. The inflammatory process is
a highly dynamic state. There is a constant change in
the volume of exudate, increased lymphatic drainage,
arrival of different cell types, and even these different cell types change during the inflammatory
process. Thus not only do we see tissue cell changes
but the arrival of polymorphs which discharge lysosomal enzymes; usually this is followed by the arrival
of mononuclear cells which transform into macrophages. These cells are actively phagocytic, they then
transform further into "activated macrophages" with
surface changes, plus increased numbers of lysosomes.
Often these cells may proliferate or fuse into giant
cells and even in this form may undergo proliferation.

Why do we need to study the inflammatory process
and to elucidate the precise role of mediators? It is
simply that our main objective is to have a better
understanding of inflammatory disease in man. We need
to answer the question is it desirable to suppress PG
formation during the inflammatory process? Should we
leave the PG's and search for other mediators and
mechanisms that are susceptible to control by new
therapeutic agents?

In the present paper we will briefly present the
results obtained using four models of inflammation.
These models have one thing in common, they have all
been produced within the pleural cavity. Another
feature common to all these reactions is that they
have been studied previously but usually as cutaneous
reaction.

One problem facing the experimental worker
studying the topic of inflammation is the inability
to collect the cells and exudate, without resorting
to artificial means such as perfusion techniques, or
merely performing histology for an approximation of
types and number of cells.

We have illustrated this talk with
 - carrageenan induced pleurisy; this is merely
that carrageenan has been widely used as an irritant
to elicit an inflammatory response.
 - a model of immediate hypersensitivity; the
reverse passive Arthus reaction, classically performed in the skin of animals, however in the present
experiments performed in the pleural cavity.
 - a model of delayed hypersensitivity once again
usually performed in the skin but here carried out in
the pleural cavity.
 -calcium pyrophosphate induced pleurisy; this
irritant has been used for the reason that crystal
deposition disease is common among the arthropathies.

We hope to show that by using these irritants in
this particular site much information may be gained
regarding the role of the prostaglandins. We have not
studied the PG intermediates; the endoperoxides, the
thromboxanes, etc. However the model systems would

equally well lend themselves to an analysis of the role
of these other factors in the inflammatory response and
we would recommend them to those devotees of the prostaglandins.

I - CARRAGEENAN PLEURISY

Carrageenan has been injected intrapleurally at a
dose of 0.1 ml per rat (1 % solution in saline) (7).

Animals were killed by exsanguination at various
time intervals after injection. The exudate was
collected into siliconised tubes at 4° C and the pleural cavity washed in medium 199. In this way exudate
volume could be calculated together with total and
differential leucocyte counts. Thus it is evident that
the pleural cavity as a site for provoking inflammation
would appear advantageous for the following reasons -
  1) Precise quantitation of various parameters such
     as quantitation of exudate volume, numbers and
     types of leucocytes.
  2) Analysis of humoral factors in exudates eg hista-
     mine, serotonin, kinins, prostaglandins, cyclic
     nucleotides, complement, macrophage inhibition
     factor, mitogenic factor(s), chemotactic factors,
     lysosomal enzymes, etc.
  3) Similar analysis may be performed on the cellu-
     lar exudate.
  4) Inflammatory leucocyte ultrastructure may be
     studied simply by collecting cells directly with
     fixative, thus avoiding distortion or artefact.
  5) Cell function may be studied with respect to
     phagocytic activity, migratory activity, mitosis,
     etc., and correlated with the above parameters.

During the pleural reaction to carrageenan it was
found that the inflammatory response was maximal around
6 hours, consisting of a mixture of polymorphs and mononuclear leucocytes. The sequential presence of the
mediators, histamine, 5-HT, and prostaglandins was confirmed. It was of interest that PGE reached higher
levels than did PGF_{α} during the early phase of the reaction (Table 1). Coincident with the peak reaction there
was a marked fall in both extracellular and intracellu-

| | Exudate collected at time (hr) | | | | | |
	1	3	6	12	24	48
Vol. fluid exudate (ml)	-	-	2.0	-	0.9	0
Total polys, X 10^6	-	-	86	-	53	30
Total monos, X 10^6	-	-	38	-	36	32
Histamine, $\mu g/10^8$ cells	9.6	4.0	3.6	5.5	4.5	3.7
5-HT, ng/10^8 cells	10.2	4.3	6.0	2.5	5.4	5.4
PGE_2, ng/10^8 cells	1.8	2.4	5.0	7.0	5.3	5.6
$PGF_{2\alpha}$, ng/10^8 cells	0	1.0	6.5	8.4	12.0	11.0

Development of the carrageenan reaction in the rat pleural cavity - analysis of mediator content in the cellular exudate fraction.

Table 1

Type of reaction	cAMP (pmoles/10^8 cells) measured at time (hr)							
	0	1	3	6	12	18	24	48
Carrageenan (rat)	180	–	65	130	–	420	670	470
Arthus (rat)	165	133	30	50	64	–	130	360
DH (guinea-pig)	95	120	–	34	–	34	24	67
CPPD (rat)	280	–	85	105	120	–	193	281

Quantitation of cAMP concentration in the cellular exudate fraction of various experimental pleurisies.

Table 2

lar (C) AMP levels (Table 2) (3).

    However as the reaction diminished there was a
corresponding rise in PGF_{α} . It may be significant that
PGF, is essentially considered as an anti-inflammatory
prostaglandin by virtue of its ability to inhibit increased vascular permeability (23).

    This would seem to support the concept of PGE
provoking inflammation and PGF_{α} inhibiting the process.
Thus, the prostaglandin family can be considered as
modulators of the inflammatory response (21).

    There are many problems surrounding the use of
carrageenan as an irritant in experimental inflammation
(6). The sequence of events following injection is
complex. Not only does carrageenan appear to be a non-
specific irritant "per se", but in addition it will
cause activation of the third component of complement,
leading to rapid activation of the complement "enzyme
cascade" system, with liberation of chemotactic factors,
formation of anaphylatoxin and the release of a host
permeability factors (10 , 24). The chronic response
to carrageenan is rather unusual, being characterised
by the development of granuloma consisting of a population of long-lived macrophages with little evidence of
mitosis or death (16).

    Another property of the different types of carra-
geenan is their varying immunogenicity. Thus repeated
injections may lead to an immune response. Carrageenan
also has anticoagulant activity.

II - REVERSE PASSIVE ARTHUS PLEURISY

    A model of immediate hypersensitivity, namely the
reverse passive Arthus reaction (RPA) was developed in
the rat pleural cavity by YAMAMOTO et Al. (28). Since
the discovery of this reaction by Maurice Arthus (1903)
this has been normally performed in the skin. The
obvious disadvantage of skin reactions is that it is
impossible to quantitate accurately the oedema and
types of infiltrating leucocytes. Usually the response
is assessed by erythema, induration and biopsy specimens of the lesion itself in order to study the cellu-

lar response.

In order to perform the RPA pleural reaction it is
necessary to use the purified IgG antibody since whole
antiserum will cause a considerable non-specific pleural reaction. This is carried out by inducing antibody
formation in rabbits following repeated sensitisation
with bovine serum albumin (BSA). The antiserum thus
produced is then fractionated into the α - globulin
fraction by sodium sulphate precipitation. This fraction
is further separated by affinity chromatography to
yield the IgG anti-BSA antibody. Passage through a
DEAE-cellulose column facilitates removal of any IgM
antibodies present.

The RPA pleurisy is induced by a single intravenous injection of antigen (BSA), followed twenty
minutes later by an intrapleural injection of antibody.
The response is rapid in onset, becoming maximal
between 4-6 hours and being characterised by a predominantly polymorphonuclear leucocyte infiltration (Table
3). Injection of the antibody intrapleurally alone
induces no exudate formation and a very small infiltration of leucocytes.

Examination of the mediators reveals that histamine
and serotonin are present during the early stages of the
reaction, their levels exceeding those for the carrageenan reaction (Table 3). The prostaglandins show a
similar pattern as for the carrageenan pleurisy.
However, the early PGE rise is more pronounced, giving
way to PGE$_x$ as the reaction wanes. Pretreatment with
cobra venom factor in order to deplete circulating
haemolytic complement caused a marked suppression of
the RPA pleurisy, with respect to both exudate volume
and cellular infiltration. The changes in (C) AMP
levels are inversely related to the reaction, initially falling towards the peak reaction, and rising to,
or above, normal levels as the reaction wanes (Table 2).

III - DELAYED HYPERSENSITIVITY PLEURISY

Again these reactions have thus far been conven-
tionally carried out in the skin. It was therefore
decided to transpose them to the pleural cavity in the

	Exudate collected at time (hr)					
	1	3	6	12	24	48
Vol. fluid exudate (ml)	-	-	1.95	0	0	0
Total polys, X 10^6	-	-	60	45	17	10
Total monos, X 10^6	-	-	20	46	40	23
Histamine, μg/10^8 cells	60.0	40.0	9.0	1.0	1.0	0
5-HT, ng/10^8 cells	50	10	15	10	11	5
PGE_2, ng/10^8 cells	0	6	4.2	2.3	3.2	0
$PGF_{2\alpha}$, ng/10^8 cells	1	2	2.7	6.0	3.2	2.0

Development of exudate and exudate intracellular mediators after intrapleural injection of purified anti-BSA antibody in the reverse passive Arthus reaction.

Table 3

guinea-pig (27). Animals were sensitised to complete
Freund's adjuvant (CFA) [heat-killed mycobacterium
tuberculosis mixed with incomplete Freund's adjuvant].
Three weeks later they were challenged intrapleurally
with 0.1 ml purified protein derivative (PPD) of tuber-
culin. In marked contraste to the RPA reaction the delayed hypersensitivity (DH) reaction was delayed in
onset, becoming maximal around 18-24 hours (Table 4).
Mononuclear cells were the dominant cell type found,
consisting of both monocytes and lymphocytes (Table 5).
Again, in contrast to RPA pleurisy, the DH reaction
was independent of the complement depletion by cobra
venom factor (27). Several types of migration inhibitory (MI) substances were also detected in the DH pleural reaction, one of which also appeared in the RPA
reaction (26, 29, 30). Thus MI factors may not always
be indicative of DH reactions as has frequently been
supposed.

The levels of prostaglandins follow the same
pattern as for the other pleural models of inflammation
suggesting the same modulation system by these mediators (Table 5). However, unlike the RPA and carrageenan reactions the DH reaction shows little evidence of
histamine or 5-HT in the pleural exudates, suggesting
the existence of other as yet unidentified mediators
of vascular permeability.

IV - CALCIUM PYROPHOSPHATE PLEURISY

Calcium pyrophosphate crystals have been isolated
from the synovial fluid of patients with "pseudo gout",
"chondrocalcinosis", or "calcium pyrophosphate crystal
deposition disease", according to the terminology one
wishes to adopt. These calcium pyrophosphate crystals
(CPPD) were injected into the pleural cavity and the
exudate removed and analysed at various time periods
(25). It is typified by an acute reaction dominated by
polymorphs, reaching maximal levels around 6 hours and
subsiding around 24 hours (Table 6). It was also found
to be independent of the complement system as shown by
prior depletion of complement with cobra venom factor
(25).

No. of animals	Exudate collected at time (hr)	Volume exudate (ml)	S.E.	^{125}I IgG (Cell-free exudate) (c.p.m.)
10	1	0.3	($\pm$ 0.2)	-
18	6	0.7	($\pm$ 0.3)	821
18	12	2.3	($\pm$ 1.5)	3750
18	18	5.0	($\pm$ 1.9)	3190
19	24	3.5	($\pm$ 1.2)	1914
16	30	2.0	($\pm$ 0.9)	-
16	48	0.5	($\pm$ 0.6)	520

Quantitation of exudate induced by intrapleural injection of PPD into CFA-sensitised guinea-pigs.

Table 4

	Exudate collected at time (hr)						
	1	3	6	12	18	24	48
Vol. fluid exudate (ml)	0	0.2	0.4	3.1	5.0	4.4	0.7
Total polys, X 10^6	-	-	11	20	43	48	15
Total monos, X 10^6	-	-	8	19	45	52	24
Histamine, ng/10^8 cells	60	100	160	122	-	64	56
5-HT, ng/10^8 cells	5	0	0	0	0	0	0
PGE$_2$, ng/10^8 cells	0	0	1.0	3.5	-	1.2	1.0
PGF$_{2\alpha}$, ng/10^8 cells	0	0	1.0	7.0	-	4.0	4.5

Development of exudate and exudate intracellular mediators after intrapleural challenge with PPD into CFA sensitised guinea-pigs.

Table 5

	Exudate collected at time (hr)					
	1	3	6	12	24	48
Vol. fluid exudate (ml)	-	-	1.9	-	0.5	0
Total polys, X 10^6	-	-	55	-	45	15
Total monos, X 10^6	-	-	20	-	32	10
Histamine, μg/10^8 cells	2.5	2.0	2.0	1.9	1.1	0
5-HT, ng/10^8 cells	1.0	1.0	0	0	0	0
PGE_2, ng/10^8 cells	0	1.0	2.7	2.7	2.3	2.2
$PGF_{2\alpha}$, ng/10^8 cells	0	0	2.6	6.8	14.3	2.1

Development of the calcium pyrophosphate dihydrate (CPPD) reaction in the rat pleural cavity – analysis of mediator content in the cellular exudate fraction.

Table 6

Examination of mediators of increased vascular permeability shows no significant participation of either histamine or 5-HT (Table 6). Thus, these results together with those of complement depletion distinguish the reaction to CPPD from that of carrageenan. There is an early rise in PGE and, as with carrageenan, this is followed by a greater rise in PGF$_{\alpha}$ as the reaction diminishes (Table 6). Thus, this reinforces the concept of modulation of the inflammatory response by the prostaglandins.

V - PROSTAGLANDINS AND CYCLIC NUCLEOTIDES

To date, there has been much speculation regarding the role of the prostaglandins in the control of the intracellular cyclic nucleotide levels. However most of this work has so far been carried out "in vitro" [see reviews by DUNN et Al., (8) and VELO and ABDULLAHI, (20)] .

Cyclic AMP may be elevated by PGE, and cyclic GMP by PGF. However, when we examine the effects of PGE and PGF we find the situation schematically in Figure 1.

Thus, an elevation of PGE, which is by itself pro-inflammatory, will cause increased vascular permeability (4), leucocyte chemotaxis (13), augment collagen biosynthesis, and sensitise receptors for mediators of increased vascular permeability (9). Yet this prostaglandin increases intracellular (C) AMP, which has been found to be essentially anti-inflammatory, viz inhibits "in vitro" lymphocyte transformation and cytotoxicity (11, 17, 18); inhibits release of mediators during "in vitro" anaphylaxis (1, 14, 15, 19); inhibits platelet aggregation "in vitro"; inhibits lysosomal enzyme secretion from leucocytes "in vitro" (2, 12, 22, 31).

PGF is basically anti-inflammatory and will inhibit increased vascular permeability (23). However, PGF causes an elevation of cyclic GMP "in vitro" which is itself pro-inflammatory, "stimulating" the above-listed processes, and thus opposing the action of cyclic AMP. It must be emphasised, however, that these

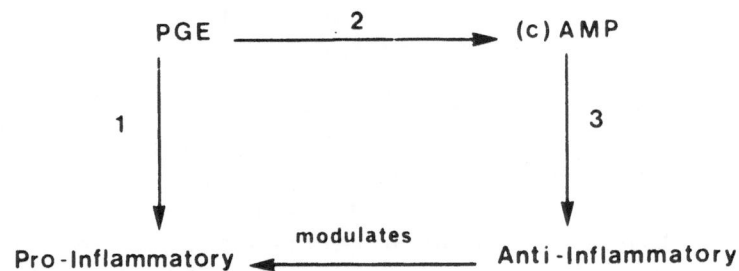

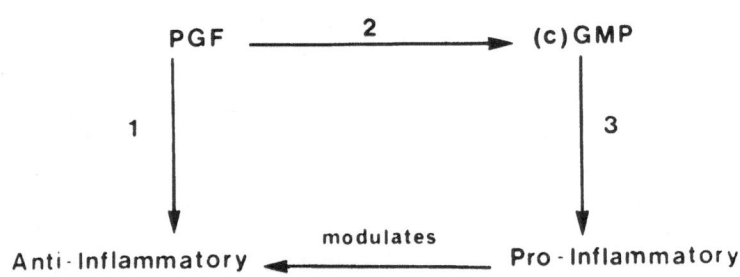

Figure 1

results have all come from "in vitro" experiments.
Only recently has it been possible to test "in vivo"
this concept of the balance between cyclic AMP and
cyclic GMP. Thus, DEPORTER et Al. (5) have found a
fall in intracellular cyclic AMP coincident with the
peak reaction followed by an increase in cyclic AMP as
the reaction waned. For cyclic GMP the reverse was
observed. Intracellular cyclic GMP levels followed the
course of the reaction and were thus inversely related
to cyclic AMP. Preliminary results indicate that a
similar inverse pattern of cyclic AMP/cyclic GMP
exists for the other models of pleural inflammation
described (ie. carrageenan, RPA, DH).

CONCLUSION

So what of the question posed at the beginning of
this paper? What is the role of the PG's in inflammation? Certainly many of the properties of certain of
the PG's are pro-inflammatory. Others are anti-inflammatory. Evidence has been produced that certain of
those with pro-inflammatory activity are present during
the development of inflammatory reactions.

Similarly as the reaction wanes so the anti-inflammatory PG's rise to higher concentrations, exceeding
that of the pro-inflammatory PG's. This by itself is
suggestive of a modulatory role of the PG's within their
own family, thus one member may elicit the response only
to be suppressed by another member of the family with
opposing actions. Yet the story becomes more complex
when one considers their relative actions on the
cyclic nucleotides. One can observe the PGE playing an
anti-inflammatory role by elevating cyclic AMP and
conversely PGF acting as pro-inflammatory agent by
elevating cyclic GMP.

Therefore one could suggest they are self modulating. It would be over naive to speculate solely on the
role of PG's in the inflammatory process. They represent merely one small piece of a vast jig saw with a
host of alternate mechanisms and pathways which may be
activated.

We have attempted to show "some" of the factors and

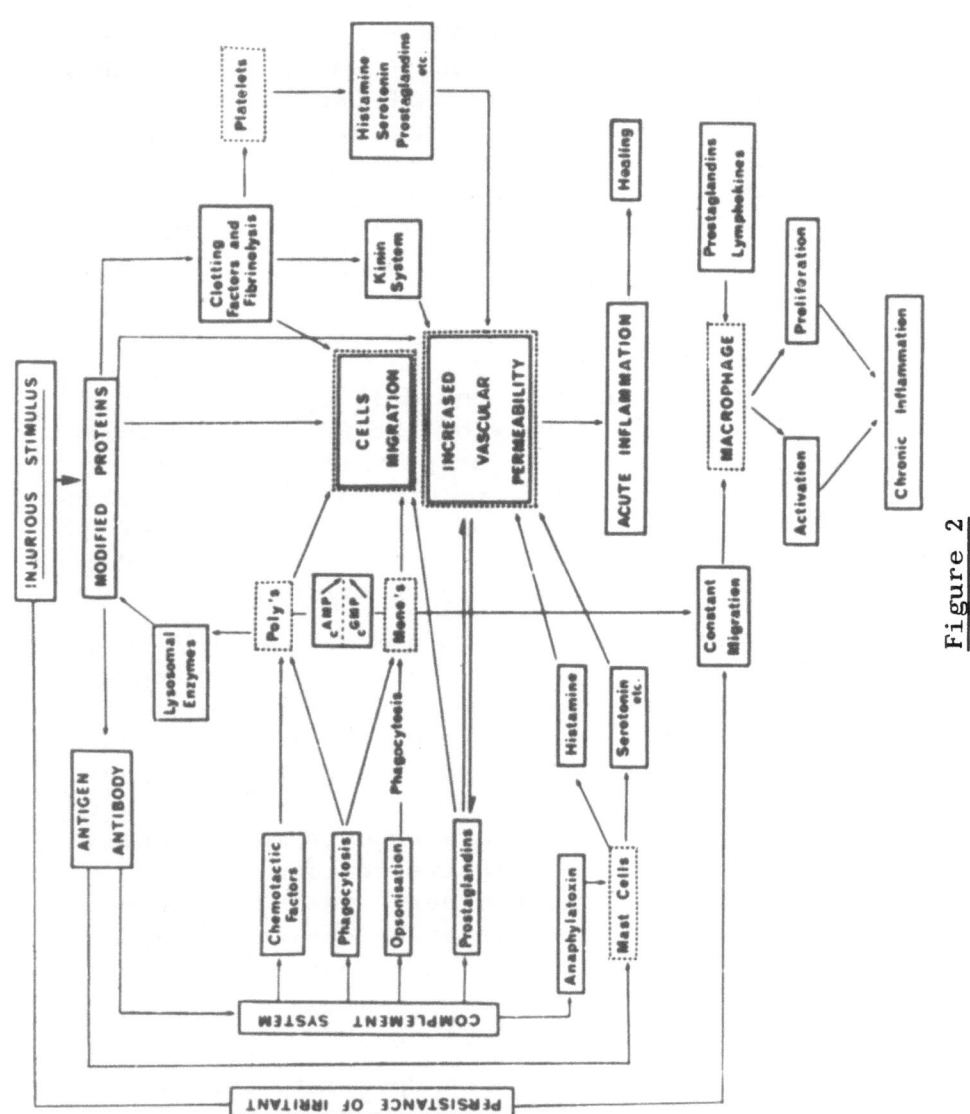

Figure 2

we stress only "some" that may be involved in the
story of acute and chronic inflammation (Fig. 2).

In conclusion we feel that the PG's are within the
entire ubiquitous family modulators of the inflammatory
response by the interrelationship from one member to
another. Secondly via the cyclic nucleotides they are
themselves self modulating. The many concepts regarding
the mode of action of the non-steroidal anti-inflammatory drugs by way of the PG's was convenient and stimulated much research. Such hypothesis must be regarded
constantly against the complex interactions shown in
Figure 2. It was many years ago that the great pharmacologist Sir Henry Dale stated that the mere presence
of a substance at a site does not necessarily implicate
that substance as a mediator.

These conclusions may appear to be pessimistic
regarding the role of the PG's in inflammation; it is
not intended to discuss them, on the contrary we await
the developments of the possible role of intermediates.
We do however stress that one must maintain a sense of
perspective when discussing any particular mediator.
No one family of mediators will explain adequately the
complexity of the varied forms of the inflammatory
response.

Financial support was given by the European Biological Research Association, C.N.R.S. (France), the
Wellcome Trust, The Arthritis and Rheumatism Council.

REFERENCES

1-AUSTEN K.F., LEWIS R.A., STECHSCHULTE D.J., WASSER-
  MAN S.I., LEID R.W. and GOETZL E.J.-
  "Generation and release of chemical mediators of
  immediate hypersensitivity".
  In Progr. in Immunol., 11, 2, 61, 1973.
  Eds. L. BRENT and J. HOLBOROW,
  North Holland Publishing Co Ltd.

2-BRUNO J.J., TAYLOR L.A., DROLLER M.J.-
  "Effects of PGE_2 on human platelet adenyl cyclase
  and aggregation".
  Nature, 251, 721, 1974.

3-CAPASSO F., DUNN C.J., YAMAMOTO S., WILLOUGHBY D.A.
  and GIROUD J.P.-
  "Further studies of carrageenan-induced pleurisy in
  rats".
  J. Path., 116, 117, 1975.

4-CRUNKHORN Pearl and WILLIS A.L.-
  "Cutaneous reactions to intradermal prostaglandins
  Brit. J. Pharmacol., 41 (1), 49, 1971.

5-DEPORTER D.A., DIEPPE P. and WILLOUGHBY D.A.-
  "Pyrophosphate-induced inflammation: an in vivo
  study of the interrelationship of intracellular
  cyclic AMP and cyclic GMP".
  Agents and Actions, 6, 476, 1976.

6-DI ROSA M.-
   In Future Trends in Inflammation.
  Eds. VELO G.P., WILLOUGHBY D.A. and GIROUD J.P.
  Piccin Medical Books, Padua, p.153, 1974.

7-DI ROSA M., GIROUD J.P. and WILLOUGHBY D.A.-
  "Studies of the mediators of the acute inflammatory
  response induced in rats in different sites by
  carrageenan and turpentine".
  J. Path., 104, 15, 1971.

8-DUNN C.J., WILLOUGHBY D.A., GIROUD J.P. and
  YAMAMOTO S.-
  "An appraisal of the interrelationship between
  prostaglandins and cyclic nucleotides in inflammation".

Biomedicine, 24, n° 4, 214, 1976.

9-FERREIRA S.H. and VANE J.R.-
"Inhibition of prostaglandin biosynthesis and the
mechanism of action of non-steroidal anti-
inflammatory agents".
In Future Trends in Inflammation.
Eds. VELO G.P., WILLOUGHBY D.A. and GIROUD J.P.
Piccin Medical Books, Padua, p. 171, 1974.

10-FONTAGNE J. and GIROUD J.P.-
"Pathopharmacologie expérimentale de la réaction
inflammatoire aigue".
Revue des Sciences Médicales,   218, 7, 1976.

11-HIRSCHHORN R., GROSSMAN J. and WEISSMANN G.-
"Effect of cyclic 3'5' -adenosine monophosphate and
theophylline on lymphocyte transformation".
Proc. Soc. Exp. Biol. Med., 133, 1361, 1970.

12-IGNARRO L.J.-
"Regulation of lysosomal enzyme secretion: Role in
inflammation".
Agents and Actions, 4, (4), 241, 1974.

13-KALEY G. and WEINER R.-
"Effect of PGE_1 on leucocyte migration".
Nature New Biol., 234, 114, 1971.

14-KALINER M. and AUSTEN K.F.-
"Cyclic AMP, ATP and reversed anaphylactic histamine
release from rat mast cells".
J. Immunol., 112, (2), 664, 1974.

15-KALINER M. and AUSTEN K.F.-
"Immunologic release of chemical mediators from
human tissues".
Ann. Rev. Pharmacol., 15, 177, 1975.

16-RYAN G.B. and SPECTOR W.G.-
"Macrophages turnover in inflamed connective tissue".
Proc. Roy. Soc. 175, 269, 1970.

17-SMITH J.W., STEINER A.L. and PARKER C.W.-
    "Human lymphocyte metabolism. Effects of cyclic and
    non-cyclic nucleotides on stimulation by
    phytohaemagglutinin".
    J. Clin. Invest., 50, 442, 1971.

18-STROM T.B., CARPENTER C.B., GAROVOY M.R., AUSTEN K.F.,
    MERRILL J.P. and KALINER M.-
    "The modulating influence of cyclic nucleotides upon
    lymphocyte mediated cytotoxicity".
    J. Exp. Med., 138, 381, 1973.

19-TAUBER A.I., KALINER M., STECHSCHULTE D.J. and
    AUSTEN K.F.-
    "Immunologic release of histamine and slow-reacting
    substance of anaphylaxis from human lung. V - Effects
    of prostaglandins on release of histamine".
    J. Immunol., 111, 27, 1973.

20-VELO G.P. and ABDULLAHI S.E.-
    "General concepts of inflammation".
    In Inflammatory Arthropathies.
    Eds. HUSKISSON E.C. and VELO G.P.
    Excerpta Medica, Amsterdam and Oxford, p. 3, 1976.

21-VELO G.P., DUNN C., GIROUD J.P., TIMSIT J. and
    WILLOUGHBY D.A.-
    "Distribution of prostaglandins in inflammatory
    exudate".
    J. Path., 111, 149, 1973.

22-VIGDAHL R.L., MARQUIS N.R. and TAVORMINA P.A.-
    "Platelet aggregation: Adenyl cyclase, PGE_1 and
    calcium".
    Biochem. Biophys., 37, 409, 1969.

23-WILLOUGHBY D.A.-
    "Effects of prostaglandins $PGF_{2\alpha}$ and PGE_1 on
    vascular permeability".
    J. Path., 96, 381, 1968.

24-WILLOUGHBY D.A. and DI ROSA M.-
Proc. Symp. Internat.
Inflammation Club. Immunopathology of Inflammation.
Exc. Med. Intern. Congr., 229, p. 28, 1970.

25-WILLOUGHBY D.A., DUNN C.J., YAMAMOTO S., CAPASSO F.,
DEPORTER D.A. and GIROUD J.P.-
"Calcium pyrophosphate-induced pleurisy in rats: a
new model of acute inflammation".
Agents and Actions, 5, 35, 1975.

26-YAMAMOTO S.-
"Studies on mechanisms of immobilization of mononu-
clear cells in the delayed hypersensitivity (DH)
reaction".
In Future Trends in Inflammation II.
Eds. GIROUD J.P., WILLOUGHBY D.A. and VELO G.P.
Birkhäuser Verlag, Basel, p. 16, 1975.

27-YAMAMOTO S., DUNN C.J., CAPASSO F., DEPORTER D.A.
and WILLOUGHBY D.A.-
"Quantitative studies on cell-mediated immunity in
the pleural cavity of guinea-pigs".
J. Path., 117, 65, 1975.

28-YAMAMOTO S., DUNN C.J., DEPORTER D.A., CAPASSO F.,
WILLOUGHBY D.A. and HUSKISSON E.C.-
"A model for the quantitative study of Arthus
(Immunologic) hypersensitivity in rats".
Agents and Actions, 5, 374, 1975.

29-YAMAMOTO S., DUNN C.J. and WILLOUGHBY D.A.-
"Studies on delayed hypersensitivity pleural exudates
in guinea-pigs. I - Demonstration of substances in
the cell-free exudate which cause inhibition of
mononuclear cell migration in vitro".
Immunol., 30, 505, 1976.

30-YAMAMOTO S., DUNN C.J. and WILLOUGHBY D.A.-
"Studies on delayed hypersensitivity pleural exudates
in guinea-pigs. II - The interrelationship of mono-
cytic and lymphocytic cells with respect to migration activity".
Immunol., 30, 513, 1976.

31-ZURIER R.B., WEISSMANN G., HOFFSTEIN S., KAMMERMAN S.,
    HSIN HSIUNG THAI-
    "Mechanisms of lysosomal enzyme release from human
    leucocytes. II - Effects of cyclic AMP, cyclic GMP,
    autonomic agonists and agents which affect microtu-
    bule function".
    J. Clin. Invest., $\underline{53}$, 297, 1974.

PROSTAGLANDINS AND MAST CELL HISTAMINE RELEASE

B. J. R. WHITTLE

DEPARTMENT OF PHARMACOLOGY, INSTITUTE OF BASIC MEDICAL

SCIENCES, ROYAL COLLEGE OF SURGEONS, LONDON, WC2A 3PN

Prostaglandin E_1 (PGE_1) increases cyclic AMP levels in rat mast cells and inhibits the release of histamine by antigen (Kaliner and Austen, 1974). In the present study, the actions of prostaglandins and drugs affecting prostaglandin metabolism on histamine release following chemical or immunological challenge have been further investigated.

Histamine release from mast cells

Mast cells were obtained by lavage of the rat peritoneal cavity with a modified buffer solution (12 ml, pH 7) containing bovine serum albumin (0.1% w/v). The solution was withdrawn, centrifuged, and the cell pellet re-suspended in the buffer (2.5 ml). Aliquots (0.1 ml) of the cell suspension were pre-incubated (5-30 min) with each substance under investigation (0.2 ml) prior to the addition of the histamine liberator (0.1 ml), and then further incubated. The incubation was terminated by addition of cold buffer (0.5 ml, 4°C). The histamine release from the mast cells was assayed fluorometrically (Shore et al, 1959) or on the superfused guinea-pig ileum.

Effects of prostaglandins on histamine release

PGE_2 reduced submaximal histamine release induced by antigen (either egg albumin or horse serum) from the mast cells of rats sensitised 14 days prior to the experiment. Likewise, the present study confirms that submaximal histamine release induced by compound 48/80 was reduced by PGE_2 (Loeffler et al, 1971; Johnson et al, 1974), and we have also observed an inhibition of ATP-induced

histamine release. Although the dose of PGE_2 causing 50% inhibition of the control histamine release (ID_{50}) was relatively high
(10.5, 16 and 7 µg/ml for antigen, 48/80 and ATP respectively),
the dose-inhibition curve was shallow (Fig. 1) and thus lower
concentrations still gave a small but significant inhibition.
$PGF_2\alpha$ also was found to inhibit histamine release, but was less
potent than PGE_2 $(ID_{50} > 20$ µg/ml$)$.

The relatively low inhibition of histamine release by the
prostaglandins did not appear to be a consequence of metabolism by
15-hydroxy prostaglandin dehydrogenase (PGDH) since two analogues
of PGE_2, (15S)-15 methyl E_2 and 16, 16 dimethyl E_2, which are
resistant to inactivation by this enzyme, were only of comparable
potency. With regard to this enzyme, both polyphloretin phosphate
and di-phloretin phosphate, which inhibit PGDH (Crutchley and
Piper, 1974), caused a marked inhibition of histamine release
induced by compound 48/80, ATP and antigen (Table 1). However,
PPP was the more potent (Fig. 2), unlike the relative potency of
these compounds as either inhibitors of prostaglandin inactivation
or as prostaglandin antagonists suggesting that they act by other
mechanisms. Furthermore, two other inhibitors of PGDH, probenecid
and frusemide $(10^{-5} - 10^{-6}$ M$)$ had no effect on 48/80-induced
histamine release.

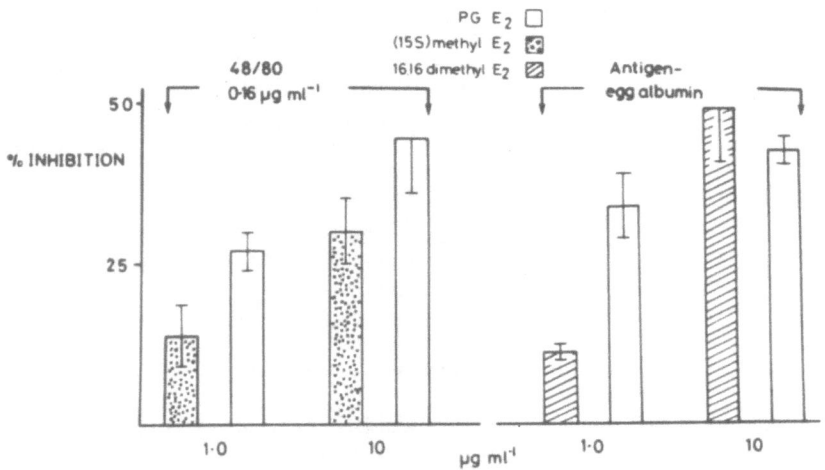

Figure 1. Inhibition of compound 48/80- and antigen-induced
histamine release from rat mast cells by prostaglandin E_2 and its
methyl analogues. Results are mean-s.e.mean of at least 5
experiments.

Prostaglandins $(PGE_1, E_2, A_1$ and $F_2\alpha)$ can also inhibit anaphylactic histamine release from human lung in vitro (Walker, 1973). Whether this ability of exogenous prostaglandins to reduce mast cell histamine release in vitro could reflect a patho-physiological modulator role in vivo during anaphylaxis requires further investigation.

Inhibition of prostaglandin biosynthesis

Low concentrations of non-steroid anti-inflammatory agents, which inhibit prostaglandin formation (Vane, 1971) can augment the release of histamine and SRS-A from guinea-pig isolated per-fused lungs following challenge with antigen (Engineer et al, 1976). This suggested that the prostaglandins released during anaphylaxis could modulate the release of these other substances.

Comparable experiments on rat mast cells have been reported (Thomas and Whittle, 1976). In contrast, similar low concentra-tions of these aspirin-like drugs had no effect on histamine release, suggesting that endogenous prostaglandins are not invol-ved with the histamine release process of the mast cell. It is possible that in the perfused lung, prostaglandins are released from cells other than those which release histamine or SRS-A. Indeed, these latter substances which are released during challenge, may lead to the prostaglandin generation, which could then act through a 'negative feed-back' mechanism to modulate the further release of the mediators.

Effects of anti-inflammatory agents

Non-steroid anti-inflammatory drugs were not thought to alter histamine release in vivo during the inflammatory process. There-fore, an unexpected finding was that, like the primary prostaglan-dins tested, the aspirin-like drugs inhibited histamine release from mast cells in vitro. The doses which reduced histamine release were greater than those which should be required to inhibit prostaglandin biosynthesis and the effects are probably not related to effects on prostaglandin formation. The findings support the observations that indomethacin reduces rat mast cell degranulation (Taylor et al, 1974) and histamine release from human lung in vitro (Walker, 1973).

In the present study, indomethacin (10 - 80 $\mu g/ml$) gave a parallel displacement of the dose-response curve for the release of histamine by compound 48/80 (0.1 - 1 $\mu g/ml$). The dose causing 50% inhibition of submaximal histamine release (ID_{50}) was 3.5, 9.0 and 21.5 $\mu g/ml$ for sodium meclofenamate, flufenamic acid and indomethacin, respectively. These drugs also inhibited histamine liberation induced by antigen (either horse serum or egg albumin)

from sensitised mast cells, or that induced by adenosine triphos-
phate or crude phospholipase-A (Table 1). Although these aspirin-
like drugs had no consistent effect on non-specific histamine
liberation with the surface-active agents Triton X-100 (0.01%) or
n- decylamine (25 µg/ml), the release of histamine following incubation of mast cells with a calcium ionophore (A23187) was markedly reduced. In contrast, the anti-inflammatory steroid, betamethasone (100 µg/ml), only slightly inhibited (18-25%) antigen-
or 48/80-induced histamine release.

Although the inhibition of histamine release by these aspirin-
like drugs could result from an elevation of mast cell cyclic AMP
levels via phosphodiesterase inhibition, histamine release stimulated by the calcium ionophore (A23187) was also inhibited, suggesting that other inhibitory mechanisms may also operate; an
increase in cyclic AMP levels is not thought to prevent ionophore-
induced histamine release (Garland and Mongar, 1976). Thus, the
possibilities that non-steroid anti-inflammatory drugs affect the
mobilization of calcium ions, or the processes of oxidative metabolism involved in histamine release have therefore been investigated.

Effects on oxidative metabolism

Submaximal histamine release induced by compound 48/80 or
ionophore-A23187 was inhibited by 2, 4 dinitrophenol (50-100 µg/ml)
which uncouples oxidative phosphorylation, and by antimycin-A
(0.02-0.5 µg/ml) which acts on the cytochrome system. This inhibition was prevented by pre-incubation with glucose (5 mM)-contain-
ing medium, as shown by others (see Diamant, 1975). In contrast,
the inhibition of histamine release by indomethacin or sodium
meclofenamate was not significantly altered by glucose incubation.

Table 1. Inhibition of histamine release from rat mast cells

Histamine liberator	(µg/ml)	Indomethacin	Meclofenamate	PPP
Compound 48/80	(0.16)	21.5	3.5	0.7
Phospholipase-A	(5.0)	13.5	2.0	–
Adenosine triphosphate	(200)	10.0	3.0	2.5
Antigen-egg albumin	(1000)	13.0	6.5	12.0

The concentration of inhibitor (µg/ml) giving 50% reduction of
histamine release (ID_{50}), following 5 min pre-incubation with mast
cells, was determined from at least 5 experiments. Control histamine release, with the concentration of each liberator shown, was
30-50% of total histamine content of the cells.

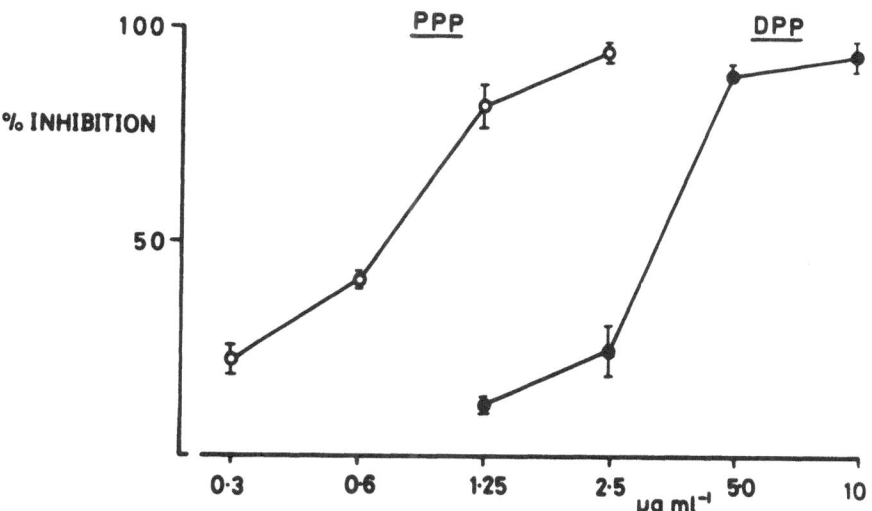

Figure 2. Inhibition of compound 48/80-induced histamine release from rat mast cells by polyphloretin phosphate (PPP) and diphloretin phosphate (DPP).

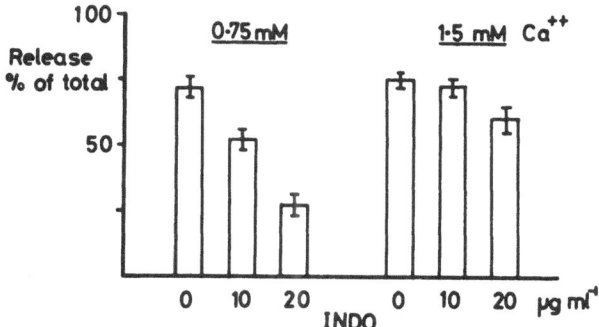

Figure 3. Effect of calcium concentration on the inhibition, by indomethacin (10-20 μg/ml), of submaximal histamine release from rat mast cells induced by the calcium ionophore A23187 (0.33 μg/ml).

These results show that whereas the inhibition of histamine release from rat mast cells, by drugs known to interfere with energy production from oxidative metabolism, can be reversed by incubation with glucose (by promoting anaerobic glycolysis), glucose-incubation did not prevent the inhibition of release produced by indomethacin or meclofenamate. Thus these aspirin-like drugs do not appear to inhibit by preventing oxidative metabolism.

Effects of calcium

Histamine release induced by the ionophore - A23187 was dependent on the dose (0.1–0.5 μg/ml) and on the calcium concentration of the incubation medium. Optimal histamine release, with a submaximal dose of the ionophore (0.33 μg/ml), was obtained with a calcium concentration of 0.75 mM; an increase in the calcium concentration up to 3 mM did not elevate the release further. The dose-dependent inhibition by indomethacin (10–40 μg/ml) or meclofenamate (1–5 μg/ml) of histamine release stimulated by this ionophore was related to the calcium concentration of the incubation medium (Fig. 3). In four experiments, the marked inhibition of ionophore-induced release by meclofenamate (5 μg/ml) was significantly reduced (from $81.4 \pm 2.9\%$ to $7.6 \pm 3.9\%$ inhibition, mean-s.e. mean: P<0.001) when the calcium concentration was increased from 0.75 to 3 mM. In contrast, the inhibition of ionophore-induced release by antimycin-A was not reversed by an increase in the calcium concentration.

The observation that an increase in the calcium concentration abolished the inhibition of ionophore-stimulated release suggests an action on a calcium-dependent stage in the release process. It has been shown that several aspirin-like drugs can interfere with the uptake and binding of calcium in various tissues (Northover, 1973). Thus, the inhibition of histamine release by non-steroid anti-inflammatory drugs in vitro could reflect actions on calcium influx into the mast cell, or on calcium mobilization or utilization within the mast cell.

It is not known whether such actions could occur in vivo with anti-inflammatory doses of aspirin-like drugs. However, it may be expedient to be aware of such a potential interaction between non-steroid anti-inflammatory drugs and calcium when interpreting the actions of such drugs in vitro on any calcium-sensitive system.

Acknowledgements
    This work was supported by a grant to Professor G. P. Lewis from the Medical Research Council. Prostaglandins were kindly supplied by the Upjohn Company, Kalamazoo and the ionophore by the Lilly Research Centre Ltd. Technical assistance was provided by Mr. R. U. Thomas.

References

CRUTCHLEY, D. J. & PIPER, PRISCILLA J. (1974).
    Prostaglandin inactivation by guinea-pig lung and its inhibi-
    tion. Br. J. Pharmac., 52, 197-203.

DIAMANT, B. (1975).
    Energy production in rat mast cells and its role for hista-
    mine release. Int. Archs. Allergy appl. Immun., 49, 155-171.

ENGINEER, DINAZ, PIPER, PRISCILLA J. & SIROIS, P. (1976).
    Interaction between the release of SRS-A and prostaglandins.
    Br. J. Pharmac., 57, 460-461P.

GARLAND, L. G. & MONGAR, J. L. (1976).
    Differential histamine release by dextran and the ionophore
    A23187: the actions of inhibitors. Int. Archs. Allergy appl.
    Immun., 50, 27-42.

JOHNSON, A. R., MORAN, N. C. & MAYER, S. E. (1974).
    Cyclic AMP content and histamine release in rat mast cells.
    J. Immun., 112, 511-519.

KALINER, M. & AUSTEN, K. F. (1974).
    Cyclic AMP, ATP and reversed anaphylactic histamine release
    from rat mast cells. J. Immun., 112, 664-672.

LOEFFLER, L. J., LOVENBERG, W. & SJOERDSMA, A. (1971).
    Effects of dibutyryl-3', 5'-cyclic adenosine monophosphate,
    phosphodiesterase inhibitors and prostaglandin E_1 on compound
    48/80-induced histamine release from rat peritoneal mast
    cells in vitro. Biochem. Pharmac., 20, 2287-2297.

NORTHOVER, B. J. (1973).
    Effect of anti-inflammatory drugs on the binding of calcium
    to cellular membranes in various human and guinea-pig tissues.
    Br. J. Pharmac., 48, 496-504.

SHORE, P. A., BURKHALTER, A. & COHN, V. H. (1959).
    A method for the fluorometric assay of histamine in tissues.
    J. Pharmac. exp. Ther., 127, 182-186.

TAYLOR, W. A., FRANCIS, D. H., SHELDON, D. & ROITT, I. M. (1974).
    Anti-allergic actions of disodium cromoglycate and other
    drugs known to inhibit cyclic 3', 5'-nucleotide phosphodies-
    terase. Int. Archs. Allergy appl. Immun., 47, 175-193.

THOMAS, R. U. & WHITTLE, B. J. R. (1976).
    Prostaglandins and the release of histamine from rat periton-
    eal mast cells. Br. J. Pharmac., 57, 474-475P.

VANE, J. R. (1971).
    Inhibition of prostaglandin synthesis as a mechanism of
    action for aspirin-like drugs. Nature, New Biol., 231,
    232-235.

WALKER, JOYCE L. (1973).
    The regulatory role of prostaglandin in the release of
    histamine and SRS-A from passively sensitized human lung
    tissue. In : Advances in the Biosciences, vol. 9, 235-240,
    Pergamon-Vieweg: Braunschweig.

PROSTAGLANDINS AND NON-STEROIDAL ANTI-INFLAMMATORY DRUGS

S.H. FERREIRA

Department of Pharmacology, Faculty of Medicine

of Ribeirão Preto, São Paulo, Brazil

The chemical aspects of the inhibition of the synthesis of prostaglandins by non-steroidal anti-inflammatory agents have been amply discussed in other papers in this volume. We shall be mainly concerned with those aspects which are relevant to the discussion of the hypothesis which proposes the inhibition of prostaglandin synthesis as the mechanism of action of aspirin-like drugs (1).

However, in order to provide a general view of the problem, we shall first summarise the physio-pathological observations which implicate prostaglandins in inflammation: prostaglandins have been detected in inflammatory exudates (2, 3, for review 4); prostaglandins mimic inflammatory signs and symptoms at concentrations found in inflammatory exudates. Prostaglandins of the E series cause erythema (5, 6) and induce oedema (7, 8) or pain by synergizing with other mediators (9). They cause pain by lowering the pain threshold to chemical or mechanical stimulation (9, 10). Prostaglandins are the most powerful known fever-inducing substances when injected into the cerebral ventricles or anterior hypothalamus. Endogenous pyrogen causes generation of prostaglandins by nervous parenchyma in parallel to the development of fever. Non-steroidal anti-inflammatory agents do not block the direct action of prostaglandins (for review 11).

Since the demonstration that aspirin-like drugs inhibit the prostaglandin synthetases from man, dog and guinea-pig (1, 12, 13), this anti-enzyme action has been amply confirmed and demonstrated in many biological

preparations and in almost all laboratory species such
as cat, rat, gerbil, mouse, sheep, dog, rabbit, cow and
sheep.
        The potency of the non-steroidal anti-inflammatory
drugs against prostaglandin synthetase varies not only
according to the source of the enzyme preparation but
also to the experimental conditions and the way in which
the enzyme is prepared (14, 15, 16). In general, however,
the overall rank order of potency is independent of the
enzyme preparation, although there are some minor
variations (even when the same preparation is used; 14).
Variations in activity among non-steroidal anti-inflammatory drugs will be discussed below.
        Despite these variations in potency, inhibition of
prostaglandin biosynthesis is clearly a general
characteristic of aspirin-like drugs. It also seems to
be a unique characteristic, for compounds selected to
represent many other types of pharmacological activity
are inactive (< 10% inhibition at 100 µg/ml). These
include chloroquine, morphine, mepyramine, probenecid,
azathioprine, para-and meta-hydroxybenzoic acid, phenergan,
atropine, methysergide, phenoxybenzamine, propranolol,
iproniazid, dorperidol, chlorpromazine and disodium
cromoglycate (17).
        Several investigators have compared anti-inflam-
matory activity of aspirin-like drugs with their
activity against prostaglandin synthetase. A good
correlation is demonstrable, even when comparison is
based on oral administration of the drug on the one hand
and direct inhibition of an in vitro microsomal enzyme
preparation on the other. The rank order was the same
against carrageenin rat paw oedema as against spleen
synthetase, except that indomethacin was out of order
for the rat paw test (17). Even more striking is the
correlation shown by comparing pairs of enantiomers.
This has been done for naproxen and for indomethacin
analogues. In each case, the member of each pair having
anti-inflammatory activity also strongly inhibited
prostaglandin synthetase, whereas the one with weak anti-
-inflammatory activity was also weak against the synthetase
(18, 15). It is important to notes however, that the
pairs of enantiomers showed, similar activity towards
protein binding, oxydative phosphorilation, lysosomal
enzyme release and that the distribution at the intracellular level and in the body fluid is governed by the
same pharmacokinetic factors.
        In spite of the general relationship observed
between anti-inflammatory activity and the effects of
aspirin like drugs on the spleen prostaglandin synthetase

preparation, enzymes prepared from other tissues show different sensitivity to the drugs. For rabbit brain synthetase, for instance, the ratio of activity between indomethacin and aspirin is 17:1 (19) whereas for bovine seminal vesicle synthetase it is 2,140:1 (13). This important property, which may be the result of a series of iso-enzymes, can explain the variations in activity within the group of compounds. For instance, the anti--pyretic analgesic drug 4-acetamidophenol (acetaminophen or paracetamol), which is ten times less effective than aspirin on the dog spleen synthetase, has almost the same potency as aspirin on brain enzymes (19, 20). Thus, the fact paracetamol has anti-pyretic activity without anti-inflammatory activity can be explained by the differential sensitivity of the prostaglandin synthetases from different tissues. Just as the anti-inflammatory activity of aspirin-like drugs correlates well with their action against spleen enzyme, so the anti-pyretic activity correlates with their action against brain enzymes (19).

Other examples of differential enzyme sensitivity are also available. There is a thousand-fold variation in the ID 50 of indomethacin against prostaglandin synthetases from different rabbit tissues (21). In the spleen enzyme, the ID 50 was 0.05 µg/ml, and in the brain enzyme, 1.0 µg/ml (these figures are in close agreement with ref. 19). In kidney enzyme, the ID 50 was 5.0 µg/ml, in the iris-ciliary body, 18.5 µg/ml and in the retina 50 µg/ml.

One of the most important points in support of the inhibition of the prostaglandin biosynthesis as the mechanism of the anti-inflammatory action of aspirin--like drugs is that a therapeutic dose leads to a plasma concentration capable of causing a strong inhibition of prostaglandin synthetases (19). Taking indomethacin as an example, the plasma concentration in man reaches 2 µg/ml. Because of protein binding (which is a property common to many of these drugs) the free plasma concentration would be 0.2 µg/ml. However, the ID 50 for indomethacin on dog spleen synthetase is only 0.05 µg/ml.

Hamberg (22) calculated daily prostaglandin turnover from the amounts of metabolites in the urine. Men consistently produced larger amounts (50-330 µg/day) than women (20-40 µg/day) but in both sexes there was a 77-98% inhibition of prostaglandin production by therapeutic doses of indomethacin (200 mg daily), aspirin (3 g daily).

Aspirin and its congeners at concentrations as low

as those required to inhibit prostaglandin generation
also block the release of a rabbit aorta contracting
substance (RCS) from guinea-pig lungs during anaphylaxis
(23, 24). Several indications, including RCS formation
from arachidonic acid (24, 25), its appearance always
with prostaglandins, its instability, and the inhibition
of its release by aspirin-like drugs (25) suggested to
Gryglewski and Vane (26) that RCS is the cyclic endoperoxide postulated as an unstable intermediate in the
biosynthesis of prostaglandins. The cyclic endoperoxide
was isolated and found to contract rabbit aorta (27).
More recently RCS was identified as thromboxane, a
substance derived from oxidation of arachidonic acid
which is capable of aggregating platelets and contracting
rabbit aorta strips (28). The fact that aspirin like
drugs inhibit endoperoxide or thromboxane formation
indicates an interference of these agents at an early
stage of the synthesis of prostaglandins.

It has been shown that several drugs capable of
inhibiting prostaglandin synthesis in vitro failed to
show any anti-inflammatory activity (32).

Recently we described a method in which inflammatory exudates were induced in rats by subcutaneous
implantation of sponges impregnated with carrageenin (29)
with this test in was possible to correlate inhibition
in vivo of prostaglandin synthesis and anti-inflammatory
activity as shown in table I. Drugs such as phenelzine,
chlorpromazine an desipramine, in spite of causing
inhibition of prostaglandin synthesis in vitro, did not
show a similar effect in vivo.

To our knowledge there is no substance capable
of inhibiting prostaglandin synthesis in vivo that does
not shown anti-inflammatory activity. The opposite was
observed with colchicine. This agent shows anti-inflammatory activity "in vivo" but potentiates the amount of
prostaglandin generated. This clearly indicates that
its action is upon other factors or mediators which
participate in the genesis of inflammatory oedema. Table I
also shows that sodium salicylate which does not inhibit
prostaglandin synthesis in vitro is as effective as aspirin
in blocking both the prostaglandin synthesis in vivo and
carrageenin induced oedema. This confirms the observation
of Willis et al. (20) who suggested that salicilates may
be converted in vivo to the active principle.

We may then conclude that inhibition of prostaglandin biosynthesis is a general property of aspirin-
-like drugs which is demonstrable at concentrations found
in body fluids during therapy. There is a general
correlation between their anti-inflammatory activity and

TABLE I. COMPARISON OF THE EFFECTS OF DIFFERENT DRUGS ON PROSTAGLANDIN SYNTHETASE IN VITRO WITH INHIBITION OF CARRAGEENIN OEDEMA AND PG PRODUCTION IN VIVO

	PG synthetase in vitro IC_{50} (μM)	Ref.	(a) Dose (mg/kg) given 3 times	% Inhibition of oedema (no. of rats)	% Reduction of PGs in sponge (no. of rats)
Indomethacin	0.17	(17)	4.0	55 (10)	> 95 (13)
Phenylbutazone	7.25	(17)	100.0	50 (5)	> 95 (6)
Aspirin	37.0	(17)	150.0	30 (5)	85 (10)
Sodium salicylate	800.0	(30)	150.0	30 (5)	80 (10)
Paracetamol	660.0	(17)	150.0	30 (5)	40 (5)
Dexamethasone	Inactive	(17)	0.1	45 (5)	30 (7)
Prednisolone	Inactive	(31)	5.0	45 (5)	0 (6)
Hydrocortisone	Inactive	(17)	20.0	45 (5)	0 (5)
Indomethacin	3.60	(32)	4.0	55 (10)	> 95 (13)
Aspirin	150.0	(32)	150.0	30 (5)	85 (10)
Phenelzine	0.37	(32)	20.0	0 (5)	0 (5)
Desipramine	123.0	(32)	20.0	0 (5)	0 (5)
Chlorpromazine	170.0	(32)	20.0	0 (5)	0 (5)
Soidum aurothiomalate (gold) (i.m.)	10-100	(17)	1.0	20 (10)	35 (5)
Mepacrine	Inactive	(30)	150.0	20 (5)	0 (5)
Penicillamine	Inactive	(30)	30.0	0 (5)	0 (5)
Colchicine (s.c.)	Potentiates	(30)	1.5	85 (5)	+40 (6)

(a) Each drug was given orally, unless otherwise stated, to groups of 5-10 rats. Three doses were given; the first at the time of sponge implantation, the second 8 h later and the third after further 13 hours.

inhibition of prostaglandin synthetase. Synthetases prepared from different tissues show different sensitivities to non-steroid anti-inflammatory drugs. This important property, which may be due to a series of iso--enzymes, can explain the variations in activity within the group of compounds. For instance, paracetamol, which is anti-pyretic and analgesic without being anti--inflammatory, has a much greater activity on brain enzyme than on spleen enzyme.

These findings together with the actions of prostaglandin (pyresis, vasodilatation, sensitization of the pain receptors or potentiation of the vascular permeability effects of other inflammatory mediators) support our theory that this anti-enzyme effect is the mechanism of action of aspirin-like drugs.

REFERENCES

1- VANE, J.R. (1971). Inhibition of prostaglandin stnthesis as a mechanism of action for aspirin-like drugs. Nature New Biol. 231: 232.

2- WILLIS, A.L. (1969). Release of histamine, kinin and prostaglandin during carrageenin-induced inflammation in the rat. In: Prostaglandins, peptides and amines 31-38, ed. by P. Mantegazza and E.W. Horton. Academic Press, London, New York.

3- GREAVES, M.W., SØNDERGAÅRD, J. and MCDONALD-GIBSON, W. (1971). Recovery of prostaglandins in human cutaneous inflammation. Br. Med. J. 2: 258.

4- FERREIRA, S.H. and VANE, J.R. (1974). Aspirin and prostaglandins. In: The Prostaglandins II, ed. P.W. Ramwell, pp. 1-39, New York, London, Plenum Press.

5- SOLOMON, L.M., JUHLIN, L. and KIRSCHENBAUM, M.B. (1968). Prostaglandin on cutaneous vasculature. J. Inv. Derm. 51: 280.

6- CRUNKHORN, P. and WILLIS, A.L. (1971). Custaneous reactions to intradermal prostaglandins. Br. J. Pharm. 41: 49.

7- MONCADA, S., FERREIRA, S.H. and VANE, J.R. (1973). Prostaglandins, aspirin-like drugs and the oedema of inflammation. Nature 246: 217.

8- THOMAS, G. and WEST, G.B. (1973). Prostaglandins as regulators of bradykinin responses. J. Pharm. Pharmac. 25: 747.

9- FERREIRA, S.H. (1972). Prostaglandins, aspirin-like drugs and analgesia. Nature, New Biology 240: 200.

10- MONCADA, S., FERREIRA, S.H. and VANE, J.R. (1975).
    Inhibition of prostaglandin biosynthesis as the
    mechanism of analgesia of aspirin-like drugs in the
    dog knee joint. Eur. J. Pharmac. 31: 250.

11- FELDBERG, W. (1974). Fever, Prostaglandins and
    Antipyretics, in Prostaglandin Synthetase inhibitors.
    H.J. Robinson and R. Vane, Raven Press, New York.

12- FERREIRA, S.H., MONCADA, S. and VANE, J.R. (1971).
    Indomethacin and aspirin abolish prostaglandin
    release from the spleen. Nature, New Biology 231:
    237.

13- SMITH, J.B. and WILLIS, A.L. (1971). Aspirin
    selectively inhibits prostaglandin production in
    human platelets. Nature, New Biology 231: 256.

14- FLOWER, R.J., CHEUNG, H.S. and CUSHMAN, D.W. (1973).
    Quantitiative determination of prostaglandins and
    malondialdehyde formed by the arachidonate oxygenase
    system of bovine seminal vesicles. Prostaglandins
    4: 325.

15- HAM, E.A., CIRILLO, V.J., ZANETTI, M., SHEN, T.Y.
    and KUEHL, F.A. Jr.(1972). Studies on the mode of
    action of non-steroidal anti-inflammatory agents. In:
    Prostaglandins in cellular biology, ed. Ramwell,
    P.W. and Pharris, B.B. 343, Plenum Press, New York.

16- TAKEGUCHI, C. and SIH, C.J. (1972). A rapid spectro-
    phometric assay for prostaglandin synthetase
    application to the study of non-steroidal anti-
    -inflammatory agents. Prostaglandins 2: 169.

17- FLOWER, R.J., GRYGLEWSKI, R, HERBACZYNSKA, C.K. and
    VANE, J.R. (1972). The effects of anti-inflammatory
    drugs on prostaglandin biosynthesis. Nature, New
    Biology 238: 107.

18- TOMLINSON, R.V., RINGOLD, H.J., QURESHI, M.C. and
    FORCHIELLI, E. (1972). Relationship between
    inhibition of prostaglandin synthesis and drug
    efficacy: support for the current theory on mode of
    action of aspirin-like drugs. Biochem. Biophys. Res.
    Comm. 46: 552.

19- FLOWER, R.J. and VANE, J.R. (1972). Inhibition of
    prostaglandin synthetase in brain explains the anti-
    -pyretic activity of paracetamol (4-acetamido-phenol).
    Nature 240: 410.

20- WILLIS, A.L., DAVIDSON, P., RAMWELL, P.W.,
    BROCKLEHURST, W.E. and SMITH, B. (1972). Release
    and actions of prostaglandins in inflammation and
    fever: inhibition by anti-inflammatory and anti-
    -pyretic drugs. Prostaglandins and cellular Biology,
    ed. Ramwell, P.W. and Pharris, B.P., 227. Plenum
    Press, New York, London.

21- BHATTACHERJEE, P. and EAKINS, K.E. (1973). Inhibition
    of the PG synthetase system in ocular tissue by
    indomethacin. The Pharmacologist 15: 209.

22- HAMBERG, M. (1972). Inhibition of prostaglandin
    synthesis in man. Biochem. Biophys. Res. Commu.
    49: 720.

23- PIPER, P.J. and VANE, J.R. (1969). Release of
    additional factors in anaphylaxis and its antagonism
    by anti-inflammatory drugs. Nature 223: 20.

24- VARGAFTIG, B.B. and DAO HAI, N. (1971). Release of
    vasoactive substances from guinea-pig lungs by slow
    reacting substance C and arachidonic acid.
    Pharmacology 6: 99.

25- PALMER, M.A., PIPER, P.J. and VANE, J.R. (1973).
    Release of rabbit aorta contracting substance (RCS)
    and prostaglandins induced by chemical or mechanical
    stimulation of guinea-pig lungs. Br. J. Pharmac.
    49: 226.

26- GRYGLEWSKI, R. and VANE, J.R. (1972). The release
    of prostaglandins and rabbit aorta contracting
    substance (RCS) from rabbit spleen and its antagonism
    by anti-inflammatory drugs. Br. J. Pharmac. 45: 37.

27- HAMBERG, M. and SAMUELSSON, B. (1974). Prostaglandin
    endoperoxides VII. Movel transformations of
    arachidonic acid in guinea pig lungs. Biochem.
    Biophys. Res. Comm. 61: 942.

28- HAMBERG, M., SVENSSON, J. and SAMUELSSON, B. (1975).
    Thromboxanes: A new group of biologically active
    compounds derived from prostaglandin endoperoxides.
    Proc. Nat. Ac. Sci. U.S.A. 72: 2994.

29- HIGGS, G.A., HARVEY, E.A., FERREIRA, S.H. and VANE,
    J.R. (1976). The Effects of Anti-Inflammatory Drugs
    on the Production of Prostaglandins In Vivo.
    Advances in Prostaglandin and Thromboxane Research,
    1: ed. B. Samuelsson and R. Paoletti.

30- FLOWER, R.J. (1974). Drugs which inhibit prosta-
    glandin biosynthesis. Pharm. Revs. 26: 33.

31- BLACKWELL, G.J. and PARSONS, M. (1975).
    Personal Communication.

32- LEE, R.E. (1974). The influence of psychotropic
    drugs on prostaglandin biosynthesis. Prostaglandins
    5: 63.

I am grateful to FAPESP (75/0296) and to the Wellcome
Foundation for research Grants.

INFLUENCE OF ANTI-INFLAMMATORY STEROIDS

ON PROSTAGLANDIN AND THROMBOXANE RELEASE FROM TISSUES

R.J. Gryglewski, R. Korbut,
and A. Dembińska-Kieć
Department of Pharmacology, Copernicus
Academy of Medicine in Cracow
31-531 Cracow, 16 Grzegórzecka, Poland

INTRODUCTION

Natural glucocorticosteroids and synthetic anti-inflammatory steroids ("steroids" for brevity) are widely used in the treatment of allergic and inflammatory conditions however the mechanism of their immunosuppressive and anti-inflammatory actions remains unknown. Steroids do not significantly affect antibody production in man but seem to prevent macrophages from responding to lymphokines that are elaborated by T lymphocytes (Claman, 1975). Steroids potentiate stimulatory effects of catecholamines and PGE_1 on accumulation of cyclic-AMP in human leucocytes and lymphocytes (Parker et al. 1973; Mendelsohn et al., 1973), and if so, this might result in a decreased liberation of eosinophil chemotactic factor of anaphylaxis (ECF-A), slow reacting substance of anaphylaxis (SRS-A) and histamine from appropriate cells (Kaliner and Austen, 1975). Indeed, steroids depress the release of histamine from anaphylactic guinea pig lungs (Gryglewski et al., 1975) and from other tissues (Kurihara and Shibata, 1975). Early reports that steroids impair the formation of kinins has not been confirmed (Eisen et al., 1968).

Under many experimental conditions glucocorticosteroids have been shown to stabilize lysosomal membranes against a variety of injuries (Weissmann, 1964). The original finding of Weissman was confirmed in a number of laboratories and the stabilizing effect on lysosomes is the most consistent biological effect of glucocorticosteroids.

363

Steroids influence many aspects of inflammatory and allergic responses. Perhaps there is no common mechanism that could explain numerous pharmacological effects of steroids but it seemed also impossible to explain a variety of pharmacological effects of aspirin until the papers of Vane (1971) and his colleagues (Smith and Willis, 1971; Ferreira et al., 1971) were published. Inhibition by steroids of the release of prostaglandins and thromboxanes from various tissues offers a new possibility for speculation on the mechanism of pharmacological action of steroids.

GLUCOCORTICOSTEROIDS AND PROSTAGLANDIN SYNTHETASE

Vane (1971) and his colleagues have discovered that aspirin and other non-steroidal anti-inflammatory drugs inhibit prostaglandin biosynthesis in vitro and this biochemical effect has been proposed as the mechanism of therapeutic action of aspirin-like drugs. This concept was later supported by much experimental and clinical evidence as reviewed by Flower (1974). In the same three pioneering papers there are short but meaningful statements about the failure of hydrocortisone (100 µg/ml) to inhibit prostaglandin generation (or release) from guinea pig lung homogenates (Vane, 1971), from aggregating human platelets (Smith and Willis, 1971) and from adrenergically stimulated perfused dog spleen (Ferreira et al., 1971).

The effect of steroids on microsomal prostaglandin synthetase was investigated by Flower et al. (1972). Prostaglandin synthetase from dog spleen is inhibited by indomethacin and fenemates (IC_{50} = 0.03 - 0.17 µg/ml), whereas hydrocortisone, dexamethasone, triamicinolone and fludrocortisone at concentrations of 100 µg/ml are inactive. Hydrocortisone, dexamethasone and fluocinolone (100 - 300 µg/ml) do not inhibit generation of PGE_2 from arachidonic acid by bovine seminal vesicle microsomes (Gryglewski, in this book). Therefore there is little doubt that corticosteroids are neither inhibitors of cyclo-oxygenase of arachidonic acid nor inhibitors of PGE_2 isomerase.

Greaves and McDonald-Gibson (1972a,b) have reported that fluocinolone at high a concentration of 220 µM inhibits prostaglandin generation from arachidonic acid by skin homogenates. In rat abdominal skin homogenates this inhibition is 30% in respect to PGE_2 and 50% in respect to $PGF_{2\alpha}$ (Greaves and McDonald- Gibson, 1972a). In human breast skin homogenates this inhibition is 50% in respect to PGE_2 and 30% in respect to $PGF_{2\alpha}$ (Greaves and McDonald Gibson, 1972b). Hydrocortisone (280 μM) is ineffective as an inhibitor of prostaglandin biosynthesis in human skin homogenates. More recently Greaves et al. (1975) have found that five potent anti-inflammatory steroids failed to inhibit prostaglandin biosynthesis by microsomes from rat skin over a wide concentration range. Thus the inability of steroids to inhibit microsomal cyclo-oxygenase was confirmed, Greaves and his colleagues suggest that in skin homogenates steroids depend for their inhibitory effect on a cell membrane-bound factor. It well might be that the presence of few unbroken cells in skin homogenates can explain the original results of Greaves and McDonald-Gibson (1972 a,b), since steroids inhibit prostaglandin release from biological systems providing that the integrity of the cell structure is preserved.

GLUCOCORTICOSTEROIDS AND PROSTAGLANDIN RELEASE

Prostaglandins are not "stored" within cells (Piper and Vane, 1971) and therefore their biosynthesis must immediately precede their release. Non-steroidal anti-inflammatory drugs inhibit prostaglandin biosynthesis at the first step of cyclo-oxygenation of arachidonic acid and consequently these drugs are inhibitors of prostaglandin release from tissues. Anti-inflammatory steroids do not inhibit prostaglandin biosynthesis at any known enzymic stage and still these drugs do inhibit prostaglandin release from intact tissues. This implies a direct interaction of steroids with biomembranes which impairs the supply of endogenous substrates for prostaglandin biosynthesis. Therefore steroids inhibit prostaglandin synthesis not being enzymic inhibitors of prostaglandin synthetase. Alternatively steroids could hinder the transmembrane transport of prostaglandins from inside to outside the cell but then prostaglandins would have to be "stored" inside the cell.

In September 1974 Herbaczyńska-Cedro and Barczak-Staszewska (1976) have presented a communication during the Second Congress of Hungarian Pharmacological Society on the inhibition of prostaglandin release by hydrocortisone. The authors bioassayed the output of prostaglandins into the venous blood of hind legs of anaesthetized dogs during electrical stimulation of sciatic nerve. This procedure causes an increase in prostaglandin-like activity in blood up to 0.5 - 3.5 ng/ml of PGE_2 equivalents. This effect can be reduced by a close intra-arterial infusion of either indomethacin (2 μg/ml) or hydrocortisone semisuccinate (2 μg/ml) but not by aldo-

sterone. On the other hand the removal of exogenous
PGE_2 from circulation of the hindquarters is not influenced by hydrocortisone. The authors conclude that
hydrocortisone although devoid of anti-prostaglandin
synthetase activity, as well as devoid of activity on
transmembrane transport of PGE_2 may somehow interfere
with the rate of prostaglandin formation in tissues.
     Lewis and Piper (1975) have published the first
data concerning the inhibition of prostaglandin release
by steroids. In that and in the next papers (Chang et
al., 1975; Lewis and Piper, 1976 a,b) the authors investigated the influence of glucocorticosteroids on the
ACTH-induced release of prostaglandins from rabbit adipose tissue. Steroids depress the release of prostaglandins but do not lower the content of prostaglandins
in the adipose tissues. The authors propose that the
inhibitory effect of steroids on prostaglandin release
is not due to the impairment of prostaglandin formation
in tissues but it is due to the inhibition of the transmembrane transport mechanism for prostaglandins.
     Gryglewski et al. (1975) have reported that ste-
roids inhibit prostaglandin release from antigen-challen-
ged lungs of sensitized guinea pigs and from noradrenali-
ne-constricted rabbit mesenteric arteries. In perfused
lungs hydrocortisone inhibits also the release of a
rabbit aorta contracting substance (RCS). The inhibitory
action of steroids does not occur in the presence of
arachidonic acid. A concept has been put forward that
steroids impair the supply of arachidonic acid for
microsomal cyclo-oxygenase owing to their membrane-
stabilizing properties.

                              LUNGS

     Perfused lungs of sensitized guinea pigs when
challenged with antigen release slow reacting substance
of anaphylaxis (SRS-A), histamine, prostaglandins, a
rabbit aorta contracting substance (RCS) and a releasing factor for RCS (RCS-RF) (Piper and Vane, 1969,1971).
RCS from lungs comprises mainly thromboxane A_2 (Svensson
et al., 1975, see also Gryglewski et al., this book).
RCS-RF is a stable principle that injected into the
perfused lungs of healthy guinea pigs results in the
release of RCS (Flower et al., 1976). The release of
prostaglandins and RCS from challenged lungs is inhibited by non-steroidal anti-inflammatory drugs (Piper and
Vane, 1969; Palmer et al., 1973). We have shown (Gryglewski et al., 1975) that the release of prostaglandins
and RCS from shocked lungs (10 mg of fresh ovalbumen)

can be reduced also by hydrocortisone (100 μg/ml),
however, the fundamental difference between indomethacin blockade and hydrocortisone blockade is that inhibitory effect of hydrocortisone but not that indomethacin has been abolished by arachidonic acid (1 μg/ml).

An elegant series of experiments was designed by
Flower et al. (1976). The release of RCS from perfused
lungs of healthy guinea pigs was stimulated either by
RCS-RF or by arachidonic acid. The first stimulus is
supposed to activate phospholipase A_2 in lungs and
consequently to liberate endogenous arachidonic acid,
while the second stimulus is just the exogenous substrate for lung enzymes to generate prostaglandins and
thromboxane A_2 (RCS). Indomethacin (2 μg/ml) and dexamethasone (2 μg/ml) were equally efficient in blocking
the RCS-RF-induced generation of RCS, however, contrary
to indomethacin, dexamethasone was unable to inhibit
the release of RCS by exogenous arachidonic acid. The
mode of action of indomethacin implies the direct blockade of microsomal cyclo-oxygenase, and then the enzyme
converts neither exogenous nor endogenous arachidonic
acid to biologically active products. The site of action
of dexamethasone is obviously different from that of
indomethacin and should be somewhere before the cyclo-
oxygenase system, since in the presence of dexamethasone
exogenous arachidonic acid is easily converted to RCS.
The relative potencies for hydrocortisone, prednisone,
prednisolone, fludrocortisone, triamicinolone, dexamethasone and betamethasone to inhibit the RCS-RF-induced
release from lungs are very close to their relative anti-
inflammatory potencies (Dr. R.J. Flower, personal communication).

In another set of experiments Dr Flower has measured by gas chromatography the efflux of arachidonic
acid from the RCS-RF-challenged lungs. In the presence
of eicosatetraynoic acid - TYA (TYA is arachidonic acid
antimetabolite that specifically blocks its conversion
by cyclo-oxygenase) the RCS-RF-induced release of arachidonic acid was severalfeld increased. This increase
does not occur when dexamethasone and TYA are simultaneously infused to lungs before the challenge with
RCS-RF. Thus Dr Flower has obtained a direct evidence
for our concept (Gryglewski et al., 1975) that steroids
hinder the liberation of arachidonic acid from its cellular stores.

The inhibitory effect of steroids on prostaglandin
and RCS release from lungs is capricious and requires
special experimental conditions. In our hands hydrocortisone (100 μg/ml) and dexamethasone (30 μg/ml) were
unable to inhibit the release of prostaglandins from

superfused guinea pig tracheal spirals and to inhibit
the release of RCS from superfused strips of lung parenchyma when these preparations were contracted by histamine (unpublished results).

SPLEEN

Ferreira et al. (1971) have reported that hydrocortisone (2 μg/ml, one experiment; 25 μg/ml, two
experiments) when infused into isolated dog spleen does
not change the prostaglandin output stimulated by adrenaline. In this experimental set indomethacin (0.75 μg/
ml) and aspirin (40 μg/ml) completely inhibit the adre-
naline-induced release of prostaglandins. Dr S.Moncada
(personal communication) has observed that the spontaneous release of prostaglandins from the perfused cat
spleen is inhibited by pretreating animals with dexamethasone supplemented with a continuous infusion of
this steroid (2 μg/ml) into the splenic artery.

In our unpublished experiments we bioassayed
prostaglandins (rat stomach strip and rat colon superfused in cascade) in the effluent from the perfused cat
spleen. An infusion of noradrenaline (1 μg/ml during
5 min) resulted in the release of prostaglandins
(11.5 $\pm$ 0.6 ng/ml of PGE_2 equivalents, n = 6, mean $\pm$
S.E.). When noradrenaline was administered during an
infusion of hydrocortisone semisuccinate (40 μg/ml,
20 min), then the release of prostaglandins was significantly diminished (2.0 $\pm$ 0.4 ng/ml of PGE_2 equivalents,
n = 6). In the presence of arachidonic acid (0.2 μg/ml)
in Krebs bicarbonate the inhibitory effect of hydrocortisone was partially reversed (6.3 $\pm$ 0.2 ng/ml PGE_2
equivalents, n = 4). In three control experiments indomethacin (3 μg/ml) completely inhibited the noradrenaline-induced release of prostaglandins and the superimposed infusion of arachidonic acid (0.2 μg/ml) did not
reverse its inhibitory effect (2 experiments).

Superfused cat splenic strips did not generate
prostaglandins when contracted by noradrenaline and
therefore we used in our "superfusion experiments" strips
of rabbit spleen. This preparation releases large amounts
of prostaglandins (19.8 $\pm$ 4.1 ng/ml PGE_2 equivalents,
n = 6) in response to a contraction induced by noradrenaline, however, neither preincubation (4 hours) nor
superfusion (40 min) of rabbit splenic strips with
hydrocortisone semisuccinate (40 - 150 μg/ml) changed
the output of prostaglandins (20.1 $\pm$ 3.9 ng/ml PGE_2
equivalents, n = 10). Indomethacin (3 μg/ml) completely
abolished the noradrenaline-induced release of prosta-

glandins (3 experiments).

Thus hydrocortisone inhibits the release of prostaglandins from perfused lungs and spleen but it does not affect the output of prostaglandins (or RCS) from superfused strips of trachea, lung parenchyma and spleen.

MESENTERY

In another unpublished series of experiments we have used fragments of mesentery of guinea pigs that were sensitized with ovalbumen (Piper and Vane, 1969). One gram of fragmented mesentery was suspended in 3 ml of Tyrode solution and incubated aerobically at $37^{\circ}C$ for 3 min in the presence of 1% ovalbumen. Then the incubation fluid was separated from the tissue and used for bioassay of prostaglandins and histamine (Gryglewski and Korbut, 1976). In twenty six experiments a challenge with antigen resulted in the release of both autacoids in amounts of 56.3 ± 11 ng PGE_2 equivalents/g wet weight and 253 ± 39 ng histamine/g wet weight (mean $\pm$ S.E.). When hydrocortisone semisuccinate (0.1 - 30 µg/ml) was preincubated with mesentery for 10 min before a challenge then the release of both autacoids was changed in a manner presented in Fig. 1. The release of prostaglandins was depressed at a whole range of concentrations of hydrocortisone, whereas histamine release was stimulated at low concentrations of this steroid (0.1-0.5 µg/ml) and depressed by high concentrations (1 - 30 µg/ml). Desoxycorticosterone (100 µg/ml) influenced neither prostaglandin nor histamine release (6 experiments).

We consider the suspension of mesenteric slices from sensitized guinea pigs as the source of immunized mast cells that respond with a direct release of histamine following antigen-antibody interaction on the cell surface (McIntire, 1973). Prostaglandins seem to be produced in parallel to histamine during the same immunological reaction but not subsequently to histamine release. In this last case it would be difficult to explain the depression of prostaglandin release associated with the stimulation of histamine release by low concentrations of hydrocortisone (Fig. 1). We propose that hydrocortisone at low concentrations impairs prostaglandin biosynthesis (stabilizing effect on lysosomes) and has no effect on the mast cell granula membranes. The shortage of prostaglandins results in an increased release of histamine because of decrease in intracellular cyclic-AMP levels (Kaliner and Austen,

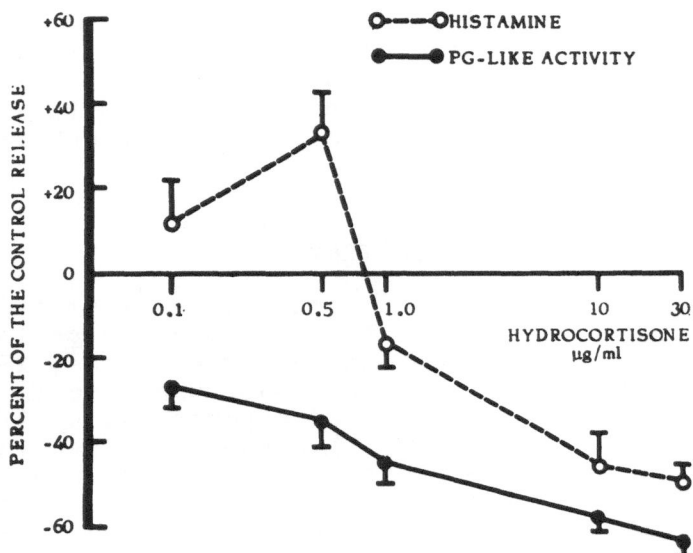

Fig. 1. The effect of various concentrations of hydrocortisone semisuccinate (abscissa) on the percent of
the control release of a prostaglandin-like substance
and histamine (ordinate) from incubated fragments of
mesentery of sensitized guinea pigs during the challenge with an antigen. The control release for a prostaglandin-like substance was 56.3 ± 11 ng PGE_2 equivalents/
/g wet tissue and for histamine 253 ± 39 ng/g wet tissue
(mean $\pm$ S.E., n = 26). Each of five concentrations of
hydrocortisone was tested in 4 - 8 experiments. Points
represent mean of percent change, bars represent S.E.

1975). Hydrocortisone at high concentrations directly
stabilizes the membranes of histamine-storage granulae
and therefore the release of histamine is also depressed, similarily as it has been observed for high concentrations of hydrocortisone in anaphylactic guinea
pig lungs (Gryglewski et al., 1975) and for dexamethasone in various tissues of sensitized guinea pigs (Kurihara and Shibata, 1975).

SYNOVIA

     Synovial membrane occupies special place in in-
vestigation of the effects of steroids on prostaglandin
release, since it is the target tissue for anti-inflammatory drugs in rheumatoid arthritis. Kantrowitz et al.

(1975) have reported that hydrocortisone at a concentration of 10 μM, prednisolone and dexamethasone at a range of concentrations of 10 nM - 1 μM inhibit significantly generation and release of PGE_2, $PGF_{2\alpha}$ and their 13,14 dihydro-15-keto metabolites by human rheumatoid synovial organ cultures. Prostaglandins in the culture media were measured by radioimmunoassay after 3 and 6 days of incubation. Additional experiments allowed the authors to conclude that steroids inhibit biosynthesis of prostaglandins by rheumatoid synovial tissue, in the absence of any detectable increase in either degradation or storage of prostaglandins in tissue explants (Kantrowitz et al., 1975). Needless to say that these data are in favour of our concept of the mechanism action of glucocorticosteroids (Gryglewski et al., 1975; Gryglewski, 1976).

The results of Floman et al. (1976) even more directly support our view that inhibitory action of steroids on prostaglandin release depends on the shortage of the endogenous substrates for prostaglandin generation. Acute arthritis was induced in rats by injection of cell-free extract of group A Streptococci into the knee joint. The inflammed synovia was incubated with corticosterone acetate, dexamethasone, aldosterone and indomethacin, each at a concentration of 100 μg/ml. Corticosterone, dexamethasone and indomethacin inhibited the release of PGs from synovial membranes by 70%, 50% and 90%, respectively, and reduced tissue content of PGEs by 40%, 40% and 70%, respectively, while aldosterone had no effect. The inhibitory effect of corticosterone was reversed by addition to the medium of arachidonic acid (2 μg/ml) ; by contrast, the inhibitory action of indomethacin was not affected by similar treatment (Floman et al., 1976).

Patrono et al. (1976) have used the superfused fragments of synovial membrane obtained from patients undergoing different kinds of orthopedic surgery and measured in the effluent the content of $PGF_{1\alpha}$ and $PGF_{2\alpha}$ by radioimmunoassay. The authors have found that hydrocortisone at a concentration of 100 μg/ml is practically inactive as an inhibitor of prostaglandin release (less than 20% inhibition), whereas non-steroidal anti-inflammatory drugs inhibit the release of $PGF_{2\alpha}$ in 50% at the following concentrations : indomethacin - 1.1 μM, fenoprofen - 2.2 μM and aspirin 172 μM. This finding is of great methodological importance since it confirms that the superfusion technique is inadequate to detect the inhibitory effect of corticosteroids on the release of prostaglandins. Thus the superfused guinea pig trachea, strips of guinea pig lungs, strips of rabbit spleen,

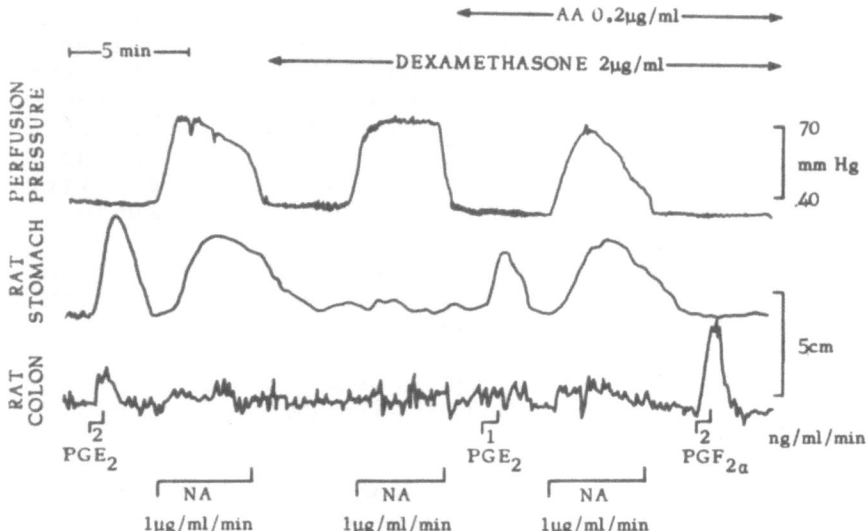

Fig. 2. The release of a prostaglandin E-like substance
(PGEs) by noradrenaline (NA) from perfused rabbit mesenteric arteries, its inhibition by dexamethasone and the
reversal of this inhibition by arachidonic acid (AA).
The effluent from perfused rabbit mesenteric arteries
superfused in cascade (2.5 ml/min) rat stomach and rat
colon which were treated with a mixture of combined
antagonists and indomethacin (1 µg/ml). The first infusion of NA (1 µg/ml/min) into mesenteric artery
caused a tachyphalactic rise in perfusion pressure
which was associated with the release of a substance
that contracted rat stomach but not rat colon. On the
basis of differential sensitivy of the assay organs to
PGE_2 and $PGF_{2\alpha}$ this substance could be spotted as PGEs.
The second infusion of NA was in the presence of dexamethasone (2 µg/ml). There was neither acute tolerance
to NA nor the release of PGEs. A response to the third
infusion of NA was much alike the first, though dexamethasone was still present in the perfusing Krebs bicarbonate, however, this time AA (0.2 µg/ml) was infused along with dexamethasone into mesenteric arteries.

and fragments of human synovia are resistent to the
inhibitory action of steroids on prostaglandin release,
whereas the same tissues when perfused through their
vascular system or incubated are sensitive to the inhibitory action of steroids.

BLOOD VESSELS

We have shown that an intravascular infusion of noradrenaline into anesthetized cats (Gryglewski and Ocetkiewicz, 1974), into perfused rabbit ear vessels (Gryglewski and Korbut, 1975) and into perfused rabbit mesenteric blood vessels (Grodzińska et al., 1976) results in the release of a PGE-like substance. An infusion of noradrenaline (1 - 3 μg/ml) into perfused rabbit mesenteric arteries is associated with the release of a substance that contracts the superfused rat stomach strip but neither affects the superfused strip of rabbit aorta nor rat colon (Fig. 2). Thus induced contractions of rat stomach strip can be matched by the PGE_2-induced contractions and then the amount of the released PGE-like substance is 3.2 $\pm$ 0.7 ng PGE_2 equivalents/ml (mean $\pm$ S.E., n = 20). This release of PGEs from mesenteric vessels is blocked by indomethacin (0.3 - 1 μg/ml), hydrocortisone (10 - 30 μg/ml) and by dexamethasone (2 - 5 μg/ml), and again steroid blockade but not indomethacin blockade can be relieved by arachidonic acid (0.2 μg/ml) (Fig. 2).

The release of PGEs from blood vessels by noradrenaline has been proposed as a feedback mechanism that limits vasoconstriction (Gryglewski and Korbut, 1975). The blockade of this mechanism by steroids can explain their vasoconstrictor action on skin vessels (Greeson et al., 1973) and their potentiating effect on the vasoconstrictor action of catecholamines (Fig.2).

OTHER TISSUES

Tashijan et al. (1975) have reported that hydrocortisone at a range of concentrations of 5 - 500 nM inhibitis in a dose-dependent manner the accummulation of PGE_2 in the medium of $HSIM_1C_1$ mouse fibrosarcoma cells. The inhibitory effect of hydrocortisone is reversible and can be detected as early as after 4 hours after its addition to the culture medium. There is no concomitant increase in intracellular PGE_2, in fact, a decrease in intracellular PGE_2 is usually observed. The acceleration of degration of PGE_2 by hydrocortisone in cultures has been also excluded. The authors propose that hydrocortisone inhibits PGE_2 accumulation in the culture medium by intererefing with prostaglandin biosynthesis. The explanation is offered that hydrocortisone binds to membrane structures and prevents activation of the prostaglandin synthesis system of the intact cell. Needless to say that these results and their

interpretation support our concept of the mechanism of
the action of glucocorticosteroids (Gryglewski et al.,
1975, Fig. 3).

Floman et al. (1976) have induced acute uveitis
in rabbits by intravitreal injection of 10 µg of endotoxin of E. coli. Inflamed iris and ciliary body have
been removed and incubated with hydrocortisone acetate
dexamethasone, aldosterone or indomethacin. All drugs
except for aldosterone inhibited the release of PGEs
from ocular tissues. The incubation with hydrocortisone
(100 µg/ml) reduced by 75% the tissue content of PGEs.
Floman et al. (1976) have also shown the inhibitory
effect of steroids on prostaglandin generation in vivo.
Paracentesis was performed on the right eye of heparini-
zed rabbit. Sixty minutes later 50 µl of the secondary
aqueous was withdrawn for determination of PGEs level.
The same procedure was performed on the left eye, but
60 min after a subconjunctival injection of 0.1 ml triamicinolone acetonide (40 mg/ml). The treatment with
this steroid reduced the release of PGEs into the aqueous following paracentesis from 1.9 ± 0.3 to 0.2 ± 0.05
ng/ml/hour (mean $\pm$ S.E., n = 8).

Another interesting evidence for the in vivo mechanism of steroid action has been recently presented
by Bonilla and Dupont (1976). The authors have investigated composition of fatty acids and prostaglandins in
left ventricular myocardium of dogs with severe low
output syndrome (LOS). LOS causes an increase in content
of arachidonic acid in myocardium, whereas the administration of dexamethasone (8 mg/kg) nullified the LOS-
-induced increase in both arachidonic acid and prostaglandin concentrations.

HYPOTHESES ON THE MECHANISM OF ACTION
OF GLUCOCORTICOSTEROIDS

There are two hypothese on the mechanism of inhibitory action of glucocorticosteroids on the release
of prostaglandins from tissues. Lewis and Piper (1975)
have postulated that steroids do not inhibit formation
of prostaglandins in tissues but do impair the active
transmembrane transport of prostaglandins from inside
to outside the cell. This concept is based on the observation that ACTH-induced release of prostaglandins (and
an unidentified chick rectum contracting substance) from
blood-perfused epigastric fat of rabbits is inhibited
by hydrocortisone. At the same time hydrocortisone does
not reduce the ACTH-induced increase in prostaglandin
content in adipose tissue (Lewis and Piper, 1975,1976a,

b). In fact in chopped adipose tissue the pretreatment
with steroids is followed by an increase in prostaglandin levels during stimulation with ACTH (Chang et al.,
1975).

Gryglewski et al. (1975) have put forward the
hypothesis that glucocorticosteroids impair generation
of prostaglandins by hindering liberation of arachidonic
acid from membrane phospholipids. This concept was o-
riginally based on the observation that inhibitory
effects of steroids on the release of prostaglandins
and RCS from guinea pig lungs and from rabbit blood
vessels are abolished by an extra supply with exogenous
arachidonic acid. Further evidence for this concept were
presented later (Gryglewski et al., 1976, this paper).

The "active transport inhibition" hypothesis of
Lewis and Piper (1975) is well documented for rat tissue,
however, the proposed mechanism does not seem to operate
in lungs (Gryglewski et al., 1975; Flower et al., 1976),
in blood vessels (Gryglewski, 1975), in synovial membranes (Kantrowitz et al., 1975; Floman et al., 1976), in
cultured fibrosarcoma cells (Tashjian et al., 1975), in
iris and ciliary body (Floman et al., 1976), in myocardium (Bonilla and Dupont, 1976) and in spleen (this paper). On the other hand the above reports support our
concept of the mechanism of steroid action (Gryglewski
et al., 1975). It is feasible to assume that adipocytes
differ from other cells in responsiveness to inhibitory
action of steroids on prostaglandin release. It might
be reminded that adenyl cyclase of adipocytes responds
to PGE_1 differently than adenyl cyclases of most other
cells [1] (see Paoletti and Puglisi, 1973).

How can glucocorticosteroids impair the supply of
arachidonic acid for prostaglandin synthetase? Flower
and Blackwell (1976) have proved that during mechanical
or immunological challenges of a tissue arachidonic acid
released as the substrate for prostaglandin biosynthesis
originates solely from phospholipids. Phospholipase A_2
is therefore the key enzyme which mobilizes the substrate for cyclo-oxygenase. Phospholipase A_2 is inhibited
by mepacrine (see Vargaftig, 1974). Having no experimental evidence for this type of steroid action we rather
believe that the well known membrane-stabilizing effect
of glucocorticosteroids (Weissmann, 1964, 1973) is responsible for their inhibitory action on prostaglandin
and thromboxane release. According to Weissmann (1973)
the tissue distortion (in our case a stimulus releasing
prostaglandins or thromboxanes) results in an escape of
lysosomal enzymes through channels created by virtue of
the incomplete fusion of lysosomal membrane and plasma
membrane. Glucocorticosteroids prevent this fusion.

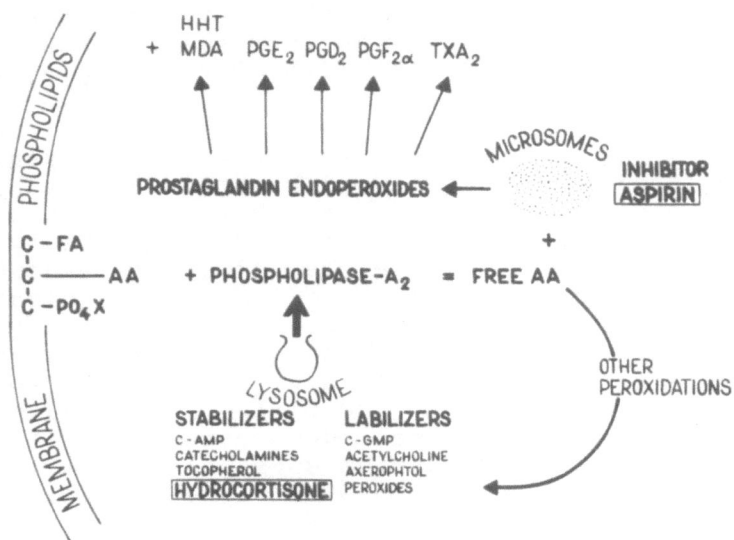

Fig. 3. Hypothetical mechanism by which glucocorticosteroids inhibit the formation and the release of prostaglandins and thromboxane A_2 from tissues. Glucocorticosteroids stabilize lysosomal membranes thus preventing
the leakage of phospholipase A_2 in response to a stimulus that usually initiates prostaglandin biosynthesis.
In the absence of phospholipase A_2 arachidonic acid cannot be liberated from membrane phospholipids. The activity of microsomal cyclo-oxygenase is supressed because
of shortage of the substrate. This causes a decreased
formation of prostaglandins, thromboxane A_2 and possibly
other products of transformation of cyclic endoperoxides.
Alternatively glucocorticosteroids may inhibit activity
of phospholipase A_2 or impair activation of phospholipase A_2. This would lead to the same final result.

Membrane-bound and soluble acid phospholipases A_2 have
been found in lysosomes of bovine adrenal medulla (Smith
and Winkler, 1968), in rat liver lysosomes (Rahman et
al., 1970) and in rat testicular lysosomes (LeGrande et
al., 1975), although phospholipases A_2 have been also
described in mitochondria in microsomes, in plasma membranes and in serum (see Vargaftig, 1974). There is
convincing evidence for the role of lysosomal phospholipases in the formation of prostaglandins during carrageenin-induced inflammation in rats (Anderson et al.,
1971). Therefore the simplest explanation for the mechanism of inhibitory action of glucocorticosteroids on

prostaglandin release is an assumption that steroids by stabilizing lysosomal membranes prevent the leakage of phospholipase A_2 and consequently prevent the liberation of arachidonic acid from membrane phospholipids (Fig. 3). Alternatively steroids can prevent activation of phospholipase A_2 (whatever this expression means) or being bound to plasma membranes impair the enzymatic cleavage of membrane phospholipids. The final outcome of either of these actions will be the shortage of sub- strates for microsomal cyclo-oxygenase resulting in in- hibition of prostaglandin and thromboxane formation that is <u>followed</u> by a decrease in prostaglandin and thromboxane release from cells. It is obvious that the integrity of the cell structure is an indispensable requirement for the effectiveness of corticosteroids.

CONCLUSION

The inhibitory effect of steroidal anti-inflam- matory drugs on the formation and/or the release of prostaglandins and/or thromboxanes A_2 have been shown in lungs, spleen, synovial membrane, fibrosarcoma cells, mesenteric mast cells, fat tissue, iris and ciliary body, myocardium and blood vessels. This effect may be of im- portance for the therapeutic and toxic actions of ste- roidal anti-inflammatory drugs. Inhibition of prosta- glandin formation may contribute to anti-inflammatory and vasoconstrictor properties of steroids as well as to their ulcerogenic action in gastric mucosa. A decre- ase in arachidonic acid and prostaglandin levels in myocardium which is induced by steroids can explain their beneficial effect on the survival rate in animals with myocardial insufficiency. It well might be that anti-shock activity of steroids is related to inhibition of prostaglandin release from vessels and thromboxane A_2 from lungs. The inhibition of thromboxane A_2 forma- tion in lungs may contribute to the effectiveness of steroids in the treatment of bronchial asthma and ana- phylactic shock, since thromboxane A_2 is a potent bronchoconstrictor, pulmonary vasoconstrictor and pro- aggregatory agent. A realistic appraisal on relevance of the proposed mechanism of steroid action must await a greater accumulation of basic experimental data.

ACKNOWLEDGEMENTS

This study was supported by The Wellcome Trust London.

REFERENCES

Anderson, A.J., Brocklehurst, W.E., and Willis, A.L.,1971,
    Evidence for the role of lysosomes in the formation
    of prostaglandins during carrageenin-induced in-
    flammation in rat. Pharmacol. Res. Commun. 3: 13.
Bonilla, C.A., and Dupont, J., 1976, Fatty acid and
    prostaglandin composition of left ventricular myo-
    cardium from dexamethasone-treated dogs with severe
    low output syndrom. Prostaglandins 11: 935.
Chang, J., Lewis, G.P., and Piper, P.J., 1975, The
    effects of anti-inflammatory steroids on levels of
    prostaglandins in adipose tissue in vitro.Proc. Br.
    Pharmacol. Soc. 17 - 19 December, p. 342P.
Claman, H.N., 1975, How corticosteroids work. J. Allergy
    Clin. Immunol. 55: 145.
Eisen, V., Greenbaum, L., and Lewis, G.P., 1968, Kinins
    and anti-inflammatory steroids. Br. J. Pharmac. 34:
    169.
Ferreira, S.H., Moncada, S., and Vane, J.R., 1971, In-
    domethacin and aspirin abolish prostaglandin release
    from the spleen. Nature New Biol. 231: 237.
Floman, Y., Floman, N., and Zor, U., 1976, Inhibition
    of prostaglandin E release by anti-inflammatory
    steroids. Prostaglandins 11: 591.
Flower, R.J., 1974, Drugs which inhibit prostaglandin
    biosynthesis. Pharmac. Rev. 26: 33.
Flower, R., Gryglewski, R., Herbaczyńska-Cedro, K., and
    Vane, J.R., 1972, Effects of anti-inflammatory drugs
    on prostaglandin biosynthesis. Nature New Biology
    238: 104.
Flower, R.J., and Blackwell, G.J., 1976, The importance
    of phospholipase A$_2$ in prostaglandin biosynthesis.
    Biochem. Pharmacol. 25: 285.
Flower, R.J., Harvey, E.A., Moncada, S., Nijkamp, F.,
    and Vane, J.R., A communication during the Meeting
    of Br. Pharmacol. Soc. 1-2 April, 1976, in Chelsea
    College. Proc. Br. Pharmacol. Soc. p.47.
Greaves, M.W., Kingston, W.P., and Pretty, K., 1975,
    Action of a series of non-steroid and steroid anti-
    inflammatory drugs on prostaglandin synthesis by
    microsomal fraction of rat skin. Brit. J. Pharmacol.
    53: 470P.
Greaves, M.W., and McDonald-Gibson, W., 1972a, Inhibition
    of prostaglandin biosynthesis by corticosteroids.
    Br. Med. J. 1: 83.
Greaves, M.W., and McDonald-Gibson, W., 1972b, Prosta-
    glandin biosynthesis by human skin and its inhibi-
    tion by corticosteroids. Br. J. Pharmac. 46: 172.

Greeson, T.P., Levan, N.E., and Freedman, R.I., 1973, Corticosteroid-induced vasoconstriction studied by xenon-133 clereance. J. Invest. Dermatol. 61: 242.

Grodzińska, L., Panczenko, B., and Gryglewski, R.J., 1976, Release of prostaglandin-like material from perfused mesenteric blood vessels of rabbits. J. Pharm. Pharmac. 28: 40.

Gryglewski, R.J., 1976, Steroid hormones anti-inflammatory steroids and prostaglandins. Pharmacol. Res. Commun. 8: 337.

Gryglewski, R.J., and Korbut, R., 1975, Prostaglandin feedback mechanism limits vasoconstrictor action of norepinephrine in perfused rabbit ear. Experientia 31: 89.

Gryglewski, R.J., and Korbut, R., 1976, Bioassay of histamine in the presence of prostaglandins. Br. J. Pharmac. 56: 39.

Gryglewski, R.J., Panczenko, B., Korbut, R., Grodzińska, L., and Ocetkiewicz, A., 1975, Corticosteroids inhibit prostaglandin release from perfused mesenteric blood vessels of rabbit and from perfused lungs of sensitized guinea pig. Prostaglandins 10: 343.

Gryglewski, R.J., and Ocetkiewicz, A., 1974, A release of prostaglandins may be responsible for acute tolerance to norepinephrine infusions. Prostaglandins 8: 31.

Herbaczyńska-Cedro, K., and Staszewska-Barczak, J., 1976, Adrenocortical hormones and the release of prostaglandin-like substances, in the symposium on prostaglandins, Second Congress of Hungarian Pharmacological Society (J. Knoll and K. Keleman, eds.), p. 157, Akademiai Kiado, Budapest.

Kaliner, M., and Austen, K.F., 1975, Immunologic release of chemical mediators from human tissues in: Annual Review of Pharmacology, (H.W. Elliot, R. George and R.Okum, eds.), p. 177. Annual Review Inc. California.

Kantrowitz, F., Robinson, D.R., McQuire, M.B., and Levine, L., 1975, Corticosteroids inhibit prostaglandin production by rheumatoid synovia. Nature 258: 737.

Kurihara, N., and Shibata, K., 1975, Effects of corticosteroids on the release of mediators in anaphylaxis of sensitized guinea pig tissues. Japan J. Pharmacol. 25: 181.

Legrande, C.E., Sorenson, K., and Buhrley, L.E., 1975, Mechanism and interactions in testicular steroidogenesis and prostaglandin synthesis. J. Steroidal. Biochem. 6: 1081.

Lewis, G.P., and Piper, P.J., 1975, Inhibition of re-
    lease of prostaglandins as an explanation of some
    of the actions of anti-inflammatory corticosteroids.
    Nature 254: 308.

Lewis, G.P., and Piper, P.J., 1976a, Inhibition of
    prostaglandin release by anti-inflammatory steroids,
    in: Advances in Prostaglandin and Thromboxane Rese-
    arch, Vol. 1 (B. Samuelsson and R. Paoletti, eds.),
    p. 121, Raven Press, New York.

Lewis, G.P., and Piper, P.J., 1976b, The action of
    corticosteroids on the prostaglandin system, in:
    The role of prostaglandins in inflammation. Procee-
    dings of a workshop held during the VIIIth European
    Rheumatology Congress, Helsinki 1975, p. 148. (G.
    P. Lewis, ed.), Hans Huber Pub. Bern-Stuttgard-
    Vienna.

McIntrie, F.C., 1973, Histamine release by antigen-
    antibody reactions, in: Histamine and Antihistami-
    nes, Vol. 1 (M. Schachter, ed.), p. 45, Pergamon
    Press, Oxford-New York-Toronto.

Mendelsohn, J., Mutler, M.M., and Boone, R.F., 1973,
    Enhanced effects of prostaglandin E_1 and dibutyryl
    cyclic AMP on human lymphocytes in the presence of
    cortisol. J. Clin. Invest. 52: 2129.

Palmer, M.A., Piper, P.J., and Vane, J.R., 1973, Release
    of rabbit aorta contracting substance (RCS) and
    prostaglandins induced by chemical or mechanical
    stimulation of guinea pig lungs. Br. J. Pharmac.
    49: 226.

Paoletti, R., and Puglisi, L., 1973, Lipid metabolism,
    in: The Prostaglandins, Vol. 1 (P.W. Ramwell, ed.),
    p. 317, Plenum Press, New York-London.

Parker, C.W., Huber, M.G., and Baumann, M.L., 1973,
    Alterations in cyclic AMP metabolism in human bron-
    chial asthma. III. Leukocyte and lymphocyte respon-
    ses to steroids. J. Clin. Invest. 52: 1342.

Patrono, C., Ciabattoni, G., Greco, F., and Grossi-
    Belloni, D., 1976, Comparative evaluation of the
    inhibitory effects of aspirin-like drugs on prosta-
    glandin production by human platelets and synovial
    tissue, in: Advances in Prostaglandin and Thrombo-
    xane Research, Vol. 1 (B. Samuelsson and R. Paole-
    tti, eds.), p.125, Raven Press,New York.

Piper, P.J., and Vane, J.R., 1969, Release of additio-
    nal factors in anaphylaxis and its antagonism by
    anti-inflammatory drugs. Nature 233: 29.

Piper, P.J., and Vane, J.R., 1971, The release of
    prostaglandins from lung and other tissues. N.Y.
    Acad. Sci. 180: 363.

Rahman, Y.E., Verhagen, J., and van der Wiel, D.F.M., 1970, Evidence of a membrane-bound phospholipase A in rat liver lysosomes. Biochim. Biophys. Res. Commun. 38: 670.

Smith, A.D., and Winkler, H., 1968, Lysosomal phospholipases A_1 and A_2 of bovine adrenal medulla. Biochem. J. 108: 867.

Smith, J.B., and Willis, A.L., 1971, Aspirin selectively inhibits prostaglandin production in human platelets. Nature New Biol. 231: 235.

Svensson, J., Hamberg, M., and Samuelsson, B., 1975, Prostaglandin endoperoxides IX. Characterization of rabbit aorta contracting substance (RCS) from guinea pig lung and human platelets. Acta Physiol. Scand. 94: 222.

Tashjian, A.H., Voelkel, E.F., McDonough, J., and Levine, L., 1975, Hydrocortisone inhibits prostaglandin production by mouse fibrosarcoma cells. Nature 258: 739.

Vane, J.R., 1971, Inhibition of prostaglandin synthesis as a mechanism of action for aspirin-like drugs. Nature New Biol. 231: 232.

Vargaftig, B.B., 1974, Search for common mechanisms underlying the various effects of putative inflammatory mediators, in: The Prostaglandins, Vol. 2. (P.W. Ramwell, ed.), p. 205. Plenum Press, New York -London.

Weissmann, G., 1973, Effect of corticosteroids on the stability and fusion of biomembranes, in: Asthma (A. Austen and L.M. Lichtenstein, eds.), p. 221. Academic Press, New York and London.

Weissmann, G., 1964, Labilization and stabilization of lysosomes. Fed. Proc. 23: 1038.

INFLUENCE OF PROSTAGLANDINS ON CENTRAL FUNCTIONS

R. Fumagalli, G.C. Folco and D. Longiave

Institute of Pharmacology and Pharmacognosy

University of Milano, 20129 Milano, Italy

Biologically active compounds have been described in brain and cerebrospinal fluid some twenty years ago. These substances behaved like unsaturated acidic lipids, displayed direct effects on a number of isolated tissue preparations and influenced the action of known agonists on some classical experimental model. In the early sixties some of these compounds were chemically characterized, their presence in the Central Nervous System (CNS) definitely assessed, and found to correspond to prostaglandins (PGs). The function of prostaglandins in CNS has, since then, become a stimulating goal. The recent discovery that brain cortex is able to form thromboxanes from endogenous precursors provides new impetus to research in this field and opens new perspectives to the understanding of the possible roles played by prostaglandins in the Central Nervous System. The aim of this chapter is to update (end of June 1976) current knowledge on prostaglandins as related to CNS, previously discussed and reported (1-3).

PROSTAGLANDINS IN BRAIN

The first specific demonstration that PGs are present in brain dates back to 1964, when Samuelsson found that ox brain contains $PGF_{2\alpha}$ (4). Thereafter it was reported that brain of various species contains prostaglandins of the E and F series. $PGF_{2\alpha}$ was the most common prostaglandin found, followed by PGE_2, while the occurrence of

$PGF_{1\alpha}$ and PGE_1 seems to be less frequent (5-9). These
data should be considered mainly from a qualitative point
of view: in fact it is now well recognized that the biosynthesis of these compounds is a rather rapid process
brought about by a variety of stimuli. This consideration
has some bearing on the rather widespread quantitative
value reported which might, at least in part, be somehow
artifactual. In fact they might simply reflect endogenous
biosynthesis occurring during animal sacrifice and tissue
manipulation, rather than the actual tissue concentration.
Furthermore the identification of prostaglandins was not
highly specific, with the exception of the work of
Samuelsson, where the identity of $PGF_{2\alpha}$ in ox brain was
based on gas chromatographic-mass spectrometric (GC-MS)
techniques. For these reasons the observations that
prostaglandins are uniformly distributed throughout the
brain and specifically localized in a synaptosomal fraction should be considered with caution (7,10). The same
holds true for the presence of PGE_1 and $PGF_{1\alpha}$ in brain:
indeed their precursor (n-6 eicosatrienoic acid) has
neither been found in brain of some species (11-13) nor
characterized for the position of double bonds in others
(14). The use of highly specific analytical techniques
coupled with particular precautions in animal sacrifice
and tissue handling allows a more correct quantitation
of the prostaglandin content in brain. When the brain is
immediately frozen or the animal sacrificed with focussed
microwave radiation, the neoformation of prostaglandins
from endogenous precursors is minimized and the quantitative values found are nearest to the actual content
(15,16). Table 1 shows how a few minute interval between
the animal sacrifice and the beginning of incubation is
sufficient to increase the PGs content of the brain. No
data on subcellular localization of PGs in the CNS obtained with these techniques have been reported. These observations also suggest that spontaneous and evoked release
of prostaglandins occurring from various parts of the CNS
in different species and in a variety of experimental
conditions (17-27) merely reflect endogenous formation
of these compounds elicited during the experiments.
In fact since the available evidence is against the existence of storing granules, it follows that PGs diffusing
out of CNS are the result of stimulated synthesis brought
about by the stimulus itself. In these studies the release
of prostaglandins appears to be related with neuronal
activity as influenced by electrical or pharmacological
means, and the pattern of released PGs varies in that
prostaglandins of the E type are predominant at rest and
the F type after stimulation. This might mean that

TABLE 1

LEVELS OF ENDOGENOUS PROSTAGLANDINS IN BRAIN (RAT CEREBRAL CORTEX) EXPRESSED AS NANOGRAMS/100 mg FRESH TISSUE WEIGHT

	PROSTAGLANDIN $F_{2\alpha}$	PROSTAGLANDIN E_2
Cerebral cortex immediately frozen in liquid nitrogen	1.1 ± 0.3 (6)[+]	N.D.[+]
Animals sacrificed by focussed microwave radiation	1.5 ± 0.5 (6)[++]	1.4 ± 0.5 (6)[++]
Tissue slices prepared to study endogenous biosynthesis of prostaglandins	8.6 ± 2.6 (6)[+] [+++] 11.5 ± 1.2 (6)[+++]	4.2 (2)[+] 4.5 ± 0.5 (6)[+++]

N.D. = Not detectable

In parenthesis the number of determinations $\pm$ S.E. or S.D.

Prostaglandins were measured by gas-chromatographic-mass spectrometric techniques.

+ Data from Wolfe et al. (15)

++ Data from Bosisio et al. (31)

+++ Data from Nicosia and G. Galli (16)

different cell types are concerned or that the end product of endogenous free arachidonic acid metabolism is affected by the interference of the stimulus with the enzymatic activities involved. The possibility exists that the stimulus enhances the 9-keto reductase activity transforming PGEs into PGFs: this activity has already been demonstrated in some peripheral tissues (28,29) and recently observed in brain cortex slices (15).

BIOSYNTHESIS AND CATABOLISM

A salient feature of prostaglandin formation in brain is the poor conversion of exogenous labeled arachidonic acid into primary PGs. Possible explanations are the dilution of the exogenous precursor by endogenous pool(s) or its incorporation into complex lipids. The former possibility is underlined by the observation that the endogenously available free arachidonic acid saturates brain cyclooxygenase (15). For this reason the biosynthetic capacity of brain in respect to prostaglandin formation has to be evaluated without addition of free arachidonic acid. Using this experimental approach it has been found that rat brain cortex, either sliced or as homogenate, can form both $PGF_{2\alpha}$ and PGE_2, with a ratio of the former to the latter of about 3 (16,30). Homogenates appear to be more active, in this respect, at least during the first hour of incubation (15). In rat brain cortical slices no feed-back regulation has been observed (3). During incubation of sliced tissue, newly formed prostaglandins diffuse out into the medium without reaching, however, a complete equilibration. It is possible that this occurs for an unspecific partition into tissue components or for the presence of specific binding site(s) : Wolfe and coworkers (15) report that a plot of tissue $PGF_{2\alpha}$ vs concentrations of this prostaglandin in the medium suggests the presence of saturable binding site(s).

Regional differences have been observed in endogenous biosynthetic capacity to form prostaglandins in rat and cat. The formation of either PGE_2 and $PGF_{2\alpha}$ in cerebral cortex of rats far exceeds that observed in the cerebellum of the same animal species (31): the endogenous biosynthesis of $PGF_{2\alpha}$ is about three fold that of PGE_2 in cerebral cortex, while this ratio approaches unity in the cerebellum. In the cat brain four regions have been explored and display different capacities to form prostaglandins from endogenous precursors (3,15). As in the rat, the amounts of newly formed prostaglandins are higher in cerebral cortex than in cerebellum, where again the ratio $PGF_{2\alpha}/PGE_2$ is about unity; caudate nucleus and hypothalamus synthesize as much $PGF_{2\alpha}$ as cerebellum, whereas PGE_2 formation is one half of that in cerebral cortex and cerebellum. Species differences have also been reported concerning rat, cat and human cerebral cortexes (15). The values of endogenously formed prostaglandins are highest in cat and lowest in man, with a higher ratio of $PGF_{2\alpha}$ versus PGE_2 in the human specimen. However, as far as regional and species differences are concerned, the available data are scanty and no definitive conclusions can be drawn.

Numerous substances are known to interfere with the biosynthesis of prostaglandins, but this property has generally been evaluated in peripheral tissues using labeled exogenous arachidonic acid as precursor. For the reasons previously discussed this is not a suitable tool for brain and the pharmacological interference with PGs biosynthesis in this tissue should be evaluated on the endogenous formation of these compounds. It is a merit of Wolfe and associates to have studied the effect of a number of substances on this parameter in suitable experimental conditions (15). These Authors have found that the formation of prostaglandins by cerebral cortex slices of rat is clearly enhanced by dopamine, norepinephrine and adrenochrome with a preferential stimulation of the biosynthesis of $PGF_{2\alpha}$, as reflected by the increased values of $F_{2\alpha}/E_2$. They have also shown that serotonine, L-Dopa, glutathione, hydroquinone, ATP, dibutyryl-cyclic AMP, histamine, Cu^{++} and bradykinin have no effect on prostaglandin biosynthesis in this experimental model. The capacity of norepinephrine to increase PGs formation in brain was formerly considered secondary to an enhanced availability of precursor due to a stimulation of lipase activity (32): however, Wolfe et al. have shown that this interpretation is not correct since the catecholamine does not affect the pool of free arachidonic acid in brain slices. Also morphine and apomorphine stimulate prostaglandin biosynthesis in rabbit brain homogenate (33).

Drugs known to inhibit prostaglandin formation in peripheral tissues are active also on endogenous biosynthesis of PGs in rat brain (15). This study has been carried out in slices and homogenates of rat cerebral cortex and a number of non steroidal antiinflammatory drugs, along with eicosatetraynoic acid, has been tested. Indomethacin and ketoprofen have the lowest ID_{50} in both preparations, prodolic acid appears to be more effective on PGE_2 than on $PGF_{2\alpha}$ formation, and eicosatetraynoate is likely to have problems in cellular membrane permeation. No data are available on psychotropic drugs and PGs formation in brain. However drugs such as reserpine, chlorpromazine, monoamine oxidase inhibitors, tricyclic antidepressant, diazepam, meprobamate and cannabis reportedly depress prostaglandin synthesis in peripheral tissues (34-36).

Wolfe and coworkers (37) have recently found that slices and homogenates of guinea-pig brain cortex and rat cortical homogenates are able to form thromboxane B_2 from endogenous precursors. Thromboxane B_2 (TxB_2), an almost biologically inactive but stable compound derives from PGs endoperoxides through an unstable biologically very active intermediate, named thromboxane A_2. Guinea pig cortex slices and homogenate form TxB_2 to a greater extent than primary prostaglandins, whereas this is not the case in the rat. In the former species the production of TxB_2 is enhanced by norepinephrine and blocked by indomethacin. This finding might represent a turning point in this field of research and opens new ways of speculation on the possible roles of arachidonic acid metabolites in the function of CNS. Available data on the capacity of brain to catabolize prostaglandins are at variance. The presence of 15-hydroxy-dehydrogenase and Δ^{13}-reductase activities have been described in pig brain, mainly in a high-speed supernatant (38), while rat and dog brains and cerebella are practically unable to break down these compounds (39). Using a different experimental approach Siggins et al. (40) have shown that prostaglandin-dehydrogenase appears to be present almost exclusively in cerebellar cortex of the rat, particularly in the Purkinje cell layer. Proper investigations with the use of GC-MS techniques have revealed that the prostaglandin-degrading capacity of cerebral cortex of rat and man is rather low: after 1 hr of incubation, rat and human cerebral cortexes transform about 1-3 nanograms/100 mg tissue of newly formed $PGF_{2\alpha}$ into the corresponding 15-Keto-13,14-dihydro metabolite (15). However, a 9-Keto-reductase activity has been demonstrated, allowing some conversion of PGE_2 into $PGF_{2\alpha}$ as previously mentioned. The poor capacity of brain to degrade prostaglandins is accepted, but it should be considered that this does not necessarily mean that catabolic break down of these compounds is low throughout the brain. It might well be that in some specific neuronal pathways, where PGs may play a role in the occurring biochemical events, there is the capacity to degrade them and that this capacity can be overshadowed when discrete areas or whole brain are evaluated in this respect.

PROSTAGLANDINS AND THE CEREBROSPINAL FLUID

The presence of prostaglandin-like material in cerebrospinal fluid (CSF) has been recognized more than a decade ago (41) but research on their chemical characterization and quantitative evaluation has appeared only recently (3,15). Work in this respect has received great impetus since the observation that prostaglandins appear to be involved in pyrogen fever (42-44); this subject will however be dealt with in another chapter of this book.

The content of $PGF_{2\alpha}$ in human CSF has been recently evaluated during diagnostic pneumoencephalography, and identification and quantitation of this prostaglandin was assessed by GC-MS methods (15). $PGF_{2\alpha}$ is fairly detectable in this fluid and the concentration varies between 50 and 100 picograms/ml in subjects with no demonstrable central pathology. These concentrations are greater than in plasma where PGs are rapidly degraded, but agree with the low capacity of brain to metabolize prostaglandins. What appears to be the normal content of $PGF_{2\alpha}$ in CSF is clearly enhanced in pathological situations such as epilepsy, meningoencephalitis and stroke. The increased $PGF_{2\alpha}$ content in CSF of epileptic patients is in agreement with the rise of this prostaglandin observed during stimulation of CNS (19). Similar results have been reported by La Torre et al. (45) using a radioimmunoassay technique: the CSF concentration of $PGF_{2\alpha}$ was higher in patients with brain hemorrhage. No data are available concerning CSF content of PGE_2 in pathological conditions of central origin other than febrile diseases.

An interesting point to make is the possible role of CSF in the egression of endogenous PGs from the Central Nervous System. Since the accumulation of prostaglandins in cerebrospinal fluid and hence in the extracellular fluid of brain can have profound effects on the CNS, these compounds should be metabolically degraded and/or rapidly removed. Available evidence, however, indicates that the capacity of brain to inactivate PGs is rather low. It has been therefore postulated that low concentrations of prostaglandins in brain fluids may be maintained via a rapid clearance through blood-brain and blood-CSF barriers (46,47). The easiest way to achieve this would be a simple diffusion process across such

barriers provided that PGs are able to cross cell
membranes. If it is assumed that red blood cell membrane
is a model for a basic cell membrane in general, the
inability of $PGF_{2\alpha}$ or PGE_1 to penetrate these cells in
rabbit would reflect the general incapacity of prostaglandins to freely permeate cell membranes (48). Bito
and coworkers have reported that chemically unaltered
PGs are accumulated against a concentration gradient in
choroid plexuses (49) suggesting that an active transport
mechanism might play a role in prostaglandin removal from
CNS. "In vivo" experiments consisting in ventricular-
cysternal perfusion in rabbit further support this
possibility and demonstrate a facilitated transport of
$PGF_{2\alpha}$ across blood-brain barriers: this mechanism appears
to be specific and probably saturable (50).   Although
no "in vivo" data are available for PGE_2, its behavior
in choroid plexuses suggests a similar possibility (49).
Anyhow, since the plasma concentration of circulating
prostaglandins is lower in respect to CSF, a facilitated
downhill diffusion, along with bulk flow removal, may
contribute to the elimination of PGs from central fluids.

ACTION OF PROSTAGLANDINS ON CNS

Behavior

The original pharmacological observations on the
central actions of prostaglandins were first made by
Horton and his group (51). Prostaglandins of the E series
have been shown to produce sedation and loss of the
righting reflex in chicks when given intravenously (51)
and in adult fowls after intraventricular injection (52).
Subsequent studies by Holmes and Horton (6) in the mouse
have demonstrated that PGE_1 injected either subcutaneously
or intravenously at the dose of 1 mg/kg, caused diminished
general motor activity and ptosis lasting 30-60 minutes.
In addition the same Author has also reported that both
PGE_1 and $PGF_{2\alpha}$ potentiated hexobarbital sleeping time in
mice at doses ranging between 0.5-1 mg/kg i.p. (6).
PGEs exert marked behavioral effects also in the rat:
intraperitoneal administration of PGE_1 induces sedation
and a state resembling persistent paradoxical sleep
accompanied by diminished muscular tone, and increases
the turnover rate of brain 5-hydroxytryptamine. These
results are compatible with the possibility that PGE_1
may work out its effects through an action on 5-hydroxy-
tryptamine metabolism in brain. Moreover PGE_2 injected

subcutaneously or intraperitoneally at various doses
causes a significant depression of conditioned avoidance but
not escape responses in both naïve and trained rats (54).
These Authors conclude that PGE_2 has specific tranqui-
lizing effects rather than general sedative effects.
The doses used to induce these actions are relatively
large and could lead to unphysiological concentrations
within the brain or strongly affect the cardiovascular
system resulting in a decreased cerebral regional blood
flow (55). However PGs are actively metabolized by lungs
and liver (56) so that the amounts actually reaching the
brain after an intraperitoneal injection might well be
rather small (57).

The most intense and long lasting effects were
obtained injecting PGE_1 through a chronically implanted
cannula into a lateral ventricle of an unanaesthetized
cat. Decreased motor activity and definite signs of
stupor and catatonia developed within 20 minutes after
injections, together with moderate dilatation of the
pupils. The threshold dose of PGE_1 which produce
behavioral effects was approximately 3 μg/kg and effects
lasted up to 48 hours. Prostaglandin E_2 (12 μg/kg) given
through the same route, developed central actions resembling those obtained after an intraventricular injection
of 7 μg/kg of PGE_1. Conversely, prostaglandins of the F
type show weak sedative actions in cats and mice. In the
latter animal species PGF_2 unlike PGE_1 does not antagonize convulsions induced by pentamethylenetetrazole (PMT)
but like amphetamine potentiate the effects of sub-
convulsant doses of PMT (6,58). Extensive studies by
Horton and Main (58,59) have shown that neither $PGF_{1\alpha}$
nor $PGF_{2\alpha}$ cause sedation in the chick. On the other hand
intravenous administration of $PGF_{2\alpha}$ in young chicks
causes marked limb extension with some dorsiflection of
the neck. This effect lasting 10-30 minutes is still
present under light urethane anaesthesia but disappears
in deeply anaesthetized animals.

The site and mechanism of action of the sedative
effect of PGEs is unknown. The observations that PGs are
widely distributed in the CNS indicate that these
compounds may influence various regions of this system.
The hypothalamus (2) and the brain-stem pathways (60)
seem to be favourite areas in this respect although no
definite mechanism of action has yet been proposed. The
vasoconstrictor properties of prostaglandins on cerebral
vessels (55) might play a significant role, although a
modulatory action at central adrenergic nerve endings

could also be postulated. As recently pointed out by
Horton (61) the fact that prostaglandins are active
throughout the CNS, could simply speak in favour of their
pharmacological activities without casting any light upon
their role (physiological or pathological) in this system.
The recent discovery by Wolfe (37) that thromboxanes are
actively synthetized in the CNS poses new problems to
investigators trying to elucidate the biological role of
this class of compounds.

Motor Pathways

The effects of prostaglandins on motor functions
have been investigated rather extensively and at the
present time there is an ever-growing literature dealing
with their influence on experimental convulsions and on
spinal mechanisms. The administration of prostaglandins
of the E type to the experimental animals displays some
protection against pentamethylenetetrazole induced
convulsions at the dose of 0.5 mg/kg i.p. and exhibits
almost complete protection at the dose of 1 mg/kg i.p.
(6). On the other hand PGEs fail to prevent convulsions
induced by picrotoxin (1), partially protect against
electroshock-induced seizures and the results obtained
in strychnine convulsions are conflicting (3,62). The
involvement of cerebellar cyclic 3',5'-adenosine-mono-
phosphate (cAMP) and cyclic 3',5'-guanosine-mono-phosphate
(cGMP) in experimental convulsions has been recently
reported (63), and the existence of a strict correlation
between the increase in cerebellar cGMP and the onset of
convulsions induced by PMT has been demonstrated (64).
This, however, does not exclude the possibility that
changes in cerebellar cGMP levels represent a defensive
response of the body versus the effects of convulsants.
Experiments carried out to ascertain whether the anticonvulsant properties of PGE_2 might be reflected by
changes in cerebellar cyclic nucleotides, have shown that
this prostaglandin completely prevents the increase in
cerebellar cGMP which is likely to be the primary
biochemical modification triggering the convulsive phase.
Basal levels of cerebellar cGMP are not modified by PGE_2.
Cyclic AMP variations invariably follow the appearance
of convulsions. As a matter of fact the state of brain
anoxia induced by seizures might be responsible for the
increase of this nucleotide, as pointed out by Sattin
(65).

Concerning the mode of action of PGE_2, an indirect

effect cannot be ruled out. As already pointed out prostaglandins are known to strongly affect the cardio-vascular system, and Yamamoto et al. (55) have noted a decreased cerebral regional blood flow following intra-carotid injection of PGs of the E and F types. Consider-ing the dose of PGE_2 used in the experiment mentioned above this possibility should be considered. However $PGF_{2\alpha}$, which like PGE_2 decreases cerebral regional blood flow, completely lacks anticonvulsant properties and does not influence the concentrations of cerebellar cyclic nucleotides.

More selective experiments aimed to better elucidate the anticonvulsant properties of PGEs, has been recently carried out by using harmaline, an alkaloid of Peganum Armala, which induces tremors of cerebellar origin (66, 67). These tremors are rapid in onset and long lasting and are characterized by a marked increase of cerebellar cyclic GMP : cyclic AMP levels remain unaffected (64,68). PGE_2 (1 mg/kg i.p.) is active against both tremors and cGMP rise produced by the alkaloid, although the duration of its protection is relatively short lasting (20-30 minutes). These results suggest a possible localization of the site of action of exogenous PGE_2. Since harmaline fails to induce tremors and the concomitant increase of cerebellar cGMP in newborn rats, where climbing fibers do not yet synapse with Purkinje cells (69,70), and since mice with selective degeneration of Purkinje cells show very low basal levels of cGMP and tremors are facilitated (70), it is likely that PGE_2 acts at the synaptic link between climbing fibers and Purkinje cells. The putative stimulatory neurotransmitter involved at this level seems to be glutamic acid (71,72), found to increase the levels of cGMP in slices of mouse cerebellum (73). Since this effect is Ca^{++} dependent, it is also tempting to speculate that the antitremorogenic action of PGE_2 might be related to this ionic mechanism (73). Prostaglandins of the F type were completely ineffective in this respect. On the other hand, altered levels of cyclic GMP might simply mean involvement of Purkinje cells to the overall process, hence a defensive mechanism, since these cells are inhibitory in nature. Quesney et al. (74) have carried out studies on prostaglandins and penicillin-induced epilepsy in the cat. These investigators have tested $PGF_{2\alpha}$ and 15(S)-15 methyl PGE_2 methyl ester against generalized epileptic activity induced by this antibiotic. Their results clearly indicate that non toxic doses of the PGE_2 analogue and $PGF_{2\alpha}$ markedly reduce epileptic activity in their experimental model.

Recently a report by Zatz and Roth (75) has demonstrated that electroconvulsive shock (ECS) raises $PGF_{2\alpha}$
in rat cerebral cortex and this increase is abolished
by indomethacin without blocking convulsions. It seems
therefore unlikely that the convulsions result from an
ECS-induced rise in $PGF_{2\alpha}$ levels. These results are of
interest, but the absolute values of prostaglandin content
reported by these Authors should be considered with
caution because of the well known stimulation in PGs
synthesis by anoxia following decapitation (31) and tissue
trauma (76).

Recent clinical studies have been reported by Lyneham
et al. (77) on epileptic seizures and atypical electro-
encephalograms (E.E.G.) occurring in patients aborted
by intraamniotic instillation of $PGF_{2\alpha}$. This report has
casted considerable concern on the potential use of
prostaglandins as oxytocic, but reports by other
laboratories (78,79) are at variance with this statement.
To draw definite conclusions about the role of PGs in
convulsive state from the above cited observations is
definitely overoptimistic. Further studies are necessary
in order to elucidate the complex nature of the events
leading to a convulsive state and the possible relationships between PGs and brain neurotransmitters.

Prostaglandins are active stimulant of spinal
reflexes but their actions are complex and difficult to
interpret. $PGF_{2\alpha}$ given i.v. at the dose of 2 to 100 μg/kg
to anaesthetized or unanaesthetized chicks increases
gastrocnemius muscle tension (59). This contraction is
abolished by sciatic nerve section and tachyphylaxis
occurs very often. In addition to this $PGF_{2\alpha}$ increases
gastrocnemius muscle tension also in decapitated and
spinal chicks (59). In sectioned sciatic nerves, small
doses of urethane abolish reflex contraction to electrically induced stimulation of the cut central stump of the
controlateral sciatic nerve, while $PGF_{2\alpha}$ still elicits
contraction even in presence of very high doses of
urethane. These observations led Horton and Main to
postulate an action of $PGF_{2\alpha}$ directly on the chick spinal
cord. This conclusion has been supported by other observations done by the same Authors. $PGF_{2\alpha}$ cannot potentiate
contractions of the gastrocnemius muscle elicited by
electrical stimulation of its own motor nerve even in
doses which contract the ipsilateral muscle; again a
proof that $PGF_{2\alpha}$ potentiate spinal reflexes by central
and not peripheral actions (59). The studies of Coceani
(23,80) in the isolated superfused frog spinal cord are

in agreement with this view. PGEs effects are much less consistent. The cross extensor reflex is inhibited in the slightly anaesthetized chick by PGE_1. On the other hand in the spinal chick, PGE_1 potentiate the cross extensor reflex (59). These observations suggest the presence of at least two sites of action of PGE_1, one at spinal cord level resulting in an excitation of the reflex, and one located at higher centres causing inhibition of the extensor reflex.

Increase in gastrocnemius muscle tension by PGE_1 has been found with both decerebrate and spinal cats. This increase is blocked by section of ventral routes of spinal nerves but not of dorsal roots, and this is consistent with a spinal localization of the site of action of PGE_1. These effects are long lasting and give way to tachyphylaxis. Furthermore it is likely that they are due to a facilitated firing of the α-motoneurones (81,82), by stimulation of direct excitatory pathways.

Conclusions similar to those discussed for chicks, can account also for the cat. The following statement is taken from Horton (1): ".... in the spinal cat as in the spinal chick, PGE_1 appears to have a facilitatory action on spinal reflexes. When higher centres are present, this facilitation is not observed presumably because PGE_1 acts upon supraspinal centres with descending pathways which impinge upon neurones of the spinal reflex pathways".

Hypothalamus

A) Fever and thermoregulation. This subject is discussed in a further chapter of this book.

B) Food intake. Data have appeared concerning a possible role of prostaglandins on the regulatory mechanisms of food intake. When these compounds are administered subcutaneously or directly injected into various hypothalamic areas, a decreased food intake is observed in rat and sheep (83-86). In this respect PGEs are more active than PGFs and PGA_1, while PGB_1 appears devoid of any action. However when prostaglandin E_1 is injected in the anterior hypothalamus of the sheep, its action on food intake is reversed (86). In a further report PGE_1 did not affect eating in the rat when injected in hypothalamus, hyppocampus and amigdala, exactly in the same regions where norepinephrine stimulated food intake

(87). From these results it is not possible to draw any
conclusion on a role, if any, played by prostaglandins
on ingestive behavior.

C) Release of pituitary hormones. There is increasing
evidence that prostaglandins have pharmacological actions
suggesting a possible role in the balance of physiological
mechanisms underlying hormone release processes at the
hypothalamus-pituitary axis. A number of "in vitro" and
"in vivo" studies have shown that exogenous prostaglandins
increase the secretion of GH, ACTH, LH, FSH, Prolactin
and TSH (88-101). These experiments were based on addition
of PGs to emipituitaries, injections of PGs by parenteral
routes or in hypothalamic regions and in the anterior
pituitary, and on the use of prostaglandin synthetase inhibitors. Different prostaglandins are involved in these
events: for instance PGEs are more active than PGAs, PGBs
and PGFs in releasing GH from cultured anterior pituitary
cells (102), and PGE_1 but not PGE_2 and PGFs is able to
stimulate prolactin release when injected into the 3rd
ventricle of rats (99). Interestingly, only growth hormone
appears to have its secretion altered by a direct pituitary effect of prostaglandins (89,90). For the other
hormones the predominant effect of prostaglandins appears
to be exerted at the hypothalamic level. In fact PGs raise
ACTH secretion when stereotaxically applied to the median
eminence region of the hypothalamus, but fail to do so if
injected in the anterior pituitary (93,103). Again, female
rats respond to i.v. PGE_2 with a rise in plasma LH levels,
but a previous injection of an anti-LHRH serum prevents
the rise induced by this prostaglandin, indicating that
the effect of PGE_2 is secondary to LHRH release from the
hypothalamus and not displayed at the pituitary level
(104). However prostaglandins appear to modulate, although
in an opposite way, the secretion of TSH and ACTH elicited
in the anterior hypophysis by the corresponding releasing
factors (102,104). That endogenous PGs are involved in the
release of pituitary hormones is suggested by the use of
synthetase inhibitors. Pretreatment with indomethacin
decreases normal serum LH levels in rats, and this
decrease recovers after administration of $PGF_{2\alpha}$ (105);
the same antiinflammatory drug significantly inhibits
the TSH response to exogenously administered TRH (106).

It is generally agreed that cyclic adenosine monophosphate is involved in pituitary secretion of hormones
as influenced by the specific factors released from the
hypothalamus. The stimulatory action of prostaglandin E_1
on GH release from anterior pituitary "in vitro",

paralleled by a rise of cAMP, is significantly reduced
by the prostaglandin antagonist 7-oxa-13-prostynoic acid
with a concomitant drop of the cyclic AMP content in this
tissue (90). In addition, since PGs raise and somatostatin
reduces adenyl-cyclase activity, the increased release of
growth hormone brought about by prostaglandins appears to
be mediated via this cyclic nucleotide (107-109). Another
example relates to LH and FSH release: addition of LHRH
leads to stimulation of cyclic AMP accumulation in rat
anterior pituitary gland "in vitro" (110-112), and a close
correlation is observed between rates of LH and FSH
release and changes in intracellular cyclic AMP concentrations (113). This implies a cause-effect relationship
between the releasing hormone and cAMP, as pointed out
by the finding that theophylline and the dibutyryl-
derivative of the cyclic nucleotide have a stimulatory
effect on LH output (104). Not all reports are however
consonant with the intermediary role of cAMP between PGs
and hormone release. For instance the observation that
cAMP is implicated as the mediator of TSH secretion by
the anterior pituitary (115) is at variance with a more
recent report. While TRH added "in vitro" markedly
increased TSH release, it failed to enhance cyclic AMP
accumulation. PGE_1, on the other hand, was ineffective
on TSH liberation at concentrations which were maximally
stimulatory of cyclic AMP levels (116).

It is known that various neurotransmitters can
influence the release of pituitary gonadotropins (117)
and that prostaglandins may be implicated in the mechanism
of action of some of these transmitters (118). On this
basis, it is tempting to speculate that the effects of
PGs on pituitary release of hormones might, at least in
part, be mediated by neurotransmitters. Although this
interaction deserves further attention, it is worth
mentioning a recent report indicating that the effect of
prostaglandin E_2 on LH release is not mediated by
adrenergic, dopaminergic, serotoninergic or cholinergic
receptors (119). Prostaglandins might play some role also
in the hormonal secretion by neurohypophysis. PGE_2 enhances
urine osmolarity and decreases urine flow in rats (120),
suggesting an effect on ADH output, while $PGF_{2\alpha}$ appears
to be involved in oxytocin release (121).

Brain Stem

Prostaglandins affect brain stem neurones as originally assessed by Kaplan (122), who found that $PGF_{2\alpha}$
induces an increase in systemic arterial blood pressure
which is mediated by central neurones. Moreover elegant
works by Lavery et al. (123), have shown that in the dog
$PGF_{2\alpha}$ given into the vertebral artery causes augmented
pressor response, due to an increase of cardiac output.
Since these effects are no longer present when the drug
is given intravenously, the authors attribute the vascular
changes to an activation of "cardioregulatory centres
within the territory of distribution of the vertebral
artery". Similar conclusions have been attained by
Rinchuse and Deuben (124) who had observed that PGE_1, A_1
and A_2 elevated arterial blood pressure in the rat when
infused into the carotid artery. Infusions of identical
doses into the femoral vein failed to cause a pressor
response. Experiments by McQueen and Ungar (125), in dogs
exibiting marked sinus arrythmia, point out a selective
action of PGE_1 on brain stem neurones. In fact intra-
carotid administration of PGE_1 reduces or suppresses sinus
arrythmia and the effect is not altered by denervation
of the carotid sinus region. Conclusions drawn by these
authors suggest an inhibitory role of PGE_1 on brain stem
involving respiratory and vasomotor pathways. Interesting
studies have been carried out by Carlson et al. (126),
on PGE_1 infusion in humans. They report an increased
lung ventilation with a concomitant drop in arterial pCO_2
and respiratory alkalosis. The primary event leading to
changes in ventilation, might well be a direct stimulation
by PGE_1 of the brain stem respiratory centres.

Local superfusion of $PGF_{2\alpha}$ to the surface of the
"medulla oblungata" causes reduction in synaptic transmission through the cuneate nucleus, likely due to an
action on postsynaptic sites (23,127). These results
obtained by Coceani et al., together with other studies
by the same Author underline the fact that prostaglandins
can stimulate or depress synaptic events in brain stem
under a variety of experimental conditions, and propose
a role of PGs as modulators of transmitter action at
receptor level.

Cerebral Circulation

As previously pointed out (55), prostaglandins exert
powerful actions on cerebral vessels, and a number of

recent studies has attempted to elucidate the mechanisms
through which PGs act in this respect. Pennink and
associates (128) have performed an angiographic study of
experimental cerebral vasospasm injecting blood mixed
with $PGF_{2\alpha}$ into the chiasmatic cistern of the dog and have
found a statistically significant difference when compared
with injections of blood alone. In addition $PGF_{2\alpha}$ injected
alone caused cerebral vasospasm whereas PGE_1 had no effect.
In line with these observations are the results obtained
by Denton and coworkers (129), who have found that $PGF_{2\alpha}$
evoked selective increase in cerebrovascular tone in dogs.
This effect is mimicked by serotonine. PGE_1 tested in the
dog selectively reduced cerebrovascular tone, but was
without effect in the monkey. Moreover Yamamoto et al.
(55), examining the effects of intracarotid infusion of
PGE_1 and $PGF_{2\alpha}$ on dog brain circulation by different
methods have shown that PGE_1 constricted the epicerebral
arteries and this effect was abolished when 0.08% ethanol
was added. $PGF_{2\alpha}$ constricted epicerebral arterial vessels
and lenghtened the cerebral circulation time. These
results speak in favour of the role of prostaglandins as
factors regulating the cerebral blood flow and involved
in the mechanism of cerebral vasospasm. La Torre et al.
(45) found no correlation between $PGF_{2\alpha}$ levels in CSF and
cerebral vasospasm.

Rosemblum (130), has recently suggested that among
the causes of cerebral vasospasm, one should keep in mind
not only the actions of single agents, but also the combined effects exerted by prostaglandins together with
brain neurotransmitters (norepinephrine, 5-HT), that are
active on cerebral vessels. For instance the constrictor
action of $PGF_{2\alpha}$ on pial arterioles is enhanced by norepinephrine. At variance with these observations is a
report by Welch et al. (131) demonstrating that intra-
carotid PGE_1 markedly increase external carotid flow and
reduce internal carotid flow in the monkey. $PGF_{2\alpha}$
constricted both arteries. These Authors hypothize that
the cranial flow changes, which are typical of migraine
headache, may be due to a single biochemical agent which
acts differently on internal and external carotid flows.
This would cause a concomitant dilation of the external
carotid vasculature and a "steal" of blood from the
internal carotid vessels.

MICROIONTOPHORETIC APPLICATION

The first report stating a direct effect of prostaglandins on central neurones has been presented by
Avanzino et al. (132), who demonstrated that PGs have a
selective excitatory or inhibitory action when applied
in very small amounts to brain stem neurones. In more
extensive investigations Hoffer et al. (133) have shown that
PGEs and PGFs exert a highly variable direct action on
the spontaneous firing of Purkinje cells, and a potent
antagonism to norepinephrine (NE) responses (PGE_1, PGE_2)
without affecting that of exogenous cAMP. The latter
effect is present regardless of whether Purkinje cells
firing inhibition is obtained through microiontophoretic
application of NE (40), or by stimulation of noradrenergic
inhibitory pathways to the cerebellar cortex from the
"locus coeruleus" (134). These Authors place therefore
the site of action of PGEs on these cells at the level of
cyclic AMP formation. These observations, together with
the fact that PGEs are present in the cerebellar cortex
(20) and are released from there upon nerve stimulation
(20) suggest an endogenous modulatory role on this adrenergic synapse. Recently it was shown (135) that ionto-
phoretic application of PGE_2 or $PGF_{2\alpha}$ to hypothalamic
neurones cause a long lasting activation of their firing
rate. These Authors suggest that PGs might have the
capacity to modulate firing of the hypothalamic neurones
implicated in the control of gonadotropin secretions.
Moreover Traber and coworkers (136) have studied the
effects of PGE_1 on membrane potentials of neuroblastoma
x glioma hybrid cells, and have found that this prostaglandin cause hyperpolarization and increase cAMP levels.
On the other hand norepinephrine and dopamine antagonize
the rise of cAMP induced by this prostaglandin and exert
opposite effects on transmembrane potentials. These
results further emphasize the exhistence of a possible
interplay among PGs, neurotransmitters and second
messenger(s).

PROSTAGLANDINS AND CYCLIC NUCLEOTIDES

It is now well established that prostaglandins of the
E series stimulate, with a few exceptions, cyclic AMP
formation in several organs and tissues. The influence
on this "second messenger" might be the basis of the
wide spectrum of pharmacological actions of these
compounds. Besides the interplay between PGs and cyclic
AMP at the level of hormone release in the hypothalamus-

pituitary axis, the influence of prostaglandins on the cAMP system takes place also in other parts of the Central Nervous System, and it has been demonstrated both "in vivo" and "in vitro". Wellman and Schwabe reported a modest but significant increase of cyclic adenosine monophosphate in whole brain of rats and mice given PGEs i.v., while $PGF_{2\alpha}$ was almost devoid of action in this respect (137). Time-course experiments indicated that this action is maximal one minute after injection and rather long-lasting. The elevation of the cyclic nucleotide levels in response to PGE_2 was highest in the cerebral cortex, intermediate in the thalamus and lowest in cerebellum and brain stem. The increased cAMP content correlates with sedation and stupor observed in the animals. Research by various investigators have shown that PGEs, but not PGFs, stimulate cyclic AMP formation in brain "in vitro". As found in our laboratory, this response is species-dependent (138,139), in that rat but not rabbit, guinea pig and human cerebral cortex respond to PGEs with an increased formation of cyclic AMP. Similar results have been reported by others laboratories in murine neuroblastoma cells (140) and in cultured cells of fetal rat brain (141). This interaction occurs in other brain regions as well, such as striatum, diencephalic tissues, hippocampus and cerebellum, the lowest response being observerd in medulla oblungata (142). Since rat neocortex increases cAMP formation when challanged either with PGE_2 or norepinephrine, the possibility that these two agonists are acting on the same site was considered. However the findings that the actions of PGE_2 and nor-epinephrine are synergistic (139), and that cyclic AMP formation is stimulated at birth by prostaglandin E_2 but not by catecholamines (143), as shown in Table 2, are against this possibility. In addition, α and β blockers partially prevent the rise of cyclic AMP elicited by nor-epinephrine but not by PGE_2 (142,144). In cultured human astrocytoma cells no synergistic effects on cAMP formation was observed between PGE_1 and other putative neurotrans-mitters (145), raising the question of which type of cells (neurons or glia) are specifically affected by PGEs. Because prostaglandins of the E series enhance cyclic AMP levels in neuroblastoma cells (140), where catecholamines are ineffective, and the opposite happens in glioma cells (146), it might be inferred that prostaglandins are mainly concerned with neuronal cells. However this assumption is in contrast with reported data showing that astrocytoma cells and some clones of murine neuroblastoma respond to PGEs and norepinephrine with increased levels of cyclic AMP (145,147).

TABLE 2

EFFECT OF NOREPINEPHRINE, ADENOSINE AND PGE$_2$ ON CYCLIC AMP FORMATION IN SLICES OF WHOLE BRAIN FROM 5-DAY-OLD RATS

COMPOUND 100 μM	+ CONVERSION % ($\overline{x}$ + S.E.)	STATISTICAL EVALUATION	RATIO OF STIMULATED VERSUS CONTROLS
NONE (6)	0.09 + 0.01	----	1
NOREPINEPHRINE (6)	0.11 + 0.01	vs.control N.S.	1.2
ADENOSINE (6)	0.42 + 0.09	vs.control P<0.01	4.6
PGE$_2$ (9)	0.95 + 0.1	vs.control P<0.001	10.5

+Calculated on the percentage of conversion of prelabeled ATP into cyclic AMP.

In parenthesis number of experiments.

Fumagalli et al. (143).

The results obtained with sliced brain are of difficult interpretation because of the extreme heterogeneity
of brain as a tissue. At the moment no correlation is
possible between accumulations of the cyclic nucleotide
evoked by PGEs and specialized activities of neuronal
pathways. However, as suggested by Coceani (2), the very
rapid PGE_1-induced increase in cyclic AMP formation
occurring in murine neuroblastoma cells (140) having some
properties of differentiated neurones (148) is compatible
with a role of this interaction in neuronal events: for
instance endogenous PGEs might function as intermediate
messengers in the activation of the adenyl-cyclase by
neurohumoral stimuli.

It has become increasingly clear that cyclic adenosine
monophosphate is involved in the regulatory processes
pertaining cell growth and differentiation. Gilman and
Nirenberg (140) found that PGE_1 (and to a lesser extent
PGE_2), but not PGFs, PGAs and PGBs, has a powerful
stimulatory effect on cyclic AMP levels in murine neuroblastoma C1300 cultures and that such a prostaglandin,
alone or in presence of theophylline, highly reduced the
rate of cell growth. This finding confirmed previous
reports on cell lines of peripheral origin in which addition of exogenous cyclic AMP (149,150) or PGE_1 (151) slows
the rate of cell division. Cultured clones of murine
neuroblastoma have been thereafter largely investigated
and it was shown that dibutyryl cyclic AMP and some
prostaglandins induce morphological differentiation (152-
154). This differentiation,consisting mainly in the
formation of axons, induced by prostaglandin of the E
type appears to be irreversible (153). Moreover such
morphological changes are inhibited by vinblastine and by
cycloheximide but not by actinomycin D, suggesting that
microtubule assembly and the synthesis of new proteins,
but not of new RNA, are necessary (153). Similar results
have been reported on morphological differentiation of
cultured human glioma cells (155). That prostaglandins
may have some bearing upon cellular control mechanisms
has been realized also for peripheral tissues. It has
been claimed that an inverse relationship exists between
PGs-synthetase activity and cell proliferation (156), but
a report by Hammarström and coworkers shows that virus-
transformed fibroblast cell cultures synthesize more PGE_2
than the corresponding normal cells (157). A possible
explanation of this discrepancy is that certain tumoral
or transformed cultured cells have a reduced responsiveness, as far as cAMP formation is concerned, to pharmacological concentrations of PGEs when chronically exposed

to physiological levels of these prostaglandins, as
recently found by Leichtling et al. (158) in astrocytoma
cells. These Authors suggest that this failure is due to
a selective desensitization of PGE_1-responsive adenyl-
cyclase, although the possibility that this prostaglandin
increases phosphodiesterase activity cannot be ruled out
(159). In an update by Jaffe (160) it is summarized that
neoplastic cells are able to form prostaglandins both
"in vivo" and "in vitro" with a relative prevalence of
PGEs, whose concentrations are higher than in normal non
transformed cells.

It is at least premature to stress any possible
implications concerning the cellular events observed
(cellular growth and differentiation) and the biochemical
parameters considered (cellular concentrations of cyclic
AMP, of prostaglandins, and cAMP formation as elicited
by PGs). For instance, norepinephrine, which elevates the
cyclic AMP content in a number of cultured glioma cells,
causes only minor changes in the morphology of a human
glioma (155). The available data are often in contrast
and simply suggest that PGs and cyclic adenosine monophosphate are somehow involved in cellular replication
and differentiation (161-165). It should also be considered that these effects of prostaglandins do not appear
to be necessarily mediated by cyclic AMP. Indeed the
addition of dibutyryl-cAMP to culture media reduced cell
replication and stimulated PGs synthesis; moreover, in a
virus-induced sarcoma tumor in mice, indomethacin
decreased PGs content and the size of the tumor, without
affecting the levels of the cyclic nucleotide (160).

Further interest in the relationship between prostaglandins and cyclic AMP system has been provided by
Collier and Roy (166), reporting that opiates antagonize
the production of cAMP elicited by prostaglandins of the
E type in rat brain homogenates. This effect of morphine
is more evident with low concentrations of PGEs, and
heroin was more and methadone less active than morphine
itself. These results were confirmed by other investigators
on different experimental models (neuroblastoma and neuroblastoma x glioma hybrid cells) (167). Naloxone counter-
acts morphine also on this biochemical parameter (166,
167), but at high concentrations it increases cyclic AMP
formation and potentiate the action of PGE_1 (167). The
inhibiting effect of opiates on PGE_1-induced increase in
cyclic AMP formation appears to be stereospecific since
levorphanol, but not its biologically inactive enantiomer
dextrorphan, is active in this respect (136). Investi-

gations aimed to further explore the antagonism between morphine and prostaglandin E_1 on cyclic AMP system suggest a non-competitive type of inhibition (168). This antagonism does not apply to all cell types: it occurs in neuroblastoma and in neuroblastoma x glioma hybrid cells, but not in glioma and in glioma x fibroblast hybrid lines (167,168). The stimulatory action of prostaglandin E_1 on cyclic AMP biosynthesis in cultured cells of central origin is not only antagonized by morphine but also by adrenergic and cholinergic agonists via α-adrenergic and muscarinic receptors respectively (136,169). If cultured cells are preincubated with morphine for 15 hours, the basal levels of cellular cyclic AMP and their sensitivity to PGE_1 are doubled, but the latter event is prevented when cycloheximide is added during the preincubation step. The same picture results when long-term preincubation is performed with adrenergic and cholinergic agonists (170). All these antagonists of the PGE_1-increased cyclic AMP formation enhance the cellular levels of cyclic GMP (170). Interestingly enkephalins mimic morphine in increasing the cyclic GMP content and in antagonizing the prosta-glandin-evoked rise in cAMP in the same cultured cells, as reported at the 5th International Narcotic Research Conference held in Scotland (171).

No comprehensive interpretations of the interaction between opiates and prostaglandin E_1 on cyclic AMP generating system are available. Collier and associates (172-175) have proposed the possibility that this inter-action might be implicated in the mechanism of morphine analgesia and opiate dependence, but most likely the PGE_1-sensitive adenyl-cyclase represents a model of this enzyme responding to morphine (174). However this inter-action might be implicated in the decreased antinociceptive effect of morphine in the rat given prostaglandin E_1 by intraventricular route (176), although an opposite effect has been reported when PGE_1 is administered intraperito-neally at much higher doses (177). No data are available on the effect of prostaglandins on cyclic GMP metabolism in CNS "in vitro". We have discussed in a previous session of this chapter the effects of exogenously administered PGs on cGMP as related to convulsions and tremors in experimental animals.

REMARKS

Central Nervous System contains prostaglandins and
is able to form them from endogenous precursors. Since
PGs display a wide spectrum of pharmacological activities
also at this level, investigators have tried to seek a
biological role(s) for these compounds in this system.
It is possible that the actions of PGs of E and F types
on brain is only an aspect of their pharmacology: actually
in most experiments reported the doses and/or concentrations used are largely pharmacological. Perhaps the use
of drugs inhibiting their endogenous formation would
represent a proper tool in disclosing physiological roles
of these compounds in CNS. However the overall picture may
be complicated by effects other than inhibition of prostaglandin biosynthesis. For instance, the use of indomethacin to establishe whether endogenous PGs are involved
in TSH secretion from the pituitary might lead to
erroneous conclusions, being this drug directly impairing
thyroid function and hence increasing blood TSH levels
by a negative feed-back mechanism (106).

Prostaglandins have been proposed to act as local
hormones or as modulators. In keeping with the former role
is the luteolytic action of $PGF_{2\alpha}$ and the regulation of
intrarenal blood flow by PGEs, but no evidence seems to
support this role in nervous tissue. Newly synthesized
prostaglandins may act as intracellular messengers either
by directly affecting enzymatic activities or by influenc-
ing intracellular levels of other biologically active
substances such as cyclic nucleotides and/or calcium ions.
However it should be stressed that cyclic nucleotides not
necessarily act behind prostaglandins, as it might be the
case, for instance, in the firing of the Purkinje cells
(133). The finding that cyclic AMP regulates the formation
of prostaglandin endoperoxide in human platelets (178)
may suggest that a similar regulatory mechanism could
operate also in CNS. With regard to neurotransmission,
prostaglandins are generally viewed in the role of modulators. They might be formed in the target cells in
response to an appropriate stimulus, such as neurotransmitter impact, and then work pre- or post-junctionally
to modulate synaptic events. Indeed, serotonin elicits
release of PGs (21) and it has also been found that the
output of other transmitters can be modulated by prostaglandins (179). This suggests for instance that the model
of prostaglandin-norepinephrine interaction proposed for
peripheral sympathetic synapses (180) may hold true for
CNS as well. However some experimental evidence cast some

doubts on this model (181). A major point in seeking a role for these compounds in CNS is the recent report that brain cortex is able to form thromboxanes from endogenous precursors : this finding gains new insights to possible functions of arachidonic acid metabolites in the nervous system.

ACKNOWLEDGMENTS

This review and the experimental results of the Authors reported in it were supported in part by a grant of the C.N.R. (Consiglio Nazionale delle Ricerche, Roma) No. 75.00522.

The Authors are indebted to Mrs. L. Rossoni Moriggi for her invaluable assistance.

REFERENCES

1. HORTON, E.W., (1972). Prostaglandins. In: Monographs on Endocrinology, vol. 7, Springer-Verlag, New York
2. COCEANI, F., (1974). Prostaglandins and the Central Nervous System. Arch. Intern. Med. 133, 119-132
3. WOLFE, L.S., (1975). Possible roles of prostaglandins in the Nervous System. In: Advances in Neurochemistry. vol. 1, Eds. B.W. Agranoff and M.H. Aprison, Plenum Publ. Corp., New York, pp. 1-49
4. SAMUELSSON, B., (1964). Identification of a smooth muscle-stimulating factor in bovine brain. Biochim. Biophys. Acta, 84, 218-219
5. AMBACHE, N., BRUMMER, H.C., ROSE, J.G. and WHITING, J., (1966). Thin-layer chromatography of spasmogenic unsaturated hydroxy-acids from various tissues. J. Physiol. (Lond.), 185, 77-78P
6. HOLMES, S.W. and HORTON, E.W. (1968). Prostaglandins and the Central Nervous System. In: Worcester Symposium on Prostaglandins, Eds. P.W. Ramwell and J.E. Shaw, John Wiley & Sons Inc., New York, pp. 21-36
7. HOLMES, S.W. and HORTON, E.W. (1968). The identification of four prostaglandins in dog brain and their regional distribution in the Central Nervous System. J. Physiol. (Lond.), 195, 731-741
8. WOLFE, L.S., COCEANI, F. and PACE-ASCIAK, C. (1967). Brain prostaglandins and studies of the action of prostaglandins on the isolated rat stomach. In: Nobel Symposium 2 on Prostaglandins, Eds. S. Bergström and B. Samuelsson, Almqvist & Wiksell Publ., Stockholm, pp. 265-275

9. HORTON, E.W. and MAIN, I.H.M. (1967). Identification
    of prostaglandins in central nervous tissues of the
    cat and chicken. Br. J. Pharmacol. 30, 582-602.
10. KATAOKA, K., RAMWELL, P.W. and JESSUP, S. (1967).
    Prostaglandins: Localization in subcellular particles
    of rat cerebral cortex. Science, 157, 1187-1189.
11. LUNT, G.G. and ROWE, C.E. (1968). The production of
    unesterified fatty acid in brain. Biochim. Biophys.
    Acta, 152, 681-693.
12. BAZAN, N.G. (1970). Effects of ischemia and electro-
    convulsive shock on free fatty acid pool in the brain.
    Biochim. Biophys. Acta, 218, 1-10.
13. BAKER, R.R. and THOMPSON, E. (1972). Positional dis-
    tribution and turnover of fatty acids in phosphatidic
    acid, phosphoinositides, phosphatidylcholine and
    phosphatidylethanolamine in rat brain in vivo.
    Biochim. Biophys. Acta, 270, 489-503.
14. HOLUB, B.J., KUKSIS, A. and THOMPSON, W. (1970).
    Molecular species of mono-, di-, and triphospho-
    inositides of bovine brain. J. Lipid Res. 11, 558-564.
15. WOLFE, L.S., PAPPIUS, H.M. and MARION J. (1976). The
    biosynthesis of prostaglandins by brain tissue "in
    vitro". In: Advances in prostaglandins and thromboxane
    research, Eds. B. Samuelsson and R. Paoletti, Raven
    Press, New York, pp. 305-312.
16. NICOSIA, S. and GALLI, G. (1975). A mass fragmento-
    graphic method for the quantitative evaluation of
    brain prostaglandin biosynthesis. Prostaglandins,
    9, 397-403.
17. RAMWELL, P.W. and SHAW, J.E. (1966). Spontaneous and
    evoked release of prostaglandins from cerebral cortex
    of anesthetized cats. Am. J. Physiol. 211, 125-134.
18. RAMWELL, P.W. and SHAW, J.E. (1967). Prostaglandin
    release from tissues by drug, nerve and hormone
    stimulation. In: Nobel Symposium 2 on Prostaglandins,
    Eds. S. Bergström and B. Samuelsson, Almqvist &
    Wiksell Publ., Stockholm, pp. 283-292.
19. BRADLEY, P.B., SAMUELS, G.M.R. and SHAW, J.E. (1969).
    Correlation of prostaglandin release from the cerebral
    cortex of cats with electro-corticogram, following
    stimulation of the reticular formation. Br. J.
    Pharmacol., 37, 151-157.
20. COCEANI, F. and WOLFE, L.S. (1965). Prostaglandins
    in brain and the release of prostaglandin-like com-
    pounds from the cat cerebellar cortex. Can. J.
    Physiol. Pharmacol., 43, 445-450.
21. RAMWELL, P.W., SHAW, J.E. and JESSUP, R. (1966).
    Spontaneous and evoked release of prostaglandins from
    frog spinal cord. Am. J. Physiol., 211, 998-1004.

22. MATSUURA, S., KAWAGUCHI, S., ICHIKI, M., SORIMACHI, M., KATAOKA, K. and INOUYE, A. (1969). Perfusion of frog's spinal cord as a convenient method for neuro-pharmacological studies. Eur. J. Pharmacol. 6, 13-16.

23. COCEANI, F., PUGLISI, L. and LAVERS, B. (1971). Prostaglandins and neuronal activity in spinal cord and cuneate nucleus. Ann. N.Y. Acad. Sci. 180, 289-301.

24. FELDBERG, W. and MYERS, R.D. (1966). Appearance of 5-hydroxytryptamine and an unidentified pharmacolog-ically active lipid acid in effluent from perfused cerebral ventricles. J. Physiol. (Lond.), 184, 837-855.

25. HOLMES, S.W. (1970). The spontaneous release of prostaglandins into the cerebral ventricles of the dog and the effect of external factors on this release. Br. J. Pharmacol. 37, 653-658.

26. BELESLIN, D.B., RADMANOVIC, B.Z. and RAKIC, M.M. (1971). Release during convulsions of an unknown sub-stance into the cerebral ventricles of the cats brain. Brain Res. 35, 625-627.

27. BELESLIN, D.B. and MYERS, R.D. (1971). Release of an unknown substance from brain structures of un-anaesthetized monkeys and cats. Neuropharmacology, 10, 121-124.

28. HAMBERG, M., ISRAELSSON, U. and SAMUELSSON, B. (1971). Metabolism of prostaglandin E_2 in guinea pig liver. Ann. N.Y. Acad. Sci. 180, 164-180.

29. LESLIE, C.A. and LEVINE, L. (1973). Evidence for the presence of a prostaglandin E_2-9-keto reductase in rat organs. Biochem. Biophys. Res. Comm., 52, 717-724.

30. PAPPIUS, H.M., ROSTWOROWSKI, K. and WOLFE, L.S. (1974). Biosynthesis of prostaglandin $F_{2\alpha}$ and E_2 by brain tissue in vitro. Trans. Amer. Soc. Neurochem., 5, 119 (Abstract), and unpublished results.

31. BOSISIO, E., GALLI, C., GALLI, G., NICOSIA, S., SPAGNUOLO, C. and TOSI, L. (1976). Correlation between release of free arachidonic acid and prostaglandin formation in brain cortex and cerebellum. Prosta-glandins, 11, 773-781.

32. PRICE, C.J. and ROWE, C.E. (1972). Stimulation of the production of unesterified fatty acids in nerve endings of guinea pig brain "in vitro" by noradrenalin and 5-hydroxytryptamine. Biochem. J., 126, 575-585.

33. COLLIER, H.O.J., McDONALD-GIBSON, W.J. and SAEED, S.A. (1974). Morphine and apomorphine stimulate prosta-glandin production by rabbit brain homogenate. Brit. J. Pharmacol., 52, 116P.

34. BURSTEIN, S. and RAZ, A. (1972). Inhibition of prosta-glandin E_2 biosynthesis by Δ^1-tetrahydrocannabinol. Prostaglandins, 2, 369-374.

35. LEE, R.E. (1974). The influence of psychotropic drugs on prostaglandin biosynthesis. Prostaglandins, 5, 63-68.

36. KUNZE, H., BOHN, E. and BAHRKE, G. (1975). Effects of psychotropic drugs on prostaglandin biosynthesis in vitro. J. Pharm. Pharmac. 27, 880-881.

37. WOLFE, L.S., ROSTWOROWSKI, K. and MARION, J. (1976). Endogenous formation of the prostaglandin endoperoxide metabolite, thromboxane B_2, by brain tissue. Biochem. Biophys. Res. Comm. 70, 907-913.

38. ÄNGGÅRD, E., LARSSON, C. and SAMUELSSON, B. (1971). The distribution of 15-hydroxy prostaglandin dehydrogenase and prostaglandin-Δ^{13}-reductase in tissues of the swine. Acta Physiol. Scand., 81, 396-404.

39. NAKANO, J., PRANCAN, A.V. and MOORE, S.E. (1972). Metabolism of prostaglandin E_1 in the cerebral cortex and cerebellum of the dog and rat. Brain Res., 39, 545-548.

40. SIGGINS, G., HOFFER, B. and BLOOM, F. (1971). Prostaglandin-norepinephrine interactions in brain: microelectrophoretic and histochemical correlates. Ann. N.Y. Acad. Sci., 180, 302-323.

41. RAMWELL, P.W. (1964). The action of cerebrospinal fluid on the frog rectus abdominis muscle and other isolated tissue preparations. J. Physiol. (London), 170, 21-38.

42. FELDBERG, W. and GUPTA, K.P. (1972). Pyrogen fever and prostaglandin-like activity in cerebrospinal fluid. J. Physiol. (London), 228, 41-53.

43. FELDBERG, W., GUPTA, K.P., MILTON, A.S. and WENDLANDT, S. (1972). Effect of bacterial pyrogen and antipyretics on prostaglandin activity in cerebro-spinal fluid of unanaesthetized cats. Br. J. Pharmacol., 46, 550-551P.

44. MILTON, A.S. (1973). Prostaglandin E_1 and endotoxin fever, and the effect of aspirin, indomethacin, and 4-acetamidophenol. In: Advances in Biosciences, Ed. S. Bergström, Pergamon Press, Oxford, England, vol. 9, pp. 495-500.

45. LATORRE, E., PATRONO, C., FORTUNA, A. and GROSSI-BELLONI, D. (1974). Role of prostaglandin $F_{2\alpha}$ in human cerebral vasospasm. J. Neurosurg., 41, 293-299.

46. BITO, L.Z. (1972). Accumulation and apparent active transport of prostaglandins by some rabbit tissues in vitro. J. Physiol., 221, 371-387.

47. BITO, L.Z. (1975). Are prostaglandins intracellular, transcellular or extracellular autacoids? Prostaglandins, 9, 851-855.

48. BITO, L.Z. and BAROODY, R. (1974). Concentrative accumulation of ^{3}H-prostaglandins by some rabbit tissues in vitro: the chemical nature of the accumulated ^{3}H-labelled substances. Prostaglandins, 7, 131-140.

49. BITO, L.Z., DAVSON, H. and SALVADOR, E.V. (1976). Inhibition of in vitro concentrative prostaglandin accumulation by prostaglandin analogues and by some inhibitors of organic anion transport. J. Physiol., 256, 257-271.

50. BITO, L.Z., DAVSON, H. and HOLLINGSWORTH, J.R. (1976). Facilitated transport of prostaglandins across the blood-cerebrospinal fluid and blood-brain barriers. J. Physiol., 253, 273-285.

51. HORTON, E.W. (1964). Actions of Prostaglandins E_1, E_2 and E_3 on the central nervous system. Br. J. Pharmac., 22, 189-192.

52. NISTICO', G. and MARLEY, E. (1973). Central effects of Prostaglandin E_1 in adult fowls. Neuropharmacology, 12, 1009-1016.

53. HAUBRICH, D.R., PEREZ-CRUET, J. and REID, W.D. (1973). Prostaglandin E_1 causes sedation and increases 5-hydroxytryptamine turnover in rat brain. Br. J. Pharmac., 48, 80-87.

54. POTTS, W.J. and EAST, P.F. (1971). The effect of Prostaglandin E_2 on conditioned avoidance response performance in rats. Arch. Int. Pharmacodyn., 191, 74-79.

55. YAMAMOTO, Y.L., FEINDEL, W., WOLFE, L.S., KATOH, H. and HODGE, C.P. (1972). Experimental vasoconstriction of cerebral arteries by Prostaglandins. Neurosurg., 37, 385-397.

56. FERREIRA, S.H. and VANE, J.R. (1967). Prostaglandins: their disappearance from and release into the circulation. Nature (Lond.), 216, 868-873.

57. HOLMES, S.W. and HORTON, E.W. (1968). The distribution of tritium labelled Prostaglandin E_1 injected in amounts sufficient to produce central nervous effects in cats and chicks. Br. J. Pharmac., 34, 32-37.

58. HORTON, E.W. and MAIN, I.H.M. (1965). Differences in the effects of Prostaglandin $F_{2\alpha}$, a costituent of cerebral tissue, and Prostaglandin E_1, on conscious cats and chicks. Int. J. Neuropharmac., 4, 65-69.

59. HORTON, E.W. and MAIN, I.H.M. (1967). Further observations on the central nervous actions of Prostaglandins $F_{2\alpha}$ and E_1. Br. J. Pharmacol., 30, 568-581.

60. DAUGHERTY, J.H., MARRAZZI, M.A. and MARRAZZI, A.S. (1974). The effects of Prostaglandin on cerebral cortical evoked potentials. Fed. Proc., 33, 286-293.

61. HORTON, E.W. (1976). Prostaglandins-mediators, modu-
    lators or metabolites? J. Pharm. Pharmac. $\underline{28}$, 389-392.
62. DURU, S. and TÜRKER, R.K. (1969). Effect of Prosta-
    glandin E_1 on the strychnine induced convulsion in
    the mouse. Experientia, $\underline{25}$, 275.
63. GOLDBERG, N.D., HADDOX, N.K., HARTLE, D.K. and
    HADDEN, J.W. (1972). The biological role of cyclic
    3'5' guanosine monophosphate. In: Pharmacology and
    future of man, Proc. 5th Intern. Congr. Pharmacology,
    San Francisco, Calif. vol. 5, Cellular Mechanisms,
    Karger, pp. 146-169.
64. FOLCO, G.C., LONGIAVE, D., BERTI, F., FUMAGALLI, R.
    and PAOLETTI, R. (1976). Prostaglandin E_2 and central
    cyclic nucleotides. In: Advances in Prostaglandins
    and Thromboxane Research, vol. 1, Eds. Samuelsson B.
    and Paoletti R., Raven Press, New York, pp. 305-312.
65. SATTIN, A. (1971). Increase in the content of adeno-
    sine 3'-5'-monophosphate in mouse forebrain during
    seizures and prevention of the increase by methyl-
    xanthines. J. Neurochem., $\underline{18}$, 1087-1096.
66. LLINAS, R. and VOLKIND, R.A. (1972). The olivo
    cerebellar system: functional properties as revealed
    by harmaline induced tremor. Exp. Brain Res., $\underline{18}$,
    69-87.
67. LLINAS, R. and VOLKIND, R.A. (1972). Repetitive
    climbing fiber activation of Purkinje cells in the
    cat cerebellum following administration of harmaline.
    Fed. Proc., $\underline{31}$, 377.
68. MAO, C.C., GUIDOTTI, A. and COSTA, E. (1975). Inhi-
    bition by diazepam of the tremor and the increase of
    cerebellar cGMP content elicited by harmaline.
    Brain Res., $\underline{83}$, 516-519.
69. HENDERSON, G.L. and WOOLLEY, D.E. (1970). Ontogenesis
    of drug-induced tremor in the rat. J. Pharmacol.
    Exptl. Ther. $\underline{175}$, 113-120.
70. SPANO, P.F., KUMAKURA, K., GOVONI, S. and TRABUCCHI,
    M. (1975). Postnatal development and regulation of
    cerebellar cyclic guanosine monophosphate system.
    Pharmacol. Res. Comm., $\underline{7}$, n. 3, 223-237.
71. MAO, C.C., GUIDOTTI, A. and COSTA, E. (1974). The
    regulation of cyclic guanosine monophosphate in rat
    cerebellum: possible involvement of putative amino-
    acid neurotransmitters. Brain Res., $\underline{79}$, 510-514.
72. YANG, A.B., OSTER-GRANITE, M.L., HERNDON, R.M. and
    SNYDER, S.H. (1974). Glutamic acid: selective deple-
    tion by viral induced granule cell loss in Hamster
    cerebellum. Brain Res., $\underline{73}$, 1-13.
73. FERRENDELLI, J.A., CHANG, M.M. and KINSCHERF, D.A.
    (1974). Elevation of cyclic GMP levels in central

nervous system by excitatory and inhibitory amino-acids. J. Neurochem., 22, 535-540.

74. QUESNEY, L.F., GLOOR, P., WOLFE, L.S. and JOZSEF, S. (1976). Effect of $PGF_{2\alpha}$ and 15(5)-15-methyl PGE_2 methyl esther on feline generalized penicillin epilepsy. In: Advances in Prostaglandins and Thromboxane Research, vol. 2, Ed. Samuelsson B. and Paoletti R., Raven Press, New York, pp. 387-390.

75. ZATZ, M. and ROTH, R.H. (1975). Electroconvulsive shock raises prostaglandins F in rat cerebral cortex. Biochem. Pharmacol., 24, 2101-2103.

76. PIPER, P. and VANE, J. (1971). The release of Prostaglandins from lung and other tissues. Ann. N.Y. Acad. Sci., 180, 363-385.

77. LYNEHAM, R.C., McLEOD, J.G., SMITH, I.D., LOW, P.A., SHEARMAN, R.P. and KORDA, A.R. (1973). Convulsions and electroencephalogram abnormalities after intra-amniotic prostaglandin $F_{2\alpha}$. Lancet, ii, 1003-1005.

78. CRAFT, J. (1973). Prostaglandins and convulsions. Lancet, ii, 1389.

79. THIERY, M., AMY, J.J., DE HEMPTINNE, D. and YO LE SIAN (1974). Lancet, i, 918.

80. COCEANI, F. and VITI, A. (1973). Actions of Prostaglandin E_1 on spinal neurones in the frog. Adv. Biosci., 9, 481-487.

81. DUDA, P., HORTON, E.W. and McPHERSON, A. (1968). The effects of Prostaglandins E_1, $F_{1\alpha}$ and $F_{2\alpha}$ on monosynaptic reflexes. J. Physiol., 196, 151-162.

82. HORTON, E.W. and MAIN, I.H.M. (1969). Actions of Prostaglandin E_1 on spinal reflexes in the cat. In: Prostaglandins peptides and amines, Ed. Mantegazza P. and Horton E.W., Academic Press, London, pp. 121-122.

83. SCARAMUZZI, O.E., BAILE, C.A. and MAJER, J. (1971). Prostaglandins and food intake of rats. Experientia, 27, 256-257.

84. BAILE, C.A., BEAN, S.M., SIMPSON, C.W. and JACOBS, H.L. (1971). Feeding effects of hypothalamic injections of prostaglandins. Fed. Proc., 30, 375.

85. BAILE, C.A., SIMPSON, C.W., BEAN, S.M., McLAUGHLIN, C.L. and JACOBS, H.L. (1973). Prostaglandins and food intake of rats: a component of energy balance regulation? Physiol. Behav., 10, 1077-1086.

86. MARTIN, F.H. and BAILE, C.A. (1973). Feeding elicited in sheep by intrahypothalamic injections of PGE_1. Experientia, 29, 306-307.

87. WHISHAW, I.Q. and VEALE, W.L. (1974). Comparison of the effect of Prostaglandin E_1 and Norepinephrine injected into the brain on ingestive behavior in the rat. Pharmacol. Biochem. Behavior, 2, 421-425.

88. KRAGT, C.L. and BERGSTROM, K.K. (1975). Interactions of Prostaglandin E_1 (PGE_1) and LRH on anterior pituitary function. Prostaglandins 10, 833-851.

89. MacLEOD, R. and LEHMEYER, J.E. (1970). Release of pituitary growth hormone by Prostaglandins and dibutyryl Adenosine cyclic 3'-5' monophosphate in the absence of protein synthesis. Proc. Natl. Acad. Sci. U.S.A., 67, 1172-1179.

90. RATNER, A., WILSON, M.C. and PEAKE, G.T. (1973). Antagonism of prostaglandin-promoted pituitary cyclic AMP accumulation and growth hormone secretion in vitro by 7-OXA-13-Prostynoic acid. Prostaglandins, 3, 413-418.

91. de WIED, D., WITTER, A., VERSTEEG, D.H.G. and MULDER, A.H. (1969). Release of ACTH by substances of central nervous system origin. Endocrinology, 85, 561-569.

92. PENG, T.C., SIX, K.M. and MUNSON, P.L. (1970). Effects of Prostaglandin E_1 on the hypothalamo-hypophyseal-adrenocortical Axis in rats. Endocrinology, 86, 202-206.

93. HEDGE, G.A. and HANSON, S.D. (1972). The effects of Prostaglandins on ACTH Secretion. Endocrinology, 91, 925-933.

94. LABHSETWAR, A.P. (1973). Neuroendocrine basis of ovulation in hamsters treated with Prostaglandin $F_{2\alpha}$. Endocrinology, 92, 606-610.

95. RATNER, A., WILSON, M.C., SRIVASTAVA, L. and PEAKE, G.T. (1974). Stimulatory effects of prostaglandin E_1 on rat anterior pituitary cyclic AMP and luteinizing hormone release. Prostaglandins, 5, n. 2, 165-170.

96. SATO, T., HIRONO, M., JYUJO, T., IESAKA, T., TAYA, K. and IGARASHI, M. (1975). Direct action of Prostaglandins on the rat pituitary. Endocrinology, 96, 45-49.

97. HARMS, P.G., OJEDA, S.R. and McCANN, S.M. (1974). Prostaglandin-induced release of pituitary gonadotropins: Central Nervous System and pituitary sites of action. Endocrinology, 84, 1459-1464.

98. SATO, T., JYUJO, T. IESAKA, T., ISHIKAWA, J. and IGARASHI, M. (1974). Follicle stimulating hormone and prolactin release induced by prostaglandins in rat. Prostaglandins, 5, n. 5, 483-490.

99. OJEDA, S.R., HARMS, P.G. and McCANN, S.M. (1974). Central effect of Prostaglandin E_1 (PGE_1) on Prolactin release. Endocrinology, 85, 613-618.

100. BROWN, M.R. and HEDGE, G.A. (1974). In vivo effects of prostaglandins on TRH-induced TSH secretion. Endocrinology, 85, 1392-1397.

101. BATTA, S., FIORINDO, R.P., JUSTO, G., MOTTA, M., SIMONOVIC, I., ZANISI, M. and MARTINI, L. (1974). Role of cholinergic mechanism and of Prostaglandins

in the control of LH and FSH secretion. In: Neuroendocrine control of fertility, Int. Symp., Simla,
Ed. Anand Kumar T.C., pp. 155-168.

102. DROUIN, J. and LABRIE, F. (1976). Specificity of the
stimulatory effect of prostaglandins on hormone
release in rat anterior pituitary cells in culture.
Prostaglandins, 11, 355-364.

103. HEDGE, G.A. (1976). Hypothalamic and pituitary effects of prostaglandins on ACTH secretion. Prostaglandins, 11, 293-301.

104. DROUIN, J., FERLAND, L., BERNARD, J. and LABRIE, F.
(1976). Site of the in vivo stimulatory effect of
Prostaglandins on LH release. Prostaglandins, 11,
367-375.

105. SATO, T., JYUJO, T., HIRONO, M. and IESAKA, T. (1975).
Effects of Indomethacin, an inhibitor of prostaglandin synthesis, on the hypothalamic-pituitary
system in rats. J. Endocr. 64, 395-396.

106. THOMPSON, M.E. and HEDGE, G.A. (1976). Suppression
of Thyrotropic hormone secretion by prostaglandin
synthesis inhibitors. Endocrinology, 98, n. 3,
787-793.

107. BORGEAT, P., LABRIE, F. and GARNEAU, P. (1975).
Characteristics of action of prostaglandins on cyclic AMP accumulation in rat anterior pituitary
gland. Can. J. Biochem., 53, 455.

108. ZOR, U., KANEKO, T., SCHNEIDER, H.P.G., McCANN, S.M.,
LOWE, I.P., BLOOM, S., BORLAND, B. and FIELD, J.B.
(1969). Stimulation of anterior pituitary adenyl
cyclase activity and adenosine 3'-5' cyclic phosphate
by hypothalamic extract and prostaglandin E_1. Proc.
Nat. Acad. Sci., 63, 918-925.

109. BORGEAT, P., LABRIE, F., DROUIN, J., BELANGER, A.,
IMMER, H., SESTANJ, K., NELSON, V., GOTZ, M.,
SCHALLY, A.V., COY, D.H. and COY, E.J. (1971).
Inhibition of adenosine 3'-5' monophosphate accumu-
lation in anterior pituitary gland in vitro by
growth hormone-release inhibiting hormone. Biochem.
Biophys. Res. Comm., 56, 1052-1059.

110. BORGEAT, P., CHAVANCY, G., DUPONT, A., LABRIE, F.,
ARIMURA, A. and SCHALLY, A.V. (1972). Stimulation
of adenosine 3'-5' monophosphate accumulation in
anterior pituitary gland in vitro by synthetic luteinizing hormone-releasing-hormone/follicle stimulating hormone-releasing hormone (LH-RH/FSH-RH). Proc.
Natl. Acad. Sci. U.S.A., 69, 2677-2681.

111. KANEKO, T., SAITO, S., OKA, H., ODA, T. and
YANAIHARA, N. (1973). Effects of synthetic LH-RH
and its analogs on rat anterior pituitary cyclic AMP
and LH and FSH release. Metabolism, 22, 77-78.

112. MAKINO, T. (1973). Study on the intracellular mecha-
     nism of LH release in the anterior pituitary gland.
     Am. J. Obstet. Gynaecol., 115, 606-614.
113. BORGEAT, P., LABRIE, F., COTE, J., RUEL, F., SCHALLY,
     A.V., COY, D.H., COY, E.J. and YANAIHARA, N. (1974).
     Parallel stimulation of cyclic AMP accumulation and
     LH and FSH release by analogs of LH-RH in vitro.
     J. Mol. Cell Endocrinol., 1, 7-20.
114. RATNER, A. (1970). Stimulation of luteinizing hormone
     release in vitro by dibutyryl-cyclic AMP and theo-
     phylline. Life Sci., 9, 1221-1226.
115. STEINER, A.L., PEAKE, G.T., UTIGER, R.D., KARL, I.E.
     and KIPNIS, D.M. (1970). Hypothalamic stimulation
     of growth hormone and thyreotropin release in vitro
     and pituitary 3'-5' adenosine cyclic monophosphate.
     Endocr., 86, 1354-1360.
116. TAL, E., SZABO, M. and BURKE, G. (1974). TRH and
     prostaglandin action on rat anterior pituitary:
     dissociation between cyclic AMP levels and TSH
     release. Prostaglandins, 5, n. 2, 175-182.
117. de WIED, D. and de JONG, W. (1974). Drug effects
     and hypothalamic-anterior pituitary function. Ann.
     Rev. Pharmacol., 14, 389-412.
118. HINMAN, J.W. (1972). Prostaglandins. Ann. Rev.
     Pharmacol., 41, 161-178.
119. HARMS, P.G., OJEDA, S.R. and McCANN, S.M. (1976).
     Failure of monoaminergic and cholinergic receptor
     blockers to prevent prostaglandin E_2-induced lutei-
     nizing hormone release. Endocrinology, 98, 318-323.
120. VILHART, H. and HEDQVIST, P. (1970). A possible
     role of prostaglandin E_2 in the regulation of vaso-
     pressin secretion in rats. Life Sci., 9, 825-830.
121. COBO, E.C., RODRIGUEZ, A. and VILLAMIZAR, M. (1974).
     Milk enjecting activity induced by prostaglandin $F_{2\alpha}$.
     Amer. J. Obst. Gynaecol., 118, 831-836.
122. KAPLAN, H.R., GREGA, G.J., SHERMAN, G.P. and BUCKLEY,
     J.P. (1969). Central and reflexogenic cardiovascular
     actions of prostaglandins E_1. Intern. J. Neuro-
     pharmacol., 8, 15-24.
123. LAVERY, H.A., LOWE, R.D. and SCROOP, G.C. (1970).
     Cardiovascular effects of prostaglandins mediated
     by the central nervous system of the dog. Br. J.
     Pharmac., 39, 511-519.
124. RINCHUSE, D.J. and DEUBEN, R.R. (1976). Central
     mediated pressor effect by prostaglandins in the rat.
     Prostaglandins, 11, n. 3, 523-530.
125. McQUEEN, D.S. and UNGAR, A. (1969). The modification
     by Prostaglandin E_1 of central nervous interactions
     between respiratory and cardio-inhibitory pathways.

In: Prostaglandins, Peptides and Amides, Eds. Mantegazza P. and Horton E.W., Academic Press, New York,
pp. 123-124.

126. CARLSON, L.A., EKELUND, L.G. and ORÖ, L. (1969).
Circulatory and respiratory effects of different
doses of prostaglandin E_1 in man. Acta Physiol.
Scand., 75, 161-169.

127. COCEANI, F., DREIFUSS, J.J., PUGLISI, L. and WOLFE,
L.S. (1969). Prostaglandins and membrane function.
In: Prostaglandins, Peptides and Amides, Eds. Mantegazza P. and Horton E.W., Academic Press, New York,
pp. 73-84.

128. PENNINK, M., WHITE, R.P., CROCKARELL, J.R. and
ROBERTSON, J.T. (1972). Role of prostaglandin $F_{2\alpha}$
in the genesis of experimental cerebral vasospasm.
Angiografic study in dogs. J. Neurosurg., 37, 398-406.

129. DENTON, I.C., WHITE, R.P. and ROBERTSON, J.T. (1972).
The effects of prostaglandins E_1, A_1 and $F_{2\alpha}$ on the
cerebral circulation of dogs and monkeys. J. Neurosurg., 36, 34-42.

130. ROSENBLUM, W.I. (1975). Effects of prostaglandins
on cerebral blood vessels: interaction with vasoactive amines. Neurology, 25, 1169-1171.

131. WELCH, K.M.A., SPIRA, P.J., KNOWLES, L. and LANCE,
J.W. (1974). Effects of prostaglandins on the
internal and external carotid blood flow in the
monkey. Neurology, 24, 705-710.

132. AVANZINO, G.L., BRADLEY, P.B. and WOLSTENCROFT, J.H.
(1966). Actions of Prostaglandins E_1, E_2 and $F_{2\alpha}$
on brain stem neurones. Br. J. Pharmac., 27, 157-163.

133. HOFFER, B., SIGGINS, G. and BLOOM, F. (1970). Possible
cyclic AMP mediated adrenergic synapse to rat cerebellar Purkinje cells: combined structural physiological and pharmacological analysis. In: Role of
cyclic AMP in cell function, Ed. Greengard P. and
Costa E., Advances in Biochemical Psychopharmacology,
vol. 3, Raven Press, New York, pp. 349-370.

134. SIGGINS, G., HOFFER, B. and BLOOM, F. (1971). Studies
on norepinephrine containing afferents to Purkinje
cells of rat cerebellum: III. Evidence for mediation
of norepinephrine effects by cyclic 3'-5' adenosine
monophosphate. Brain Res., 25, 535-553.

135. POULAIN, P. and CARETTE, B. (1974). Iontophoresis
of Prostaglandins on hypothalamic neurones. Brain
Res., 79, 311-314.

136. TRABER, J., REISER, G., FISCHER, K. and HAMPRECHT,
B. (1975). Measurements of adenosine 3'-5' monophosphate and membrane potential in neuroblastoma
x glioma hybrid cells: opiates and adrenergic agonists cause effects opposite to those of Prostaglandin
E. Febs Letters, 52, n. 2, 327-332.

137. WELLMANN, W. and SCHWABE, U. (1973). Effects of prostaglandins E_1, E_2 and $F_{2\alpha}$ on cyclic AMP levels in brain in vivo. Brain Res., $\underline{59}$, 371-378.

138. BERTI, F., TRABUCCHI, M., BERNAREGGI, V. and FUMA-GALLI, R. (1972). The effects of prostaglandins on cyclic AMP formation in cerebral cortex of different mammalian species. Pharmac. Res. Comm., $\underline{4}$, 253-259.

139. BERTI, F., TRABUCCHI, M., BERNAREGGI, V. and FUMA-GALLI, R. (1972). Prostaglandins on cyclic AMP formation in cerebral cortex of different mammalian species. Adv. in Biosci., $\underline{9}$, 475-480.

140. GILMAN, A.G. and NIRENBERG, M. (1971). Regulation of adenosine 3'-5' cyclic monophosphate metabolism in cultured neuroblastoma cells. Nature, $\underline{234}$, 356-358.

141. GILMAN, A.G. and SCHRIER, B.K. (1972). Adenosine cyclic 3'-5' monophosphate in fetal rat brain cell cultures. I. Effect of cathecolamines. Molecular Pharmacology, $\underline{8}$, 410-416.

142. DISMUKES, K. and DALY, J.W. (1975). Accumulation of adenosine 3'-5' monophosphate in rat brain slices: effects of prostaglandins. Life Sci., $\underline{17}$, 199-210.

143. FUMAGALLI, R., BERTI, F., FOLCO, G.C. and OMINI, C. unpublished observations.

144. PAOLETTI, R., BERTI, F., FUMAGALLI, R. and FOLCO, G.C. (1974). Some interrelations between prosta-glandins and cyclic nucleotides. In: Future Trends in Inflammation, Eds. Velo G.P., Willoughby D.A. and Giroud J.P., Piccin Medical Books, pp. 11-18.

145. PERKINS, J.P. (1973). Adenyl Cyclase. Adv. Cyclic Nucl. Res., $\underline{3}$, 2-64.

146. GILMAN, A.G. and NIRENBERG, H. (1971). Effect of catecholamines on the adenosine 3'-5' cyclic mono-phosphate concentrations of clonal satellite cells of neurons. Proc. Nat. Acad. Sci. U.S.A., $\underline{68}$, 2165-2168.

147. SAHU, S.K. and PRASAD, K.N. (1975). Effect of neuro-transmitters and prostaglandin E_1 on cyclic AMP levels in various clones of neuroblastoma cells in culture. J. Neurochem., $\underline{24}$, 1267-1269.

148. GILMAN, A.G. (1972). Regulation of cyclic AMP meta-bolism in cultured cells of the nervous system. Adv. Cyclic Nucl. Res., $\underline{1}$, 389-410.

149. RYAN, W.L. and HEIDRICK, M.L. (1968). Inhibition of cell growth in vitro by adenosine 3'-5' monophosphate. Science, $\underline{162}$, 1484-1485.

150. HEIDRICK, M.L. and RYAN, W.L. (1970). Cyclic nucleo-tides on cell growth in vitro. Cancer Res., $\underline{30}$, 376-378.

151. PASTAN, I., JOHNSON, G.S., OTTEN, J., PEERY, C.V. and D'ARMIENTO, M. (1971). Role of cyclic AMP in the abnormal growth and morphology of transformed fibroblasts. Fed. Proc., 30, 1047.

152. PRASAD, K.N. and HSIE, A.W. (1971). Morphologic differentiation of mouse neuroblastoma cells induced in vitro by dibutyryl adenosine 3'-5' cyclic monophosphate. Nature New Biology, 233, 141-142.

153. PRASAD, K.N. (1972). Morphological differentiation induced by Prostaglandin in mouse neuroblastoma cells in culture. Nature New Biology, 236, 49-52.

154. ADOLPHE, M., GIROUD, J.P., TIMSIT, J., FONTAGNE, J. and LECHAT, P. (1974). Action de la prostaglandine A_2 sur la proliferation et la differenciation morphologique d'une lignée cellulaire de neuroblastome murin. Compte. Rendus Seances Soc. Biologie, 168, 694-698.

155. EDSTRÖM, A., KANJE, M. and VALUM, E. (1974). Effects of dibutyryl cyclic AMP and prostaglandin E_1 on cultured human glioma cells. Exp. Cell Res., 85, 217-223.

156. THOMAS, D.R., PHILPOTT, G.W. and JAFFE, B.M. (1974). The relationship between concentration of prostaglandin E and rates of cell replications. Exp. Cell Res., 84, 40-46.

157. HAMMARSTRÖM, S., SAMUELSSON, B. and BJURSELL, G. (1973). Prostaglandin levels in normal and transformed baby-hamster-kidney fibroblasts. Nature New Biol., 243, 50-51.

158. LEICHTLING, B.H., DROTAR, A.M., ORTMANN, R. and PERKINS, J.P. (1976). Growth of astrocytoma cells in the presence of prostaglandin E_1: effect on the regulation of cyclic AMP metabolism. J. Cyclic Nucl. Res., 2, 89-98.

159. MAGANIELLO, V. and VAUGHAN, M. (1972). Prostaglandin E_1 effects on adenosine 3'-5' cyclic monophosphate concentration and phosphodiesterase activity in fibroblasts. Proc. Nat. Acad. Sci. U.S.A., 69, 269-273.

160. JAFFE, B.M. (1974). Prostaglandins and cancer: un update. Prostaglandins, 6, 453-461.

161. CHLAPOWSKI, F.J., KELLY, L.A. and BUTCHER, R.W. (1976). Cyclic nucleotides in cultured cells. Adv. Cyclic Nucl. Res., 6, 245-338.

162. McMANUS, J.P. and WHITFIELD, J.F. (1975). Cyclic AMP, prostaglandins, and the control of cell proliferation. Prostaglandins, 6, 245-338.

163. JACOBSON, H.I. (1974). Oncolytic action of Prostaglandins. Cancer Chemotherapy Reports, Part 1, 58, 503-511.

164. STEIN-WERBLOWSKY, R. (1974). The effect of prosta-
     glandins on tumor implantation. Experientia, $\underline{30}$,
     957-959.
165. EISENBARTH, G.S., WELLMAN, D.K. and LEBOVITZ, H.E.
     (1974). Prostaglandin A_1 inhibition of chondrosarcoma
     growth. Biochem. Biophys. Res. Comm., $\underline{60}$, 1302-1308.
166. COLLIER, H.O.J. and ROY, A.C. (1974). Morphine-like
     drugs inhibit the stimulation by E prostaglandins
     of cyclic AMP formation by rat brain homogenate.
     Nature, $\underline{248}$, 24-27.
167. TRABER, J., FISCHER, K., LATZIN, S. and HAMPRECHT,
     B. (1975). Morphine antagonizes action of prosta-
     glandin in neuroblastoma and neuroblastoma x glioma
     hybrid cell. Nature, $\underline{253}$, 120-122.
168. TRABER, J., FISCHER, K., LATZIN, S. and HAMPRECHT,
     B. (1974). Morphine antagonizes the action of prosta-
     glandin in neuroblastoma cells but not of prosta-
     glandin and noradrenaline in glioma and glioma x
     fibroblast hybrid cells. Febs Letters, $\underline{49}$, 260-263.
169. TRABER, J., FISCHER, K., BUCHEN, C. and HAMPRECHT,
     B. (1975). Muscarinic response to acetylcholine in
     neuroblastoma x glioma hybrid cells. Nature, $\underline{255}$,
     558-560.
170. TRABER, J. and HAMPRECHT, B. (1976). Action of neuro-
     hormones and opiates on neuroblastoma x glioma hybrid
     cells in culture. In: Advances in Prostaglandin and
     Thromboxane Research, vol. 1, Eds. Samuelsson B. and
     Paoletti R., Raven Press, New York, pp. 337-340.
171. IVERSEN, L. and DINGLEDINE, R. (1976). Enkephalin:
     the latest instalment. Nature, $\underline{262}$, 738-739.
172. COLLIER, H.O.J. and ROY, A.C. (1974). Hypothesis
     inhibition of E prostaglandin-sensitive adenyl cy-
     clase as the mechanism of morphine analgesia.
     Prostaglandins, $\underline{7}$, 361-376.
173. COLLIER, H.O.J., McDONALD-GIBSON, W.J. and SAEED,
     S.A. (1974). Apomorphine and morphine stimulate
     prostaglandin biosynthesis. Nature, $\underline{252}$, 56-58.
174. ROY, A.C. and COLLIER, H.O.J. (1975). Prostaglandins,
     cyclic AMP and the biochemical mechanism of opiate
     agonist action. Life Sci., $\underline{16}$, 1857-1862.
175. COLLIER, H.O.C., FRANCIS, D.L., McDONALD-GIBSON,
     W.J., ROY, A.C. and SAEED, S.A. (1975). Prostaglan-
     dins, cyclic AMP and the mechanism of opiate depen-
     dence. Life Sci., $\underline{17}$, 85-90.
176. FERRI, S., SANTAGOSTINO, A., BRAGA, P.C. and GALA-
     TULAS, I. (1974). Decreased antinociceptive effect
     of morphine in rats treated intraventricularly with
     Prostaglandin E_1. Psychopharmacologia (Berlin) $\underline{39}$,
     231-235.

177. BHATTACHARYA, S.K., REDDY, P.K.S.P., DEBNATH, P.K. and SANYAL, A.K. (1975). Potentiation of antinociceptive action of morphine by prostaglandin E_1 in albino rats. Clin. Exp. Pharmacol. Physiol., 2, 353-357.

178. MALMSTEN, C., GRANSTRÖM, E. and SAMUELSSON, B. (1976). Cyclic AMP inhibits synthesis of prostaglandins endoperoxide (PGG_2) in human platelets. Biochem. Biophys, Res. Comm., 68, n. 2, 569-576.

179. BERGSTRÖM, S., FARNEBO, L.A. and FUXE, K. (1973). Effect of prostaglandin E_2 on central and peripheral catecholamine neurones. Eur. J. Pharmacol., 21, 362-368.

180. HEDQVIST, P. (1973). Autonomic neurotransmission. In: The Prostaglandins, vol. 1, Ed. Ramwell P.W., Plenum Press, New York, pp. 101-131.

181. DUBOCOVICH, M.L. and LANGER, S.Z. (1975). Evidence against a physiological role of prostaglandins in the regulation of Noradrenaline release in the cat spleen. J. Physiol., 251, 737-762.

PROSTAGLANDINS AS MODULATORS OF AUTONOMIC NEUROEFFECTOR TRANSMISSION

Per Hedqvist

Department of Physiology, Karolinska Institutet

S-10401 Stockholm 60, Sweden

The concept of chemical transmission at autonomic synapses, originally proposed by Elliot (1), and firstly verified by Loewi (2), evidently represents the basis for our present knowledge about how nerve impulses are conveyed to effector cells. For many years it was believed that the only factor determining the release of transmitter, and hence the ensuing effector response, was the number of nerve impulses travelling down the axon. However, research over the past decade has disclosed that the transmitter quantum released per nerve impulse is not fixed, but subject to facilitation and depression processes. Moreover, the nerve terminal membrane seems to possess receptors through which a number of different substances, naturally occurring in the body, may act to increase or decrease the amount of transmitter released in response to forthcoming nerve impulses.

There is a growing body of evidence indicating that prostaglandins (PGs) represent one group of compounds which may be of importance as modulators of autonomic transmission. Thus, PGs are naturally occurring substances, which are released from autonomically innervated tissues in response to a variety of stimuli, including nerve impulses; PGs influence both the release of transmitter and effector responses to the secreted transmitter; and inhibitors of PG synthesis usually produce effects opposite to those of the PGs. The aim of this paper is to draw the attention to some basic aspects of PG action on autonomic neuroeffector transmission, but it makes no claim to be a comprehensive review of the literature within this field.

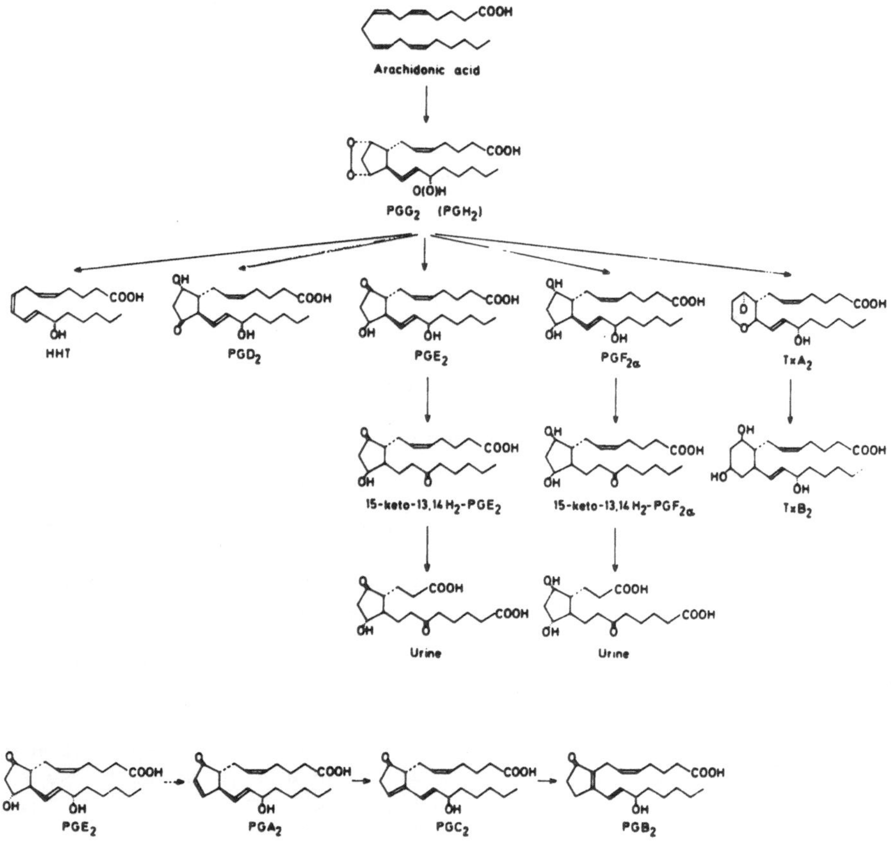

FIG.1. Pathways in prostaglandin biosynthesis and metabolism.

PROSTAGLANDIN FORMATION AND METABOLISM

It is well established that PGs are formed from certain C-20 unsaturated fatty acids of which arachidonic acid (5,8,11,14-eicosa-tetraenoic acid) is the most prevalent in mammalian tissues. As shown in Figure 1, arachidonic acid is attacked by a membrane bound multienzyme complex, termed PG synthetase to form the two endoper-oxides, PGG_2 and PGH_2. These are rapidly transformed into the primary PGs, PGE_2 and $PGF_{2\alpha}$, but also into PGD_2, HHT and Thromboxane A_2 (cf.3). At least theoretically, PGE_2 may be transformed into PGA_2 and subsequently into PGC_2 and PGB_2 (cf.4). PGs are metabolized to a variety of products. The principal sequence of reactions ultimately leading to the excretion of dioic acids in the urine is attack by 15-hydroxy-dehydrogenase and Δ^{13}-reductase, followed by β-oxdation and ω-oxidation (cf.4). It should be recalled that two other PG-precursors (8,11,14-eicosatrienoic acid and 5,8,11,14,17-eicosapentaenoic acid) give rise to PG-products differing from those formed from arachidonic acid only in the number of double bonds in the side chains.

PROSTAGLANDIN RELEASE FROM AUTONOMICALLY INNERVATED TISSUES

PGs are not stored to any significant extent in tissues. Rather a variety of different stimuli seem to trigger the immediate formation and release of PGs. Thus, stimulation of the adrenergic nerves to the dog and cat spleen causes a frequency dependent release of PGs, principally PGE_2 and to a lesser extent $PGF_{2\alpha}$ (5,6,7). Injection of adrenaline or noradrenaline (NA) also produce release of these PGs. The spleen is not unique in having a basal PG release which can be greatly increased by nerve stimulation or injection of catecholamines. Closely similar results have been obtained with a number of adrenergically innervated tissues, such as the rabbit heart (8), dog and rabbit kidney (9,10), rat and dog adipose tissue (11,12) and rat and guinea pig vas deferens (13,14). The important information provided by these studies is that adrenergic nerve impulses release a PGE, and PGEs are very effective in inhibiting the release of NA in response to forthcoming nerve impulses. In fact, the concentrations of exogenous PGEs required to significantly depress the release of NA are of the same order of magnitude as those that can be released from stimulated tissues. In this context it is worth considering that primary PGs are rapidly and effectively by one single passage through the lung circulation (6). Control of adrenergic transmission must therefore be executed by locally released rather than circulating PGs.

Several studies on spleen and kidney have revealed that α-adrenoceptor blockers abolish the release of PGs and the smooth muscle contraction elicited by nerve stimulation or catecholamines (5,6,7,15). On the other hand, catecholamine-induced release of PGs from dog spleen and rabbit heart is not greatly impaired by surgical or chemical degeneration of the nerves (7,16). These observation indicate that contraction of the effector cell is an essential trigger mechanism for PG synthesis, and have been regarded as evidence that the source of PGs released by nerve stimulation is strictly extraneuronal. However, knowing that even in densely innervated organs the nerves make up only a fraction of a per cent of the total tissue mass, it is quite concievable that a small but nevertheless important fraction of the PGs overflowing from stimulated tissues is neuronal in origin. The fact that PGs can be extracted from several types of nerve tissue would seem to support such a view.

PG formation and release readily occurs also in cholinergically innervated tissues. In the rat stomach, guinea pig ileum, rabbit heart and bovine iris there is a low basal PG release, which is markedly increased by nerve stimulation or administration of acetylcholine (ACh)(17,18,19,20). This stimulated PG release, which is blocked by atropine, seems to consist of PGE_1 and/or PGE_2 with little or no $PGF_{2\alpha}$ or other PGs.

Knowing that the release of PGs from an isolated tissue has a definite tendency to increase with time, possibly as a result of tissue deterioration, it is not unlikely that already the surgical trauma in the beginning of the experiment induces an artificial or unphysiologically high basal PG release from the tissue. On the other hand, the increased release induced by nerve stimulation must be considered relevant, which suggests that nerve impulses cause release of PGs in autonomically innervated organs even under strict in vivo conditions. However, before the precise function of the PGs in autonomic transmission can be established, it is essential to know more about the pattern of PGs released by nerve impulses, and to what extent drugs which affect autonomic transmission also influence the release of PGs.

EFFECTS OF PROSTAGLANDINS AND PROSTAGLANDIN SYNTHESIS INHIBITORS ON AUTONOMIC NEUROEFFECTOR TRANSMISSION

Adrenergic Transmission

Over the past decade much interest has been devoted to the action of PGs on the cardio-vascular system. Numerous reports have shown that PGEs are potent vasodepressors, although usually being quite modest in their direct action on the heart. Besides causing a direct vasodilatation, the PGEs affect the adrenergic neuroeffector transmission at both pre -and post-junctional levels in a great number of vascular beds and smooth muscle tissues.

Topical application of PGE_1 to mesenteric and cremasteric vessels in the rat results in a reduced responsiveness to NA, which persists long after the direct vasodilating effect has vanished (21). On the other hand, PGE_1 has been reported to potentiate responses to NA in the rabbit aorta and mesenteric artery in vitro (22), as well as vascular responses to NA in dog uterus in situ and in perfused rat kidney (23,24).

Vasoconstrictor responses to nerve stimulation are inhibited by PGE_1 and/or PGE_2 in a number of tissues, such as the cat spleen (25), cat and dog hindleg (26,27) and dog and rabbit kidney (28,29) Since in most of these tissues responses to NA are less depressed, unchanged or even enhanced, indirect evidence is provided for PGEs inhibiting the release of transmitter from the adrenergic nerve terminals in these tissues. Such an effect has been demonstrated in the cat spleen (25), rabbit heart, kidney and ear (8,30,31), isolated vessels from cat and man (32,33), guinea pig vas deferens

(34), and rat and rabbit iris (35,36). PGEs have been postulated
to have the opposite effect in some tissues, notably the rat kidney
and dog hindpaw (24,37). However, in these cases the evidence was
indirect and based on differential effects of PGEs on responses to
nerve stimulation and NA, and transmitter release, which could have
given the proper answer, was not measured.

It is apparent that PGEs may depress transmitter release from
adrenergic nerve terminals, and hence the effector response to
nerve impulses. This effect has been demonstrated in so many tissues from different animal species and from man that it seems jus-
tified to conclude that this is an essential and common action of
the PGEs. The argument is somewhat weakened because all data so far
obtained derive from experiments with isolated tissues and organs
perfused with salinic media. Measurement of transmitter release in
organs perfused with blood should be of great value.

The interaction of PGEs with locally released NA at the level
of the effector cell is more complex. Both inhibitory and stimulant
effects have been noted. Although it appears that low doses of PGEs,
which inhibit NA release, often act as postjunctional inhibitors,
there are several exceptions from this rule. It is concievable,
therefore, that PGEs, in addition to inhibiting NA release, might
serve also another function in some tissues, that of maintaining
the reactivity of the effector cell to catecholamines ((38).

Relative to the PGEs less attention has been paid to the capacity of other PGs to influence the adrenergic neuroeffector transmission. A mainly stimulant action on effector responses to nerve
stimulation occurs with $PGF_{2\alpha}$ (39), whereas PGAs appear to cause
variable effects; both inhibition and enhancement have been reported (27,40,41,42). However, in those cases where transmitter release
was measured (heart, kidney, vas deferens) $PGF_{2\alpha}$ and PGA_2 had either
no effect or caused inhibition (8,41,42). It appears therefore,
that enhancement of effector responses to adrenergic nerve impulses
by PGF and PGA is mostly, if not wholly, a postjunctional phenomenon.
PGD_2, isomer to PGE_2, and the endoperoxides, PGG_2 and PGH_2 inhibit
the stimulated release of NA in the guinea pig vas deferens, but
they all are much less effective than PGE_2 (42).

Basically three compounds with capacity to inhibit PG synthesis have been tested for effects on adrenergic neuroeffector transmission. Two of them, indomethacin and meclofenamic acid, belong
to the aspirin family, whereas the third is a substrate analogue,
the acetylenic derivative of arachidonic acid, 5,8,11,14-eicosa-
tetraynoic acid (ETA). Because PGEs, which represent the bulk of
PGs released by nerve stimulation, inhibit the release of NA from
adrenergic nerves, PG synthesis inhibitors are expected to enhance
the release of NA. Such an effect has been uniformly demonstrated
in the rabbit heart, guinea pig vas deferens, rabbit kidney, and
isolated vessels from cat and man (8,30,32,33,43), while in the cat
spleen ETA enhanced (44) and indomethacin and meclofenamic acid had
no effect (45,46). The effector responses to nerve stimulation are

are also enhanced by PG synthesis inhibitors in most of these tissues, although in the cat spleen and rabbit kidney indomethacin has been reported either to cause enhancement or to leave the responses unaffected (15,24,30,45,46).

In addition to the abovementioned in vitro and in situ experiments, there are several in vivo studies pertaining to the action of PG synthesis inhibitors on adrenergic neurotransmission. Thus, indomethacin increases urinary excretion of NA in rats, the animals being either cold-stressed or kept at room temperature (47,48). Furthermore, oral administration of indomethacin increases NA turnover rate in a number of rat tissues, such as heart, spleen, submandibular gland and adipose tissue (49).

Summarizing, the bulk of available evidence indicates that inhibitors of PG synthesis increase the release of NA per nerve impulse as a consequence of inhibition of local PGE formation. This view is supported by observations that PG synthesis inhibitors are unable to enhance NA release in the presence of exogenous PGEs, and that the NA release mechanism is hyperreactive to the inhibitory effect of exogenous PGEs when local PG synthesis is blocked (30,44, 50). However, conflicting results have been published and further investigations are required before the concept of a PGE-mediated feed-back control of adrenergic transmission can be established as physiologically relevant.

Cholinergic Transmission

PG action on autonomic cholinergic neuroeffector transmission has been studied mainly in the heart and gastro-intestinal tract. In the perfused rabbit heart and in the guinea pig heart in situ, PGE_1 and/or PGE_2 inhibit the negative chronotropic responses to vagal nerve stimulation, but not to ACh (51,52,53). Since PGE release from the heart can be induced by vagal nerve stimulation (19), it has been suggested that PGEs control the cholinergic neuroeffector transmission in the heart by restriction of the release of ACh (8), that is in a way similar to that proposed from the adrenergic system. However, at least the prejunctional target for such an action has been questioned, because it has been reported that PGEs block the chronotropic responses to both vagal nerve stimulation and ACh in guinea pig atria (54), and that they have no effect on negative chronotropic responses to electrical stimulation of the rabbit sino-atrial node preparation (55). Further careful characterization of possible pre -and post-junctional actions of PGEs on cholinergic transmission in this organ are therefore required.

In contrast to the heart, gastro-intestinal smooth muscle is usually stimulated by PGs, and it has been suggested that a continous and low PG synthesis serves the function of maintaining the inherent tone of the gut (56). In addition, it is well established that PGEs enhance the contraction response to cholinergic nerve stimulation in guinea pig and rabbit ileum (52,57,58), while the effects on contractions elicited by ACh are less clear; both enhanced and unchanged responses have been reported (57,58). On the other hand, contraction responses to nerve stimulation are markedly reduced or even abolished by indomethacin, and partially or completely restored by subsequent administration of subspasmogenic doses of PGE_1 and PGE_2 (52,58,59). The contraction response to ACh seems to be little affected by indomethacin (58). Several attempts have been made to disclose a prejunctional effect of PGEs and PG synthesis inhibitors by measuring the release of ACh from intestinal smooth muscle. However, at the present time there seems to be no indication that PGEs enhance and indomethacin inhibits ACh release induced by nerve stimulation (18,52,57). Taken together all these observations indicate that PGEs stimulate the cholinergic neuroeffector transmission in the intestinal tract, and imply that the effect is mostly, if not wholly, a postjunctional phenomenon.

In the sphincter muscle of the bovine iris PGEs and $PGF_{2\alpha}$ enhance contractions induced by cholinergic nerve stimulation (60). On the other hand, PG synthesis inhibitors relax the muscle and inhibit the contraction response to nerve stimulation, and subsequent administration of subspasmogenic doses of PGEs partially restores the contraction response. The striking similarities in the action of PGs and PG synthesis inhibitors in this tissue and the intestinal tract invites the speculation that this is a principle operating in cholinergically innervated smooth muscle organs in general.

CONCLUSION

PGEs are potent inhibitors of neurally induced transmitter release from adrenergic nerve terminals in a great number of tissues. PGEs are also released from stimulated adrenergically innervated tissues, and there is considerable evidence that inhibition of local PGE production is associated with increased NA release by nerve activity. It appears likely, therefore, that adrenergic transmission is controlled by a local PGE-mediated feed-back mechanism, which operates through restriction of NA release.

Apparently PGEs do not operate through the same mechanism in cholinergic transmission. Possibly only with the exception of the heart and gastric secretion, PGEs seem to enhance cholinergic transmission, and to do this through a postjunctional action. However, the fact that substantial amounts of PGs are released from stimulated tissues, and that PGEs and PG synthesis inhibitors produce opposite effects, merit the assumption that this system also is subject to some control by locally formed PGEs.

REFERENCES

1. Elliot, T.R., (1905): J. Physiol. (Lond.) 32:401-467.
2. Loewi, O., (1921): Arch. Ges. Physiol. 189:239-242.
3. Samuelsson, B., (1976): In "Advances in Prostaglandin and Thromboxane Research", eds B. Samuelsson and R. Paoletti, pp.1-6. Raven Press, New York.
4. Coceani, F. and Pace-Asciak, C.R., (1976): In "Prostaglandins: Physiological, Pharmacological and Pathological Aspects", ed. S.M.M. Karim, pp.1-36. MTP, Lancaster.
5. Davies, B.N., Horton, E.W. and Withrington, P.G., (1967): J. Physiol. (Lond.) 188:38-39P.
6. Ferreira, S.H. and Vane, J.R., (1967): Nature (Lond.) 216:868-873.
7. Gilmore, N., Vane, J.R. and Wyllie, J.H., (1968): Nature (Lond.) 218:1135-1140.
8. Wennmalm, Å., (1971): Acta physiol. scand. Suppl.365:1-36.
9. Dunham, E.W. and Zimmermann, B.G., (1970): Am. J. Physiol. 219: 1279-1285.
10. Davis, H.A. and Horton, E.W., (1972): Br. J. Pharmacol. 46:658-675.
11. Shaw, J.E. and Ramwell, P.W., (1968): J. Biol. Chem. 243:1498-1503.
12. Fredholm, B.B., Rosell, S. and Strandberg, K., (1970): Acta physiol. scand. 79:18A-19A.
13. Swedin, G., (1971): Acta physiol. scand. Suppl.369:1-34.
14. Hedqvist, P. and Euler, U.S.v., (1972): Nature New Biol. 236: 113-115.
15. Needleman, P., Douglas, J.R., Jakschik, B., Stocklein, P.B. and Johnson, E.M., (1974): J. Pharmacol. exp. Ther. 188:453-460.
16. Junstad, M. and Wennmalm, Å., (1973): Acta physiol. scand. 87: 573-574.
17. Bennett, A., Friedmann, C.A. and Vane J.R., (1967): Nature (Lond.) 216:873-876.
18. Botting, J.H. and Salzman, R., (1974): Br. J. Pharmacol. 50: 119-124.
19. Junstad, M. and Wennmalm, Å., (1974): Br. J. Pharmacol. 52:375-379.
20. Posner, J., (1973): Br. J. Pharmacol. 49:415-427.
21. Messina, E.J., Weiner, R. and Kaley, G., (1974): Microvasc. Res. 8:77-89.
22. Strong, C.G. and Chandler, J.T., (1972): In "Prostaglandins in Cellular Biology", eds P.W. Ramwell and B.B. Pharriss, pp.369-383. Plenum Press, New York-London.
23. Clark, K.E., Ryan, M.J. and Brody, M.J., (1973): Adv. Biosci. 9:779-782.
24. Malik, K.U. and McGiff, J.C., (1975): Circ. Res. 36:599-609.
25. Hedqvist, P., (1970): Acta physiol. scand. Suppl.345:1-40.
26. Hedqvist, P., (1972): Eur. J. Pharmacol. 17:157-162.

27. Kadowitz, P.J., (1972): Br. J. Pharmacol. 46:395-400.
28. Frame, M.H., Hedqvist, P. and Åström, A., (1974): Life Sci. 15: 239-244.
29. Lonigro, A.J., Terragno, N.A., Malik, K.U. and McGiff, J.C., (1973): Prostaglandins. 3:595-606.
30. Frame, M.H. and Hedqvist, P., (1975): Br. J. Pharmacol. 54:189-196.
31. Hadházy, P., Magyar, K., Vizi, E.S. and Knoll, J., (1976): In "Advances in Prostaglandin and Thromboxane Research", eds B. Samuelsson and R. Paoletti, pp.365-368. Raven Press, New York.
32. Hedqvist, P., (1974): In "Prostaglandin Synthetase Inhibitors", eds H.J. Robinson and J.R. Vane, pp.303-309. Raven Press, New York.
33. Stjärne, L. and Gripe, K., (1973): Naunyn-Schmiedebergs Arch. Pharmacol. 280:441-446.
34. Hedqvist, P., (1974): Acta physiol. scand. 90:86-93.
35. Bergström, S., Farnebo, L.O. and Fuxe, K., (1973): Eur. J. Pharmacol. 21:362-368.
36. Neufeld, A.H. and Page, E.D., (1975): Exp. Eye Res. 20:549-561.
37. Kadowitz, P.J., Sweet, C.S. and Brody, M.J., (1971): J. Pharmacol. Exp. Ther. 177:641-649.
38. Horrobin, D.F., Manku, M.S., Karmali, R., Nassar, B.A. and Davies, P.A., (1974): Nature (Lond.) 250:425-426.
39. Brody, M.J. and Kadowitz, P.J., (1974): Fed. Proc. 33:48-60.
40. Kadowitz, P.J., Sweet, C.S. and Brody, M.J., (1971): J. Pharmacol. Exp. Ther. 179:563-572.
41. Frame, M.H., (1976): In "Advances in Prostaglandin and Thromboxane Research", eds B. Samuelsson and R. Paoletti, pp.369-373. Raven Press, New York.
42 Hedqvist, P., (1976): In "Advances in Prostaglandin and Thromboxane Research", eds B. Samuelsson and R. Paoletti, pp.357-363.
43. Hedqvist, P., (1973): Adv. Biosci. 9:461-473.
44. Hedqvist, P., Stjärne, L. and Wennmalm, Å., (1971): Acta physiol. scand. 83:430-432.
45. Hoszowska, A. and Panczenko, B., (1974): Pol. J. Pharmacol. Pharm. 26:137-142.
46. Dubocovich, M.L. and Langer S.Z., (1975): J. Physiol. (Lond.) 251:737-762.
47. Stjärne, L., (1972): Acta physiol. scand. 86:388-397.
48. Junstad, M. and Wennmalm, Å., (1972): Acta physiol. scand. 85:573-578.
49. Fredholm, B.B. and Hedqvist, P., (1975): Br. J. Pharmacol. 54:295-300.
50. Fredholm, B.B. and Hedqvist, P., (1973): Acta physiol. scand. 87:570-572.
51. Wennmalm, Å. and Hedqvist, P., (1971): Life Sci. 10(Part 1):465-470.
52. Hall, W.J., O'Neill, P. and Sheehan, J.D., (1975): Eur. J. Pharmacol. 34:39-47.

53. Feniuk, W. and Large, B.J., (1975): Br. J. Pharmacol. 55:47-49.
54. Hadházy, P., Illés, P. and Knoll, J., (1973): Eur. J. Pharmacol. 23:251-255.
55. Park, M.K., Dyer, D.C. and Vicenzi, F.F., (1973): Prostaglandins 4:717-730.
56. Ferreira, S.H., Herman, A. and Vane J.R., (1972): Br. J. Pharmacol. 44:328P-329P
57. Illés, P., Vizi, E.S. and Knoll, J., (1974): Pol. J. Pharmacol. Pharm. 26:127-136.
58, Kadlek, O., Masek, K. and Seferna, I., (1974): Br. J. Pharmacol. 51:565-570.
59. Bennett, A., Eley, K.G. and Stockley, H.L., (1975): Prostaglandins 9:377-384.
60. Gustafsson, L., Hedqvist, P. and Lagercrantz, H., (1975): Acta physiol. scand. 95:26-33.

INFLAMMATORY PAIN AND FEVER

S.H. FERREIRA

Department of Pharmacology, Faculty of Medicine

of Ribeirão Preto, S.P., BRAZIL

The existence of an ongoing inflammatory process is generally indicated by pain or fever and these symptoms are also the first to be relieved by non-steroidal anti-inflammatory therapy. This is a brief review of the participation of prostaglandins in the genesis of these symptoms.

Prostaglandins have been detected in most inflammatory exudates and in cerebrospinal fluid (CSF) during fever (for a review see 1, 2). These findings reflect the fact that prostaglandins are generated by almost any type of cells that have been injured or submitted to intense activity. Distortion or loss of integrity of cell membrane is thought to trigger activation of phospholipases as well as cyclo-oxygenases involved in prostaglandin synthesis. Prostaglandins present in the CSF do not originate solely from the hypotalamic regions related to fever but from all parenchimal cells. However, the specific type of cell which generates prostaglandins under pyrogen stimulation has not yet been identified. In fact the problem is far from settled with respect to other tissues as well. Are local or migrating cells the major contributors to the prostaglandin content of inflammatory exudates? Are the functionally active prostaglandins generated by the reacting structures themselves? Are the prostaglandins which sensitize vessels or nerve endings independently generated? Is it possible to block prostaglandin synthesis in peripheral nerves without affecting their generation by endothelial cells? These are important but unresolved problems whose solutions are of fundamental importance to the understanding of the contribution of prostaglandins to the development of the inflammatory response. In spite of the limitations of our current understanding of the generation and release of prostaglandins the bulk of the evidence strongly implicates prostaglandins in fever and pain.

FEVER

A remarkably large variety of stimuli can produce fever: several types of endotoxins, tissue damage, antigen-antibody reaction, malignancy, etc. Usually the fever-producing agent is referred to as a pyrogen. It is important to distinguish between exogenous and endogenous pyrogens. This is illustrated by the action of endotoxins produced by gram negative bacteria. Beeson (3) and Bennett and Beeson (4) were the first to demonstrate that endogenous pyrogen could be generated by granular leucocytes and neutrophils upon stimulation exogenous pyrogen. Other cells also produce endogenous pyrogen. Both endogenous and exogenous pyrogen injected directly into the cerebral ventricles induce fever but the response to exogenous pyrogen is delayed, possibly indicating the local generation of endogenous pyrogen. Milton and Wendlant (5) first showed that the fever produced by injections of endotoxins and endogenous pyrogen into cerebral ventricles or into the pre-optic anterior hypothalamus was blocked by aspirin-like drugs. Prostaglandin E_1 is the most powerful pyretic agent known when injected either into cerebral ventricles or directly·into the anterior hypothalamus (6, 7). The hyperthermic effect is dose-dependent, almost immediate, long lasting and is not blocked by aspirin-like drugs.

There is a generation of prostaglandin E_2 like substance in the central nervous system during fever and after intravenous injection of pyrogen the concentrations in the cerebrospinal fluid rise 2.5-4.0 - fold, sometimes to as to as much as 35 ng/ml (8).

Aspirin-like drugs do not abolish either the formation of endogenous pyrogen by leucocytes (9) or the pyretic action of prostaglandins injected into the third ventricle of cats. However, they inhibit both the generation of prostaglandins in the central nervous system and the fever caused by pyrogens or 5-hydroxy-tryptamine injected into the cerebral ventricles. The 5-10 - fold increase in the amount of prostaglandin released into the cerebrospinal fluid observed at the height of endotoxin-induced fever in dogs is suppressed by the administration of indomethacin (10).

Many recent findings support the hypothesis that prostaglandins act as fever mediators. However, the

observation by Cranston et al. (11) that PG antagonist
SC19220 when given together with prostaglandins or
leucocyte pyrogen only prevented prostaglandin fever
would seen to argue against such a role for prostaglandin.
However this may only indicate that the prostaglandins
generated at the site of action are insensitive to a PG
antagonist, either because the antagonist is unable to
reach the sites were PG is being generated or because
other substances derived from arachidonic acid oxidation
such as thromboxanes may also be partially responsible
for the fever. In this late case the prostaglandin
antagonist would be ineffective.

        Although there is no direct evidence that pyrogens
stimulate prostaglandin synthetase in the brain the main
evidence for the theory is based on the following
observations: a) prostaglandins cause fever when injected
directly into the pre-optic anterior region; b) prostaglandins can be generated in these areas; c) antipyretic
which inhibit cyclo-oxygenase abolish pyrogen fever, and
there is a substantial reduction in the increased
concentration of prostaglandins in CSF.

HYPERALGESIA AND PAIN

        When we suggested that the anti-algesic effects
of prostaglandin synthetase inhibitors such as aspirin-
-like drugs could be explained by the removal of the
sensitizing action of prostaglandins upon pain receptors
several observations had already been made relating
prostaglandins and pain.

        Hyperalgesia induced by an intradermal injection
of prostaglandin had already been described (12, 13) and
its was known that intravenous prostaglandins cause
headaches, as well as pain along the veins into which
they are infused (14, 15). Prostaglandins also caused
pain when injected intramuscularly (16). When given
abdominally to mice, prostaglandins were the most
powerful writhing reflex inducer. However, at the time
of the discovery that aspirin-like drugs inhibit the
synthesis of prostaglandins (17, 18, 19) there were
conflicting views concerning the role of prostaglandins
as pain mediators. Retrospectively, I think that the
main arguments against this role for prostaglandins were:
a) the failure of prostaglandins to cause pain when
instilled on a blister base (20); b) the fact that in
an other study on the dermal vasodilatory effect of prostaglandins (21) there was no mention of pain or hyperalgesia
(21); c) the lack of a clear demonstration that no other
putative inflammatory mediators had similar hyperalgesic
effects; d) scientists at that moment were looking for

a "classical pain mediator" and not for a substance
which would just sensitize the pain receptors to the
action of chemical or mechanical stimulation.
        There were two observations which made us realize
the uniqueness of prostaglandins as "pain" mediators:
a) when infused subdermally at concentrations likely to
be found at inflamed sites, prostaglandins did not
cause overt pain however, theycaused hyperalgesia (i.e.
a state in which overt pain can be aroused by normally
painless mechanical stimulation). Other putative
me.diators of inflammation such as bradykinin or histamine
did not cause hyperalgesia when given intradernally or
infused subdermally, although they could cause a much
more pronounced erythema or oedema (22); b) the other
observation was made with regard to intra-splenic
injections of bradykinin and prostaglandin release.
        Lim and his colleagues (23, 24, 25) used
nociception induced by intra-arterial injections of
bradykinin into the spleen to show that aspirin-like
drugs act peripherally as analgesics. To test wheter
bradykinin induces pain through prostaglandin release,
we injected bradykinin into the spleen of the dog (26).
Prostaglandins were released in similar amounts, both
in vitro and in vivo by injections of adrenaline or
bradykinin. As adrenaline is a much weaker pain-
-producing substance than bradykinin in this system, it
was clear that a prostaglandin could not be mediator of
the pain-producing activity of bradykinin.
        Our experiments with subdermal infusions made in
order to mimic the continuous release of a mediador that
would occur at the site of an injury, revealed two
important features of the action of prostaglandins: (a)
prostaglandins sensitize the pain receptors to mechanical
and chemical stimulation and (b) that the effect was
cumulative (22). This cumulative effect was demonstrated
by infusing for 2 hours a concentration of prostaglandin
which was ineffective for a 30 min period. Thus, a slow
rate of release of prostaglandins during a long period
of time (as might occur in inflammation) is capable of
causing hyperalgesia. In these experiments there was no
overt pain during separate infusions of PGE_1, bradykinin,
histamine or a mixture of bradykinin with histamine.
However, when PGE_1 was added to bradykinin or histamine
(or mixture of both) strong pain developed. Furthermore,
in areas made hyperalgesic by an infusion of prostaglandins
(tested by pressure) a second infusion of either histamine
or bradykinin caused overt pain. However, at the site
where bradykinin or histamine had been previously infused
(without producing hyperalgesia) a second infusion prostaglandin caused little or no pain.

This unique ability of prostaglandins to sensitize pain receptors is also demonstrabe in the rat. Carrageenin when injected in rat paw causes a marked hyperalgesia after four hours. One hour after carrageenin administration there is no hyperalgesia. On the other hand bradykinin, histamine or 5H-Tryptamine cause litle or no hyperalgesia when given alone or together wizh carrageenin. But prostaglandins which causes a mild effect by themselves cause a marked hyperalgesia when added to carrageenin (Fig. 1).

Another important observation concerned pruritus. Neither histamine, bradykinin nor prostaglandin E_1 subdermal infusions by themselves caused itch (22). However, when prostaglandin E_1 was infused with histamine. itching was always recorded (when prostaglandin E_1 was infused with bradykinin, there was pain rather than itch). This role of prostaglandins in potentiating the effects of histamine was later confirmed by Greaves and McDonald--Gibson (27).

The sensitizing action of prostaglandins to pain induced by bradykinin was shown to occur in the dog's spleen knee joint also.

We used the reflex rise in blood pressure induced by intra-arterial bradykinin injections into the spleen of lightly anaesthetized dogs as an indication of sensory stimulation. Doses of bradykinin, which released prostaglandin from the spleen, caused a dose-dependent reflex increase in blood pressure, which was reduced by indomethacin. When prostaglandin E_1 was given together with bradykinin to the indomethacin-treated dogs, the reflex increase in blood pressure was restored, sometime to greater than control values (26).

In extending this work, we have used the same technique to study the effect of prostaglandins on the nociceptive reflex induced by bradykinin injected into dog knee joints (28). Increasing doses of bradykinin induced a dose-dependent rise in blood pressure. Local treatment with aspirin or indomethacin partially inhibited the bradykinin responses, probably by inhibiting the local formation of prostaglandins generated by the trauma caused by the manipulation of the joints. Infusion of prostaglandins produced a long lasting potentiation of the algogenic effect of bradykinin. When local generation of prostaglandins was blocked by indomethacin and the enhanced sensitivity of the joints restored by an infusion of prostaglandins, further injection of indomethacin did not alter the response to the injection of bradykinin into the joints. This fact indicates that aspirin-like drugs do not affected the sensitivity of the receptors to the action of bradykinin.

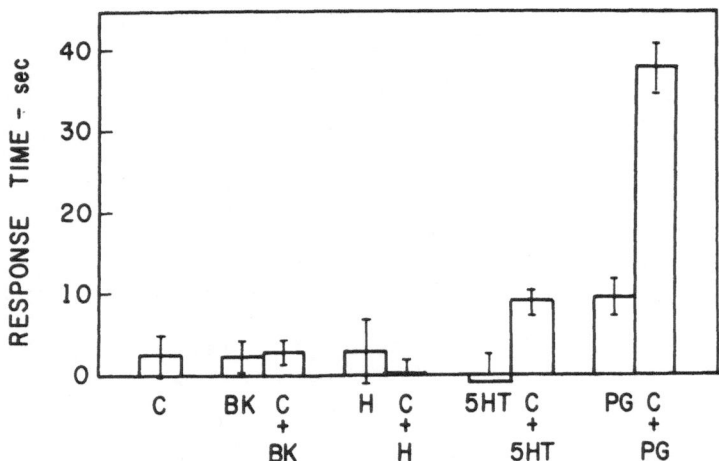

Fig. 1 - Influence of inflammatory mediator in the
rat paw hyperalgesia induced by carrageenin.
The mediators where given intraplantarly
either alone or together with carrageenin
(0.5%, W/V saline). Hyperalgesia was measured
by the Randall-Sellito method. The variation
in the response time was obtained by subtracting
the values of injuried leg from control leg.
No significant hyperalgesia was observed at
1 hour with carrageenin, bradykinin (BK, 500
ng), histamine (H, 500 ng), 5-hydroxytryptamine
(HT, 500 ng). Prostaglandin (PGE$_1$, 500 ng)
caused a mild effect. The mixture 5-Hydroxy-
tryptamine and carrageenin also caused a mild
effect when compared with the very intense
effect of prostaglandin-carrageenin mixture;
the values are mean $\pm$ EM of 5 animals.

We have also explored the possibility that fatty
acid hydroperoxides can contribute to the genesis of
pain in man (22). Intensity of the pain produced by
intradermal injections of hydroperoxides of arachidonic,
linoleic and linolenic acid was greater than that induced
by either the parent fatty acids or acetylcholine, bradykinin, histamine or prostaglandin E_1.

From all these results we can draw three conclusions: a) lipoperoxide intermediates in the prostaglandin
biosynthetic pathway may have pain-producing properties,
as do prostaglandin E_1 and E_2 in high concentrations;
depending on the intensity of activation of the prostaglandin generating system by a trauma, the generation
of the intermediate could exceed its conversion to
prostaglandin, thus causing an acute type of pain,
specially if the pain receptors are already sensitized
as is the case in a blister base. b) In low concentrations,
prostaglandin E_1 and E_2 sensitizes the pain fibres to
mechanical and chemical stimuli. c) The effect of prostaglandins E_1 and E_2 are acumulative and long-lasting.
Thus, continual generation of minute amounts of prostaglandin at a site of injury will sensitize the nerves,
so that mechanical stimulation or mediators such as
bradykinin and histamine can cause pruritus or pain. The
blockade of prostaglandin synthesis may thus explain the
anti-algesic effect of aspirin-like drugs. The reason
why some agents such paracetamol or novalgine show poor
"anti-inflammatory" activity in contrast with powerful
anti-algesic or anti-piretic activity may be due not only
to the high sensitivity of the cyclo-oxygenase of the
nervous tissues to these drugs but also to a preferential
access of these drugs to the enzymes, due to pharmacokinetic factors. Thus, prostaglandin synthetase inhibitors
which inhibit nerve or brain enzymes may do so without
affecting vessel prostaglandins, i.e., without conspicuous
"anti-inflammatory activity".

I am grateful to FAPESP (75/0296) and to the Wellcome
Foundation for research Grants.

REFERENCES

1- Ferreira, S.H. and Vane, J.R., 1974, New aspects of the
      mode of action of non-steroid anti-inflammatory
      drugs, Ann. Rev. of Pharmacol. 14: 57.

2- Feldberg, W., 1974, Fever, Prostaglandins and Anti-
      pyretics in Prostaglandin Synthetase inhibitors,
      H.J. Robinson and R. Vane, Raven Press, N.Y.

3- Beeson, P.B., 1948, Temperature elevating effect of a
      substance obtained from polymorphonuclear leuco-
      cytes, J. Clin. Invest. 27: 524.

4- Bennett, I.L. and Beeson, P.B., 1953, Studies in the
      pathogenesis of fever. II characterization of
      fever producing substances from polymorphonuclear
      leucocytes and from the fluid of sterile exudates,
      J. exp. Med. 98: 493.

5- Milton, A.S. and Wendlant, S., 1968, The effect of
      4-acetamidophenol in reducing fever produced by
      the intracerebral injection of 5-hydroxytryptamine
      and pyrogen in cóncions cat, Br. J. Pharmac. 34:
      215.

6- Milton, A.S. and Wendlant, S., 1971, Effects on body
      temperature of prostaglandins of the A, E and F
      series on injection into the third ventricle of
      unanaesthetized rats and rabbits, J. Physiol. 218:
      325.

7- Feldberg, W. and Saxena, P.N., 1971 a, Fever produced
      by prostaglandin E_1, J. Physiol. 217: 547.

8- Feldberg, W. and Gupta, K.P., 1973, Pyrogen fever and
      prostaglandin activity in cerebrospinal fluid,
      J. Physiol. 228: 41.

9- Clark, W.C. and Moyer, S.G., 1972, The effect of
      acetaminophen and sodium salicylate on the release
      and activity of leukocytic pyrogen in the cat, J.
      Pharmac. exp. Ther. 181: 183.

10-Milton, A.S., 1973, Prostaglandin release in the central
      nervous system during endotoxin induced fever. In
      Advance in Biosciences, 9: ed. S. Bergströem S.
      Bernard, Pergamon Press-Vieweg, Braunschweig, p. 495.

11-Cranston, W.I., Duff, G.W., Hellon, R.F. and Mitchell, D.,
      1976, Effect of a prostaglandin antagonist on the
      pyrixias caused by PGE_2 and leucocyte pyrogen in
      rabbits, J. Physiol. (Lond.) in press.

12- Solomon, L.M., Juhlin, L. and Kirschenbaum, M.B., 1968, Prostaglandin on cutaneous vasculature, J. Invest. Derm. $\underline{51}$: 280.

13- Michaelsson, G., 1970, Effects of antihistamines, acetylsalicylic acid and prednisone on cutaneous reactions to kallikrein and prostaglandin E_1. Acta derm. venereol. $\underline{30}$: 31.

14- Bergström, S., Duner, H., Euler, U.S., Pernow, B. and Sjovall, J., 1959, Observations on the effects of infusions of prostaglandin E in man, Acta Physiol. Scand. $\underline{45}$: 145.

15- Collier, J.G., Karim, S.M.M., Robinson, B. and Somers, K., 1972, Action of prostaglandins A_2, B_1, E_2 and $F_{2\alpha}$ on superficial hand veins of man, Br. J. Pharmac. $\underline{44}$: 374.

16- Karim, S.M.M., 1971, Action of prostaglandin in the pregnant woman, Ann. N.Y. Acad. Sc. $\underline{180}$: 483.

17- Vane, J.R., 1971, Inhibition of prostaglandin synthesis as a mechanism of action for aspirin-like drugs. Nature, New Biol. $\underline{231}$: 232.

18- Ferreira, S.H., Moncada, S. and Vane, J.R., 1971, Indomethacin and aspirin abolish prostaglandin release from the spleen, Nature, New Biol. $\underline{231}$: 237.

19- Smith, J.B. and Willis, A.L., 1971, Aspirin selectively inhibits prostaglandin production in human platelets. Nature, New Biol. $\underline{231}$: 256.

20- Horton, E.W., 1963, Action of prostaglandin E_1 on tissues which respond to bradykinin, Nature $\underline{200}$: 892.

21- Cruunkhorn, P. and Willis, A.L., 1971, Cutaneous reactions to intradermal prostaglandins, Br. J. Pharmac. $\underline{41}$: 49.

22- Ferreira, S.H., 1972, Prostaglandins, aspirin-like drugs and analgesia, Nature, New Biol. $\underline{240}$: 200.

23- Guzman, F., Braun, C. and Lim, R.K.S., 1962, Visceral pain and the pseud-affective response to intra-arterial injection of bradykinin and other algesic agents. Arch. Int. Pharm. $\underline{136}$: 353.

24- Guzman, F., Braun, C., Lim, R.K.S., Potter, G.D. and Rodgers, D.W., 1964, Narcotic and non-narcotic analgesics which block visceral pain evoked by intra-arterial injection of bradykinin and other algesic agents, Arch. Int. Pharmac. $\underline{149}$: 571.

25- Lim, R.K.S., Guzman, F., Rodgers, D.W., Goto, K., Braun, G., Dickerson, G.D. and Engle, R.J., 1964, Site of action of narcotic and non-narcotic analgesic determined by blocking bradykinin-evoked visceral pain. Arch. Int. Pharmacol. 152: 25.

26- Ferreira, S.H., Moncada, S. and Vane, J.R., 1973, Prostaglandins and the mechanism of analgesia produced by aspirin-like drugs, Brit. J. Pharmacol. 49: 86.

27- Greaves, M.W. and MacDonald-Gibson, W., 1973, Itch: Role of prostaglandins, Br. Med. J. 3: 608.

28- Moncada, S., Ferreira, S.H. and Vane, J.R., 1975, Inhibition of prostaglandin biosynthesis as the mechanism of analgesia of aspirin-like drugs in the dog knee joint, Eur. J. Pharmacol. 31: 250.

LIST OF CONTRIBUTORS

A. Bennett,
  Dept. of Surgery, King's College Hospital
  Medical School, London   SE5 8RX, UK

F. Berti,
  Institute of Pharmacology and Pharmacognosy, School
  of Pharmacy, University of Milan,   20129 Milan, Italy

M. Bygdeman,
  Dept. of Obstetrics and Gynecology, Karolinska
  Hospital, Stockholm, Sweden

P. Crabbé,
  Université Scientifique et Médicale, Grenoble, France

M. F. Cuthbert,
  Dept. of Pharmacology & Therapeutics, London Hospital
  Medical College, London E 1,   England
  &
  Chest Unit, Dept. of Medicine, King's College Hospital,
  London SE5, England

A. Dembińska-Kieć,
  Dept. of Pharmacology, Copernicus Academy of Medicine
  in Cracow, 31-531 Cracow, 16 Grzegórzecka, Poland

S.H. Ferreira,
  Dept. of Pharmacology, Faculty of Medicine of
  Ribeirão Preto, São Paulo, Brazil

G.C. Folco,
  Institute of Pharmacology and Pharmacognosy,
  University of Milan, 20129 Milano, Italy

R. Fumagalli,
  Institute of Pharmacology and Pharmacognosy,
  University of Milan, 20129 Milano, Italy

G. Galli,
  Institute of Pharmacology and Pharmacognosy, Laboratory
  of Applied Biochemistry, University of Milan,
  Milano, Italy.

J.P. Giroud,
   Département de Pharmacologie, ERA 629, C.N.R.S.
   Faculté de Médecine Cochin Port-Royal, 27 Rue du
   Faubourg Saint-Jacques, 75014 Paris, France

E. Granstrøm,
   Dept. of Chemistry, Karolinska Institutet,
   S-104 01 Stockholm 60, Sweden

L. Grodzińska,
   Dept. of Pharmacology, Copernicus Academy of Medicine
   in Cracow, 31-531 Cracow, 16 Grzegórzecka, Poland

R.J. Gryglewski,
   Dept. of Pharmacology, Copernicus Academy of Medicine
   in Cracow, 31-531 Cracow, 16 Grzegórzecka, Poland

P. Hedqvist,
   Dept. of Physiology, Karolinska Institutet,
   S-10401 Stockholm 60, Sweden

R. Korbut
   Dept. of Pharmacology, Copernicus Academy of Medicine
   in Cracow, 31-531 Cracow, 16 Grzegórzecka, Poland

D. Longiave,
   Institute of Pharmacology and Pharmacognosy,
   University of Milan, 20129 Milano, Italy

F. Maggi,
   Institute of Pharmacology and Pharmacognosy,
   University of Milan, 20129 Milano, Italy

J.C. McGiff,
   Dept. of Pharmacology and Medicine, University of
   Tennessee, Memphis, Tennessee 38163, U.S.A.

S. Nicosia,
   Institute of Pharmacology and Pharmacognosy, Laboratory
    Applied Biochemistry, University of Milan, Milan, Italy

E. Niżankowska,
   Clinical Immunology, Copernicus Academy of Medicine
   in Cracow, 31-531 Cracow, 16 Grzegórzecka, Poland

C. Omini,
   Institute of Pharmacology and Pharmacognosy, School of
   Pharmacy, University of Milan, 20129 Milan, Italy

N.L. Poyser,
   Dept. of Pharmacology, University of Edinburgh, 1 George
   Square, Edinburgh EH8 9JZ, Scotland

L. Puglisi,
   Institute of Pharmacology and Pharmacognosy, University
   of Milan, 20129 Milan, Italy

A. Robert,
   Dept. of Experimental Biology, The Upjohn Company,
   Kalamazoo, Michigan 49001, U.S.A.

B. Samuelsson,
   Dept. of Chemistry, Karolinska Institutet, S-104 01
   Stockholm 60, Sweden

T.Y. Shen,
   Research Laboratories, Merck Sharp & Dohme,
   Rahway, New Jersey 07065, U.S.A.

A. Szczeklik,
   Department of Allergology, Copernicus Academy of
   Medicine in Cracow, 31-531 Cracow, 16 Grzegórzecka,
   Poland

A. Terragno,
   Dept. of Pharmacology and Medicine, University of
   Tennessee, Memphis, Tennessee 38163, U.S.A.

G.P. Velo,
   Institute of Pharmacology, University of Padua,
   Policlinico Borgo Roma, 37100 Verona, Italy

B.J.R. Whittle,
   Royal College of Surgeons, Lincoln's Inn Fields 35/43,
   London WC2A 3PN, England

D.A. Willoughby,
   Dept. of Experimental Pathology, St. Bartholomew's
   Hospital, West Smithfield, London EC1A 7 BE, England

P. Y-K. Wong,
   Dept. of Pharmacology and Medicine, University of
   Tennessee, Memphis, Tennessee 38163, U.S.A.